T0335567

VOLUME TWO HUNDRED AND SEVEN

ADVANCES IN
IMAGING AND
ELECTRON PHYSICS

EDITOR-IN-CHIEF

Peter W. Hawkes
CEMES-CNRS
Toulouse, France

VOLUME TWO HUNDRED AND SEVEN

Advances in
IMAGING AND
ELECTRON PHYSICS

Edited by

PETER W. HAWKES
CEMES-CNRS
Toulouse, France

ACADEMIC PRESS

An imprint of Elsevier

Cover photo credit:
The cover picture is taken from Fig. 3 of the chapter by Arbouet et al. (p. 9).

Academic Press is an imprint of Elsevier
125 London Wall, London EC2Y 5AS, United Kingdom
525 B Street, Suite 1650, San Diego, CA 92101, United States
50 Hampshire Street, 5th Floor, Cambridge, MA 02139, United States
The Boulevard, Langford Lane, Kidlington, Oxford OX5 1GB, United Kingdom

ISBN: 978-0-12-815215-7
ISSN: 1076-5670

For information on all Academic Press publications
visit our website at https://www.elsevier.com/books-and-journals

Working together
to grow libraries in
developing countries

www.elsevier.com • www.bookaid.org

Publisher: Zoe Kruze
Acquisition Editor: Jason Mitchell
Editorial Project Manager: Shellie Bryant
Production Project Manager: Divya KrishnaKumar
Designer: Alan Studholme

Typeset by VTeX

CONTENTS

CONTRIBUTORS

Arnaud Arbouet
CEMES-CNRS, University of Toulouse, Toulouse, France

Giuseppe M. Caruso
CEMES-CNRS, University of Toulouse, Toulouse, France

Karl-Joseph Hanszen (deceased)
Physikalisch-Technische Bundesanstalt, Braunschweig, Germany

Florent Houdellier
CEMES-CNRS, University of Toulouse, Toulouse, France

Inder Jeet Taneja
Formerly Professor of Mathematics, Department of Mathematics, Federal University of Santa Catarina, Florianópolis, SC, Brazil

PREFACE

Among the numerous recent advances in electron microscopy, ultrafast imaging and diffraction are among the most dazzling. This volume opens with an account of the subject by A. Arbouet, G.M. Caruso, and F. Houdellier. First, the history is traced, starting with the pioneering work of Oleg Bostanjoglo, who contributed several articles to these Advances. A long section presents the instrumental features needed for high-speed image formation and recording. Several examples of information that could not have been obtained by any other technique are presented. In the closing section, endeavors to perform time-resolved holography in an ultrafast electron microscope are described.

This is followed by an account of further developments in the work of I.J. Taneja on divergence measures. Earlier essays on the subject have appeared in volumes 91, 138, and 201. Here, the generalized Gini-mean is used to establish inequalities and thence, differences of means. Readers will appreciate the fact that Taneja provides full derivations of his findings.

The volume concludes with a further chapter from *Advances in Optical and Electron Microscopy*. I am very pleased to make this work by the late Karl-Joseph Hanszen available to a wider audience. Hanszen and his colleagues were the first to put the contrast-transfer theory of electron microscope imaging on a sound foundation and this long account brings together all the early work.

As always, I am sure that readers will appreciate the authors' efforts to make their contributions accessible to a wide audience.

Peter W. Hawkes

CHAPTER ONE

Ultrafast Transmission Electron Microscopy: Historical Development, Instrumentation, and Applications

Arnaud Arbouet[1], Giuseppe M. Caruso, Florent Houdellier[1]
CEMES-CNRS, University of Toulouse, Toulouse, France
[1]Corresponding authors: e-mail addresses: arnaud.arbouet@cemes.fr, florent.houdellier@cemes.fr

Contents

Advances in Imaging and Electron Physics, Volume 207
ISSN 1076-5670
https://doi.org/10.1016/bs.aiep.2018.06.001

1. INTRODUCTION

X-ray photons and electrons are routinely used to investigate materials at the nanoscale. Whereas phase information can be retrieved in coherent diffractive imaging experiments performed with the coherent X-ray beams provided by Free Electron Lasers, Transmission Electron Microscopy is undoubtedly the most versatile and widespread technique giving access to real space imaging with nanometer spatial resolution. By enabling a direct visualization of materials down to the atomic scale, Transmission Electron Microscopes (TEM) have allowed giant steps in chemistry, biology or physics. Images, either in real or reciprocal space, as well as electron energy loss spectra (EELS) can be obtained and combined from nanoscale regions of a sample yielding structural and chemical information with unrivalled spatial resolution (Williams & Carter, 1996). Evolutions of TEM are intimately linked with the instrumental advances. For instance, the development of field emission guns delivering high brightness electron beams has opened completely new horizons in scanning TEM (STEM) where the amount of probe current is crucial (Crewe, Eggenberger, Wall, & Welter, 1968). Additionally, following the initial suggestion of Gabor in 1948, new techniques such as off-axis electron holography exploited these new sources to produce coherent electron beams and generate interferograms from which modifications of the phase of the electronic wavepacket can be detected (Gabor, 1948; Tonomura, 1999). This allowed to map the local electrostatic field, strain field or magnetic field with nanometer or sub–nanometer resolution (Dunin–Borkowski, Kasama, & Harrison, 2015). Since the seminal work of Scherzer, it is well known that the spherical and chromatic aberrations associated with the static and rotationally symmetric electromagnetic fields of magnetic lenses strongly limit the resolution of TEMs. The development of aberrations correctors using multipolar optics has allowed a giant step in spatial resolution and opened the field of analytical electron microscopy with atomic resolution (Haider et al., 1998; Krivanek et al., 2010). Considerable progress has also been made in the energy domain with the design of monochromators that can now yield sub-10 meV energy resolution (Krivanek et al., 2014). Finally, direct electron detection improved the modulation transfer function enabling higher resolution images at lower magnification and low dose imaging capabilities due to larger detective quantum efficiency. These radical innovations have revolutionized the field of Transmission Electron Microscopy and boosted the spatial and energy resolution of electron microscopes.

Despite these spectacular achievements, investigations by Transmission Electron Microscopy have been restricted until the beginning of the 1980s to systems either static or evolving on timescales compatible with the frame rate of Charge Coupled Device (CCD) cameras (typically several milliseconds per frame). Electron microscopes were capable of looking at static atoms but were not fast enough to monitor their displacement on the appropriate timescale. Yet, the observation of a physical system as it evolves in real time is essential to the understanding of its dynamics. Exactly like someone would ask a magician to slow down his movements and break them into a sequence of elementary gestures to understand a card trick, electron microscopists have long dreamed of being able to excite samples and acquire images, diffraction patterns, or energy spectra after a controllable delay to elucidate their dynamics. However, as can be seen in Fig. 1, the timescale of many important processes of the dynamics of solids lies typically in the femtosecond to nanosecond range, *i.e.* far beyond the reach of conventional *in-situ* TEMs. Following the pioneering work of Oleg Bostanjoglo at the Technical University of Berlin, spectacular progress has been made to provide TEM with a constantly improving temporal resolution. As will be discussed in this chapter, these developments rely mostly, although not exclusively, on ultrafast lasers and are largely inspired from time-resolved optical spectroscopy techniques. Indeed, the generation of femtosecond optical pulses shorter than the characteristic times of electronic relaxation or atomic motion in molecular or condensed matter systems and the ability to synchronize optical excitation and detection made possible the acquisition of "snapshots" at regular time intervals during the evolution of a system. These "pump-probe" experiments shed light on many complex relaxation paths in either chemistry or condensed matter (Shah, 1999; Mukamel, 1999; Ma, Hertel, Vardeny, Fleming, & Valkunas, 2008). However, even state-of-the-art near-field optical microscopy techniques have a spatial resolution limited to a few tens of nanometers and the information gained can only be indirectly connected to the atomic structure of the sample using models. In this context, it has soon been realized that combining electron microscopes and ultrafast lasers would provide nanometer spatial resolution and subpicosecond time resolution.

It is the purpose of this chapter to provide an introduction to the field of time-resolved TEM (Section 2), describe the major instrumental developments (Section 3), and give examples of applications in different fields (Section 4). In Section 5, the possibility of performing time-resolved electron holography experiments with ultrafast TEMs is discussed. For the sake of coherence, we intentionally restrict the scope of this chapter to time-resolved Transmission Electron Microscopy and will not address other related techniques (ultrafast scanning electron microscopy, ultrafast electron diffraction, time-resolved X-ray diffraction, and spectroscopy for instance). We deliberately chose to describe in more details investigations using ultrafast Transmission Electron Microscopes operated in stroboscopic mode which provide both femtosecond time resolution and sub–nanometer spatial resolution as the historical development of single-shot Dynamic Transmission Electron Microscopy was already the subject of several reviews. Finally, we insist on the fact that we do not consider what follows as an exhaustive review of the available literature. It is a selection of examples chosen for their ability to illustrate the richness of a dynamic research field. We apologize in advance for the inevitable omissions.

2. TIME-RESOLVED TRANSMISSION ELECTRON MICROSCOPY: TIMESCALES AND CONCEPTS

2.1 Ultrafast Dynamics in Condensed Matter: Important Timescales

As an introduction, we briefly describe here the timescales of the main physical processes governing the dynamics in atoms, molecules, and condensed matter. The motion of electrons occurs on the attosecond timescale. For instance, the revolution period of an electron in Bohr's hydrogen atom is 152 as. The timescale of the processes involving atom displacement such as chemical reactions of molecules spans the femtosecond–picosecond range. As shown in Fig. 1, vibrations and rotations in molecules typically occur on femtoseconds and picoseconds respectively. We will now focus on the dynamics of a solid excited by a femtosecond light pulse, this being the first step of the time-resolved TEM experiments described later in this review. On a timescale comparable to the laser pulse duration, the light pulse couples *coherently* to the nanostructure. In the case of metallic nanoparticles

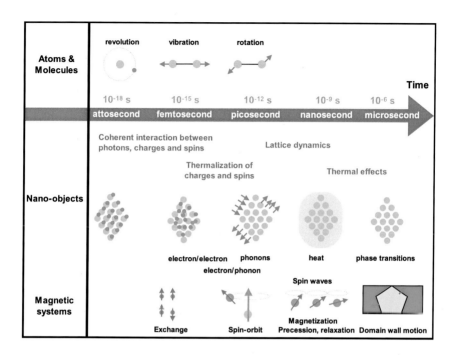

Figure 1 Timescales of the ultrafast dynamics in atoms, molecules, and nano-objects.

for instance, a collective electronic excitation such as a surface plasmon can be excited. A precise phase relation between the incident electric field and the collective charge oscillation inside the sample characterizes the coherent coupling regime. This phase relation survives until the dephasing of the plasmon in ~10 fs typically. After this initial step, the electron gas is not thermalized with only a fraction of hot electrons having gained energy from the optical excitation. The electron gas goes back to a Fermi–Dirac distribution thanks to electron–electron interactions, which redistribute the energy excess. This electronic internal thermalization occurs on a timescale of a few hundreds femtoseconds in gold and silver for instance. Coulombic interactions between the electrons and the lattice mediate the complete redistribution of the energy injected by the optical excitation on a few picoseconds.

The ultrafast energy injection places the lattice out of equilibrium and excites mechanical vibrations. The vibration period of confined acoustic modes are of the order of L/v where L is the size of the nano-object and v the speed of sound. It typically takes values in the 10–100 ps range for objects with sizes between 10–100 nm. The complete relaxation of an optically excited nano-object involves heat diffusion toward its environment. Fig. 1 summarizes the timescales of the main processes involved in the relaxation of atoms, molecules, and nano-objects. The existence of the spin degree of freedom in magnetic systems complexifies their ultrafast dynamics. The main processes involved in the magnetization dynamics are also given in Fig. 1 together with their characteristic timescale (Vaz, Bland, & Lauhoff, 2008; Bigot, Vomir, & Beaurepaire, 2009; Stohr, 2016).

It is clear from Fig. 1 that many important processes occur on a sub-nanosecond timescale. As will be detailed in the following, these processes have long remained out of reach of the time-resolved Transmission Electron Microscopes developed since the 1980s. It is only since 2005 and the development of Ultrafast Transmission Electron Microscopes operating in the stroboscopic mode that they could be investigated with the appropriate femtosecond temporal resolution. We describe in the next section the working principle of the different kinds of time-resolved TEMs.

2.2 Time-Resolved Transmission Electron Microscopy: *In-Situ* TEM, Dynamic TEM, and Ultrafast TEM

Investigating the dynamics of a physical system in the time domain inside a TEM requires (i) to be able to excite the sample, (ii) observe it with a controlled delay with respect to the excitation, and (iii) with an excitation and observation time much shorter than the characteristic duration of the physical process under scrutiny to avoid the information gained to be temporally averaged. With the variety of sample-holders available today, any kind of stimuli (electric current, magnetic field, mechanical stress, heat, etc.) can be applied to a sample directly inside the objective lens. *In-situ* TEM refers to the observation of a sample using conventional, high-resolution imaging, diffraction or spectroscopy as it evolves in real-time after being perturbed inside the electron microscope. The timescale accessible in these experiments is limited by the frame rate of the camera used for detection to typically a few milliseconds. *In-situ* TEM techniques have provided invaluable information on dynamic microstructural changes (Dehm, Howe, & Zweck, 2012).

Time-resolved TEM experiments with temporal resolution in the nanosecond range or faster rely on strategies and tools inspired from ultrafast optical spectroscopy. They are *pump-probe* experiments involving an optical pulse and a delayed electron pulse. The beam from a first pulsed laser is sent to the electron source to trigger the emission of electrons. A second beam from a pulsed laser is sent inside the objective lens of the microscope and focused on the sample. The optical **pump** pulse brings the sample out of equilibrium and the electron **probe** pulse, delayed and synchronized with respect to the excitation, is used to probe the sample during its relaxation. By systematically changing the delay between pump and probe, it is possible to record the dynamical evolution of the sample as the latter goes back to equilibrium. Controlling the pump-probe delay in the nanosecond range can be done using two independent nanosecond laser sources and a synchronization electronics. On the other hand, temporal resolutions in the femtosecond–picosecond range are almost exclusively achieved by means of mechanical translation stages. As shown in Fig. 2, both the pump and probe beams are obtained from the same laser source and the delay between pump and probe is adjusted by moving a mechanical delay stage placed on one of the optical paths: a translation of the mirrors by $D = 1$ μm yields a $2 * D/c = 6.7$ fs delay (c stands for the speed of light in vacuum). As will be discussed later, the temporal resolution depends on the laser pulse duration, electron emission characteristics (initial energy spread, number of electrons per pulse), and propagation inside the TEM (acceleration length, acceleration voltage, etc.) (Gahlmann, Park, & Zewail, 2008). Investigating dynamics on timescales longer than a few nanoseconds with mechanical translation stages poses severe constraints on the stability of the optical alignment. By getting rid of the mechanical translation stage, the technique of asynchronous optical sampling which relies on two laser sources with slightly offset repetition rates can be an interesting alternative (Bartels et al., 2007; Elzinga et al., 1987).

Time-resolved TEM experiments can be performed in two different modes. In the *single-shot* mode, only one electron pulse containing a sufficient number of electrons to get an exploitable signal (image, diffraction pattern, spectrum) is used to probe the sample with a delay with respect to the excitation of the sample by a single pump pulse (Fig. 3A). By generating a sequence of several electron pulses and displacing the image on the CCD camera with a high speed beam deflector synchronized with the laser, it is possible to acquire multiple frames at different delays with respect to the initiating optical pulse: this mode of operation is called *movie-mode DTEM* (Fig. 3B) (LaGrange, Reed, DeHope, Shuttlesworth, & Huete, 2011). In the *stroboscopic* mode, a very large number (typically $> 10^8$) of excitation-observation cycles are used to accumulate signal (Fig. 3C). This mode cannot be used to investigate irreversible processes. However, it enables TEM measurements with electron pulses containing only a few electrons per pulse. This so-called *single-electron regime* allows minimizing the

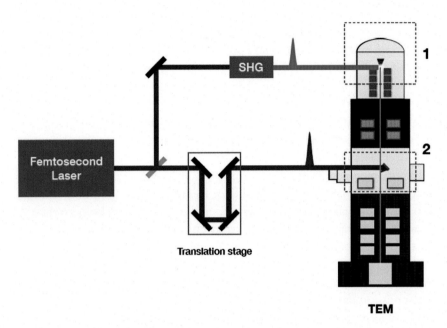

Figure 2 Principle of ultrafast pump-probe TEM experiments. The beam from a femtosecond laser source is split in two parts. The first part of the beam is sent to the electron source of the TEM where it triggers the emission of ultrashort electron pulses (1). Frequency conversion such as Second Harmonic Generation (SHG) is used to improve the emission efficiency. The second part of the beam is sent to the objective lens. It is focused on the sample and brings the sample out of equilibrium, initiating its dynamics (2). The pump-probe delay is adjusted by controlling the position of a mechanical translation stage.

Coulomb repulsion inside the electron pulse and brings the temporal resolution into the femtosecond range (Lobastov, Srinivasan, & Zewail, 2005). Time-resolved TEM in the single-shot mode is called *Dynamic Transmission Electron Microscopy (or high-speed TEM)*. Time-resolved TEM in the stroboscopic mode and single electron regime is called *Ultrafast TEM* due to its better spatio-temporal resolution.

It is evident from Fig. 2 that the conversion of a conventional TEM into a time-resolving instrument requires to modify two important parts: the *electron source* and the *objective lens*. Optical windows and focusing optics need to be added in the two regions to bring a laser beam inside the TEM column and focus it on the cathode (electron gun) or the sample (objective lens). In the case of movie-mode DTEM, it is also required to add a beam deflector to the TEM column synchronized with the laser system (LaGrange et al., 2011). Finally, it is highly desirable that a time-resolved

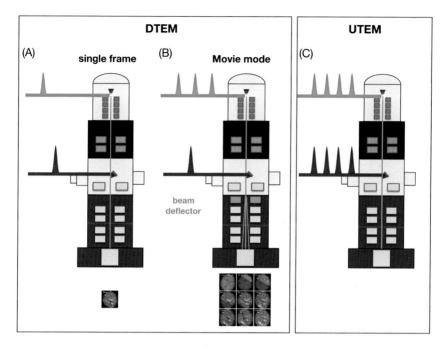

Figure 3 (A) Single-shot DTEM: a single electron pulse containing a large enough number of electrons (typically $>10^8$) is used to probe the sample after excitation by one optical pulse. (B) Movie-mode DTEM: A sequence of several electron pulses is used to probe the evolution of the sample at different delays after excitation by a single optical pulse. (C) Ultrafast TEM in stroboscopic mode: a large number of excitation/detection cycles (typically $>10^8$) is used to accumulate signal. In the single electron regime, each electron pulse only contains a few electrons therefore minimizing Coulomb repulsion and enabling TEM with sub-picosecond temporal resolution.

TEM can also be operated in conventional mode (*i.e.* with a continuous electron beam) for sample characterization prior to time-resolved experiments. Ultrafast electron sources are therefore derived from existing conventional DC electron sources. Both conventional and time-resolved TEMs have their potential largely dictated by the properties of their electron source. We therefore briefly summarize in the next section the different types of sources that can be found in conventional TEMs.

2.3 Electron Sources in Conventional TEMs

Three different types of electron sources can be found in Transmission Electron Microscopes. In a thermionic electron source, a strong electric current heats a tungsten, LaB_6, or CeB_6 cathode. At high temperature, the Fermi–

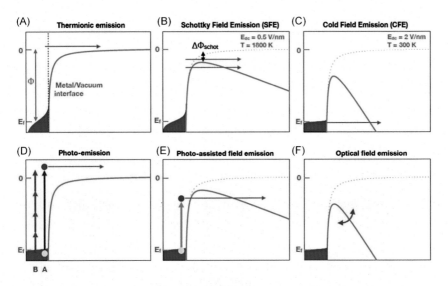

Figure 4 Conventional DC and laser-driven electron emission mechanisms. Upper row: conventional DC electron sources. Thermionic (A), Schottky Field Emission (SFE) (B), and Cold Field Emission (CFE) (C). Typical values of the DC electric field applied to the cathode and the operating temperature are given in inset. Lower row: laser-driven electron emission mechanisms. (D) One-photon photoemission (A) and multiphoton photoemission (B). (E) Photo-assisted field emission. (F) Optical field emission.

Dirac electron distribution includes an appreciable fraction of populated electronic states with energies above the vacuum level (Fig. 4A). These electrons are emitted in vacuum and then focused in a real cross–over using a Wehnelt assembly. In cold-field emission (CFE) electron sources, no heating of the cathode is required. An extraction voltage is instead applied between a conical tip, in general made of monocrystalline tungsten, and an extracting anode. This voltage, enhanced by the lightning rod effect at the tip apex lowers the potential in vacuum enough to allow for efficient tunneling of electrons from the Fermi level out of the tip (Fig. 4C). The very small size of the emitting zone is at the origin of the small virtual source size and therefore the high brightness of field–emission electron sources. As shown in Fig. 4B, Schottky field-emission guns (SFEG) are intermediate and use a field assisted thermionic emission process from a zirconia coated tungsten cathode (ZrO/W[100]). The DC electric field applied on the cathode lowers the effective material work function enabling electron emission at a lower temperature. Typical values of the DC electric field ap-

Table 1 Properties of the most common electron sources used in conventional TEMs

	W Thermionic	LaB$_6$ Thermionic	ZrO/W[100] SFEG	W[310] CFEG
Exit work function (eV)	4.5	2.4	2.92	4.25
Angular current density (μA/Sr)			100–200	10
Energy spread (eV)	1.5–3	1–2	0.7–1	0.2–0.4
Brightness (A/cm$^2 \cdot$ Sr)	$\approx 10^5$	$\approx 10^6$	10^7–10^8	10^9–10^{10}
Source size (nm)	20–$50 \cdot 10^3$	10–$20 \cdot 10^3$	10–30	2–5
Emission decay	No decay	No decay	No decay	<20% of probe current in 2 hours
Temperature (K)	2700	1700	1800	300
Vacuum pressure (Pa)	$\approx 10^{-2}$	$\approx 10^{-4}$	$\approx 10^{-8}$	$\approx 10^{-9}$
Probe current noise	<1%	<1%	<1%	<2%
Surface current density (A/cm^2)	≈ 3	≈ 30	$\approx 10^4$–10^5	$\approx 10^4$–10^6
Total emission current	$\approx$mA	$\approx$mA	$\approx$100 μA	$\approx$10 μA

plied to SFEGs and CFEGs and the temperature of the cathode are given in Fig. 4.

The most important properties of the different DC electron sources are listed in Table 1. Whereas thermionic sources yield large and very stable currents, cold field emission sources are clearly superior in terms of brightness, virtual source size, and energy spread. The latter are privileged for applications demanding high brightness and spatial coherence such as electron holography (see Section 5).

2.4 Generation of Ultrashort Electron Pulses

The high voltage required for cold field emission rules out electronic triggering of the electron emission using time-modulated voltage at high repetition rates. Similarly, the time required for temperature stabilization prevents a fast modulation of the current by thermionic or Schottky field-emission guns. Instead, it has soon been realized that driving a cathode with

laser pulses can generate electron pulses with durations in the femtosecond–
nanosecond range. We summarize in the following the processes by which
electrons can be emitted from an optically excited material.

As was first explained by A. Einstein, the absorption of one photon of
energy E_{ph} larger than the work function can trigger the emission of electrons from a materials surface (case A in Fig. 4D). In this case the emitted
current scales linearly with the light intensity. One photon photoemission is
no more possible when the photon energy is smaller than the material work
function. Instead, electron emission can arise from two different mechanisms. The dominant mechanism at low laser intensities is multiphoton
photoemission, which consists in the absorption of n photons so that the
total absorbed energy is larger than the material work function and allows
electron emission into vacuum. This process is depicted in Fig. 4D (case B)
in the case of four-photon photoemission. In the multiphoton regime, the
laser field can be considered as a small perturbation of the system. The
emission probability can be calculated using time-dependent perturbation
theory and scales as the nth power of the incident laser intensity. For larger
incident laser intensities, the laser field cannot be considered as a small perturbation anymore. Instead, it can directly modulate the potential barrier.
During a small fraction of the optical cycle, the potential barrier becomes
thin enough to allow direct tunneling from the Fermi level. The latter
process, called optical field emission, is schematized in Fig. 4F. It is clear
that optical field emission will dominate if the electrons can tunnel though
the potential barrier on a time scale shorter than the period of the laser
field. Therefore, the multiphoton and optical field emission regime can be
discriminated based on the ratio between the electron tunneling time and
the optical period. The latter is called Keldish parameter and can be writ-
ten as $\gamma = \frac{\omega\sqrt{2m\Phi}}{eE}$ in which ω is the laser circular frequency, m and e the
electron's mass and charge, E the laser peak electric field, and Φ the exit
work function of the cathode. Multiphoton photoemission is the dominant
emission mechanism for $\gamma \gg 1$ whereas optical field emission dominates
when $\gamma \ll 1$. When the material is heated or an additional static electric
field is applied to the cathode, then the emission mechanism is a combination of the previously cited mechanisms. For instance, in photo-assisted
field emission, an electron is first excited to a free electronic state below the
potential barrier by the absorption of one photon before tunneling through
the potential barrier (see Fig. 4E).

Electron emission from metallic nanotips similar to the ones used in
Schottky and cold-field emission guns (i.e. with apex radii smaller than one

micron) has first been demonstrated using nanosecond pulses (Hernandez Garcia & Brau, 2002). In 2006–2007, the generation of femtosecond electrons from nanoemitters triggered by femtosecond lasers could be demonstrated (Hommelhoff, Sortais, Aghajani-Talesh, & Kasevich, 2006; Hommelhoff, Kealhofer, & Kasevich, 2006; Ropers, Solli, Schulz, Lienau, & Elsaesser, 2007). Due to the enhancement of the laser electric field at the apex, it was possible to trigger electron emission from a small region and investigate light–matter interaction in strong optical fields (Bormann, Gulde, Weismann, Yalunin, & Ropers, 2010; Kruger, Schenk, & Hommelhoff, 2011; Schenk, Krüger, & Hommelhoff, 2010). The confinement of the emission region to strong field regions at the tip apex further yielded a brightness in the laser-driven mode similar to the conventional DC mode (Ehberger et al., 2015). The implementation of such laser-driven nanoemitters as a source for Ultrafast TEMs is discussed in Section 3 of this chapter.

However, the high spatio-temporal confinement of electron emission from laser-driven nanoemitters places an upper limit N_{max} on the number of electrons that can be emitted below the tip damage threshold. Whereas flat photocathodes can yield electron bunches having more than 10^9 electrons per pulse in the probe, the maximum number of electrons which can be photoemitted by a metallic nanotip with a single optical pulse lies typically within the range 1–1000 depending on the tip material, laser power, pulse duration, wavelength, and repetition rate. As will be further discussed in Section 3, this establishes a clear difference between UTEMs based on flat photocathodes and capable of operation both in the single pulse and stroboscopic modes and high brightness laser-driven SFEGs or CFEGs that can only be run in the stroboscopic mode. Furthermore, at 1 MHz repetition rate, the number of electrons extracted from a metallic nanotip by a laser pulse translates into a maximum current in the 0.16–160 pA range. This estimation shows that UTEMs based on laser-driven nanoemitters operate in low-dose and that useful data can only be obtained by accumulating electrons from a sufficiently large number of pulses. This comes with long acquisition times and therefore demands a careful optimization of the mechanical and optical stability of the instrument.

In Section 3, we review in chronological order the major instrumental developments in the field of time-resolved Transmission Electron Microscopy. Their applications to different fields of nanophysics will be the subject of Section 4.

3. TIME-RESOLVED TRANSMISSION ELECTRON MICROSCOPY: HISTORICAL DEVELOPMENT, INSTRUMENTAL ASPECTS

3.1 Introduction

The history of time-resolved TEM began in the Technical University of Berlin in the 1980s under the leadership of O. Bostanjoglo who developed several time-resolved electron microscopes. The Dynamic Transmission Electron Microscope (DTEM) developed in Berlin was later improved in the Lawrence Livermore National Laboratory. A major breakthrough was achieved in 2005 in the group of Nobel laureate A. Zewail at the California Institute of Technology. By using electron pulses containing only a few electrons each, Zewail and coworkers got rid of the interparticle Coulomb repulsion within the pulses that deteriorated the spatio-temporal resolution of the DTEM. Combining atomic scale spatial resolution and femtosecond time resolution, the Ultrafast Transmission Electron Microscope (UTEM) opened a completely new window on the ultrafast dynamics at the nanoscale. Recently, a lot of efforts have been put in the development of UTEMs based on high brightness ultrafast electron sources and the optimization of the temporal and spatial resolution of DTEM.

3.2 Oleg Bostanjoglo, Pioneer of High-Speed Transmission Electron Microscopy

Several modifications of a TEM aiming at gaining information in the time domain were carried out in the Technical University of Berlin before the development of the Dynamic TEM shown in Fig. 5. By using a parallel plate beam deflector synchronized with an ultrasonic oscillator located in a specimen holder, Bostanjoglo and coworkers managed to observe stroboscopically by Lorentz microscopy the magnetization dynamics induced by the mechanical stress in single and polycrystalline Ni and NiFe films (Bostanjoglo & Rosin, 1977). Electron micrographs and diffraction patterns of irreversible processes were also acquired after adding a pulsed microchannel plate and a pulsed electron beam shutter to a commercial TEM (Bostanjoglo, Tornow, & Tornow, 1987).

Four different electron microscopes, a Transmission Electron Microscope, a photoelectron microscope (Bostanjoglo & Weingartner, 1997; Weingartner & Bostanjoglo, 1998), a high energy reflection electron microscope (Bostanjoglo & Heinricht, 1990), and a mirror electron microscope

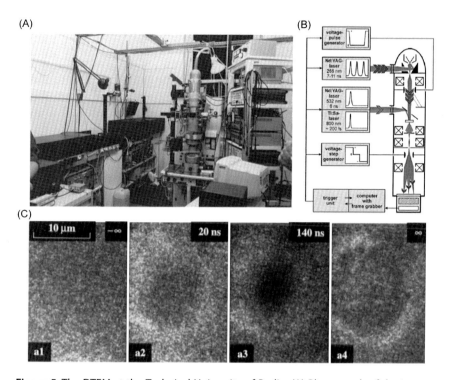

Figure 5 The DTEM at the Technical University of Berlin. (A) Photograph of the instrument. The electron microscope is a Siemens Elmiskop 1A (reprinted with permission from King, Armstrong, Bostanjoglo, & Reed, 2007). (B) Schematic diagram of the instrument (see description in the main text) (reprinted with permission from King et al., 2007). (C) Short-exposure images of a flow in a laser induced melt in an amorphous $Ni_{0.8}P_{0.2}$ film (60 nm). Exposure time was 10 ns (reprinted with permission from Bostanjoglo, 1999).

(Kleinschmidt & Bostanjoglo, 2001) were adapted for time-resolved investigations in the TU Berlin. We focus in the following on the high-speed (or dynamic) TEM (Bostanjoglo and Nink, 1997, 1996; Nink, Galbert, Mao, & Bostanjoglo, 1999). The instrument is shown in Fig. 5A. It was based on a commercial Siemens Elmiskop 1A. The laser triggering the electron emission was a Q-switched twice frequency doubled Nd:YAG laser (266 nm output wavelength). As detailed in Fig. 5B, an aluminum mirror located in the anode redirected the laser beam to the cathode tip. Focusing of the laser beam on the cathode was achieved by means of an external lens down to a spot of 20 μm ($1/e^2$ diameter). Peak currents of several milliamperes were photoemitted into a half angle of $7 \cdot 10^{-3}$ rad. Up to three frames could be recorded at different delays after the excitation of the sample by shifting the image on the camera with a beam deflector. A rectangular aperture in the intermediate lens image plane was used to avoid overlap between consecutive frames on the detector. A second Q-switched Nd:YAG laser was used to initiate the dynamics of the sample. It was synchronized with the laser source generating the electron pulses and frequency doubled to emit 5 ns or 15 ns pulses at 532 nm. A spatial resolution of $\sim$200 nm with a pulse duration <10 ns was demonstrated (Domer & Bostanjoglo, 2003).

The applications of this instrument have been extensively reviewed in several articles (Bostanjoglo, 1999, 1989; Domer & Bostanjoglo, 2003; Bostanjoglo, Elschner, Mao, Nink, & Weingartner, 2000). Special emphasis was put on non-repetitive processes related to materials processing by laser pulses such as melting, melt flow, crystalline and noncrystalline solidification, and thermal evaporation. For instance, the ablation of metal films by laser pulses (Bostanjoglo, Niedrig, & Wedel, 1994) and the laser-induced vaporization of metallic films (Bostanjoglo, Kornitzky, & Tornow, 1991) were investigated. Fig. 5C shows an example of the time-resolved imaging of a laser induced flow in an amorphous film. In these seminal studies, the low spatial and temporal resolutions were due to Coulombic repulsion in the electron pulses (space-charge effects) and to the electronics used for the electron beam deflection. It is important here to stress the fact that the developments achieved by O. Bostanjoglo and coworkers were based on vacuum and laser technologies that were far from the level of versatility that they have reached today.

3.3 Dynamic Transmission Electron Microscopy at the Lawrence Livermore National Laboratory – Movie-Mode DTEM

Following the pioneering work of O. Bostanjoglo, researchers from the Lawrence Livermore National Laboratory (LLNL) in the U.S. have improved the DTEM technology by optimizing its illumination optics and expanding its potential to acquire multiple frames (LaGrange et al., 2011; LaGrange, Reed, & Masiel, 2015).

As detailed earlier, a DTEM is well adapted to study the dynamics of irreversible phenomena with photoelectron pulses containing about $\sim$10^8–10^{10} electrons packed in $\sim$10 ns. This corresponds to a peak current in the mA regime compared to the μA of standard TEM operation mode. A conventional TEM column involves an illumination system fitted with two condenser lenses to demagnify the gun crossover located after the wehnelt in a thermoionic gun assembly. In such a design a fixed aperture stops most of the electrons after the accelerator, and a second aperture selects the paraxial electrons to be focused on the sample, removing all the electrons that suffer from condenser aberrations. In DTEM, this loss of current in the illumination system cannot be compensated by an increase of the photoemitted current without degrading the spatio-temporal resolution of the instrument. To overcome this problem, LLNL has developed a new illumination optics system (Reed et al., 2010). An extra condenser lens, called C_0, has been introduced below the accelerator and above the C_1 lens, followed by a long drift tube of approximately 20 cm with additional magnetic deflectors. The C_0 lens has a long focal length ($\approx$15 cm) to minimize the effect of geometric aberrations. Thanks to the additional C_0, the electron beam size is made very small in the C_1 lens (less than 1 mm) minimizing the contribution of the C_1 lens aberrations. It has been further shown that this additional C_0 lens does not degrade significantly the emittance, the latter being governed by the source size inside the thermionic-based electron gun. With this modified illumination system, the LLNL DTEM has succeeded to increase by a factor of 20 the current delivered in a single electron pulse (Reed et al., 2010).

The movie-mode DTEM developed at LLNL is depicted in Fig. 6A. It is based on a JEOL 2000FX (200 kV) TEM. The TEM has been modified so that both the specimen and the cathode can be driven by nanosecond laser pulses. The TEM can be used in standard mode for alignment purposes or as a DTEM. Electron emission is triggered by a laser source based on an arbitrary-waveform generator (AWG) which strikes a photocathode. The AWG temporally shapes the output of a continuous-wave fiber-laser seed with an electro-optical modulator. The modulated laser beam is then amplified in fiber and neodymium:yttrium aluminum garnet rod amplifiers and frequency-converted to ultraviolet. The AWG generates a train of electron pulses with durations ranging from 10 ns to 1 μs and delivered in any desired pattern within a window of 100 μs. A second laser focused on the sample triggers the process under scrutiny. The individual images obtained from up to 16 electron pulses at specific delays after excitation are deflected on different regions of the CCD camera using a high-speed deflector. With a separation between frames smaller than 100 ns, a sequence of images can be acquired in about 1 μs. Fig. 6B shows an example of a sequence of nine images of a moving reaction front in a 2Ti–3B nanolaminate film (LaGrange et al., 2015).

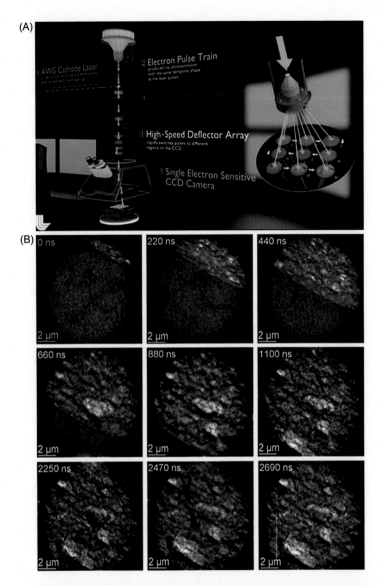

Figure 6 (A) Schematic of the Movie-Mode Dynamic Transmission Electron Microscope developed at the Lawrence Livermore National Laboratory. (B) Sequence of images of a propagating front in a reacting 2Ti–3B reactive nanolaminate film. The exposure time of each image was 17 ns. Reprinted with permission from LaGrange et al. (2015).

Movie-mode DTEM enables the observation of irreversible processes that are too fast to be captured by conventional microscopy such as the creation, motion, and interaction of individual defects, phase fronts, and chemical reaction fronts in nanoscale and microscale materials. For instance, the crystallization of phase change materials used for optical and resistive memory was studied by movie-mode DTEM: movies of laser-induced crystallization were acquired from which crystal growth rates could be measured (Santala et al., 2013). Fig. 6B gives another example in which time-resolved images of the reaction dynamics in Ti–B-based reactive nanolaminates were acquired (LaGrange et al., 2015).

3.4 Time-Resolved TEM with Atomic Scale Spatial Resolution and Femtosecond Time Resolution: The Development of Ultrafast Transmission Electron Microscopy at the California Institute of Technology

Dynamic Transmission Electron Microscopy involves electron pulses containing a large number of electrons ($>10^8$ typically) and its spatio-temporal resolution is therefore deteriorated by the Coulomb repulsion between the electrons inside each electron pulse. A major breakthrough was made in the mid-2000s at the California Institute of Technology. Inspired by their previous work in ultrafast optical spectroscopy and ultrafast electron diffraction (Lobastov et al., 2004), Zewail and coworkers developed a different kind of time-resolved TEM called Ultrafast Transmission Electron Microscope (UTEM) with a temporal resolution improved by nearly 6 orders of magnitude compared to the DTEM. In UTEM, the images, diffraction patterns, and spectra are constructed stroboscopically from billions of excitation-detection cycles, each electron pulse containing only a few electrons. The Coulomb repulsion that deteriorated the spatio-temporal resolution in DTEM is therefore minimized (Gahlmann et al., 2008) and the temporal resolution reaches the limit set by the optical pulse duration. As discussed previously, the inevitable counterpart is that UTEM cannot be used to investigate irreversible processes.

The first generation ultrafast Transmission Electron Microscope developed at Caltech, called UEM1, was based on a 120-kV G2 TWIN Tecnai TEM (FEI) (Lobastov et al., 2005). Two optical windows had been added to the TEM, one toward the photocathode and another near the sample. The electron gun assembly was a conical lanthanum hexaboride (LaB_6) cathode with a flattened surface of 300 μm at the apex. The laser system was a diode-pumped mode-locked Ti:sapphire laser oscillator, yielding sub-100-fs pulses at 800 nm with a repetition rate of 80 MHz and an average power of 1 W. Part of the beam was used to excite the sample, its power being precisely controlled by a half-wave plate ($\lambda/2$) and a Glan–Thomson polarizer. The second part of the laser beam was frequency doubled in a β-BaB$_2$O$_4$ (BBO) nonlinear crystal to yield 400-nm femtosecond pulses, redirected with computer-controlled mirrors and focused to a ~50 μm spot on the cathode. The electron source could also be operated in conventional mode by heating the cathode for sample characterization and microscope alignment. UEM1 was first used to obtain sequences of images and diffractions patterns to investigate with ultrashort temporal resolution the metal–insulator phase transition in vanadium dioxide (VO$_2$) (Grinolds, Lobastov, Weissenrieder, & Zewail, 2006) and the opening and closing of a channel in a crystalline quasi-one-dimensional semiconductor (Flannigan, Lobastov, & Zewail, 2007).

Atomic-scale information with UEM1 could only be obtained in diffraction mode. The first atomic scale resolution in real-space imaging was obtained with the second generation UEM developed at Caltech (Park, Baskin, Kwon, & Zewail, 2007). This instrument, called UEM2, is shown in Fig. 7A (Barwick, Park, Kwon, Baskin, & Zewail, 2008). It has greatly improved capabilities compared to UEM1. It is operated at 200 kV ($\lambda_{\text{de Broglie}} = 2.51$ pm) and can also be used as a Scanning Transmission Electron Microscope. Most importantly, it has been designed to be operated either as a DTEM in the single-shot mode or as a UTEM in the single-electron regime like UEM1. The instrument is based on a 200 kV TEM (Tecnai 20, FEI) modified with two optical windows, one to the photocathode and the other to the specimen. The electron source is based on a field emission gun (FEG) assembly with a lanthanum hexaboride (LaB$_6$) filament installed between suppressor and extractor electrodes. The cathode is truncated at the tip with a flat 16 μm wide area. The gun can be operated either as a thermal emission source for sample characterization or a photoemission source for time-resolved investigations.

UEM2 is equipped with a versatile laser system consisting in a diode-pumped Yb-doped fiber oscillator/amplifier, yielding ultrashort pulses of up to 10 μJ at 1030 nm with both variable repetition rate (200 kHz to 25 MHz) and pulse duration (200 fs to 10 ps). For efficient electron generation and excitation of the sample, the 1030 nm pulses are first frequency doubled (515 nm) and then tripled (343 nm) in two successive nonlinear crystals. The 2ω and 3ω harmonics are then separated from the IR fundamental beam by dichroic mirrors. The third harmonic is sent to the photocathode inside the electron gun to trigger the emission of electrons by one-photon photoemission (see Fig. 7B). As for UEM1, a fine-steering mirror mounted outside the electron gun and controlled by a computer is

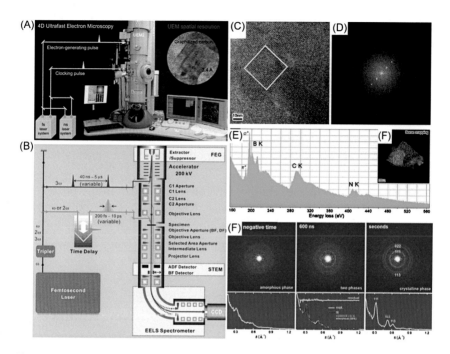

Figure 7 (A) The second generation Ultrafast Transmission Electron Microscope at the California Institute of Technology (reprinted with permission from Barwick et al., 2008). (B) Schematic of the instrument. (C) High-resolution UEM image of an organometallic crystal of chlorinated copper phthalocyanine. (D) Associated fast Fourier transform (FFT). (E) EELS spectrum of a Boron Nitride (BN) sample. Reprinted with permission from Park et al. (2007). © 2007 American Chemical Society. (F) Diffraction patterns representing three phases: before (left), during (center), and after (right) the transformation, taken from the same sample area of irradiation. Reprinted with permission from Kwon, Barwick, Park, Baskin, and Zewail (2008a).

used to align the laser beam on the photocathode. The laser beam is focused on a 30 μm spot on the cathode. Either the fundamental IR or second harmonic can be used to excite the sample inside the objective lens. The spot size on the sample is ∼30 μm. The time delay between the optical pump pulse and the electron probe pulse is adjusted by a computer-controlled mechanical translation stage.

When operated in the ultrafast mode, UEM2 delivers typically 1–100 electrons per pulse therefore getting rid of the Coulomb repulsion between particles, which temporally and spatially broadens electron pulses in DTEM. This *single electron imaging* mode allows imaging with atomic scale spatial resolution (see Fig. 7C, D) at a temporal resolution only limited by the duration of the ultrashort optical pulse. UEM2 is equipped with a Gatan Imaging Filter enabling time-resolved spectroscopic studies (Carbone, Barwick, et al., 2009; Carbone, Kwon, & Zewail, 2009; van der Veen, Penfold, & Zewail, 2015) (see Fig. 7E). As previously detailed, one of the key advantages of the UEM2 is the possibility of operating the microscope both in the single electron imaging or single shot mode. When operated in the single-shot mode, the temporal delay can be controlled with electronic synchronization and be made arbitrarily long. For instance, UEM2 was used in the single shot mode to investigate the crystallization of silicon (Kwon et al., 2008a) (see Fig. 7F). It has also been used in the single-shot mode to investigate the dynamics of the martensitic phase-transformation of iron (Park, Baskin, Barwick, Kwon, & Zewail, 2009) and the transformation of amorphous titanium dioxide nanofilms from the liquid state to the ordered crystal phase (Yoo, Kwon, Liu, Tang, & Zewail, 2015).

3.5 UTEM with High-Brightness Electron Sources: Development of a UTEM Based on a Laser-Driven Schottky Field Emission Gun in Göttingen

In conventional TEM, it is well known that electron holography, coherent nanodiffraction, or high resolution STEM experiments are extremely challenging (or nearly impossible) with a thermionic electron source but rather demand a Schottky field-emission gun (SFEG) or a cold-field emission gun (CFEG) due to their improved brightness. Similarly, whereas UTEM enabled a spectacular improvement in spatio-temporal resolution, the poor brightness of their flat photocathodes makes them inadequate for demanding applications such as ultrafast electron holography for instance.

An alternative to the flat photocathodes implemented in UTEMs appeared one decade ago when the emission of electrons from metallic nanotips similar to the ones implemented in SFEG or CFEG could be triggered using a low power, high repetition rate, ultrashort laser beam focused on the tip apex (Hommelhoff, Sortais, et al., 2006; Hommelhoff, Kealhofer, et al., 2006; Ropers et al., 2007). Due to the lightning rod effect and the possible additional contribution from plasmonic resonances in the case of noble metal nanotips, a strong enhancement of the incident laser electric field occurs at the tip apex. This enhancement of the electric field enables the investigation of processes specific of light–matter interaction in strong optical fields (Schenk et al., 2010; Bormann et al., 2010; Kruger et al., 2011; Sivis, Duwe, Abel, & Ropers, 2013). Most importantly, electron emission is confined to strong field regions at the tip apex leading to a brightness

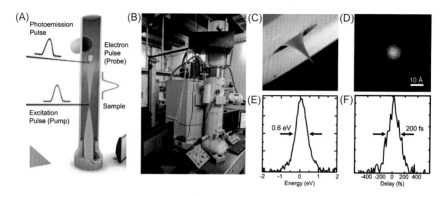

Figure 8 Schematic (A) and picture (B) of the Ultrafast Transmission Electron Microscope based on a femtosecond laser-driven SFEG developed in the University of Göttingen. (B) Schematic of the instrument. (C) SEM image of the tip assembly. (D) Focused ultra-short electron probe. (E) Energy resolution and (F) temporal resolution. Reprinted with permission from Feist et al. (2017).

in the laser-driven mode which is similar to the conventional DC mode (Ehberger et al., 2015).

Laser-driven nanoemitters in SFEGs have first been implemented in Scanning Electron Microscopes (Yang, Mohammed, & Zewail, 2010; Mohammed, Yang, Pal, & Zewail, 2011). The first implementation of a laser-driven SFEG in a Transmission Electron Microscope has been achieved in the University of Göttingen (C. Ropers, S. Schäfer). Motivated by their research activity on ultrafast photoemission from nanotips, the development of the ultrafast TEM in Göttingen began in 2009. The TEM, a JEOL 2100-FEG TEM with a Schottky electron gun, was installed in 2012 and 2013 followed by the modifications of the instrument to trigger electron emission with a femtosecond laser and numerical simulations of its performances (Paarmann et al., 2012). The first pulsed operation of the TEM was obtained in 2014 (Bormann, Strauch, Schäfer, & Ropers, 2015).

The Göttingen UTEM is a 200 kV TEM (JEOL 2100-FEG) equipped with a laser-driven Schottky emitter (Fig. 8). The Schottky field-emitter is a [100]-oriented monocrystalline W tip coated with a zirconia layer (W[100]/ZrO) widely used in conventional TEM (Swanson & Schwind, 2009). In conventional mode, the emitter is heated at a temperature of about 1800 K. The high temperature of the tip melts a droplet of zirconia coming from the reservoir resulting in the creation of a ZrO overlayer on the (100) front facet of the W tip. This additional layer reduces the work function down to 2.9 eV. Extraction fields of about 0.5–1 V/nm further reduce the work function by the Schottky effect. Emission currents of several tens μA can be obtained in the conventional mode enabling *in-situ* characterization of the sample before and after time-resolved experiments (Feist et al., 2017).

The laser system involves a regenerative Ti:sapphire amplifier delivering pulses of about 10 nJ at a tunable repetition rate of up to 800 kHz. Part of the output is frequency doubled in a β-BaB$_2$O$_4$ (BBO) nonlinear crystal and the 400-nm femtosecond pulses enter the emitter region through an optical window on the side of the electron gun. This optical access serves routinely to inspect the electron emission pattern. The second harmonic is then redirected to the tip inside the vacuum chamber of the electron source. It is focused from outside the TEM column on the cathode in a spot diameter of about 20 μm full-width-at-half-maximum (FWHM).

In the photoemission mode, the temperature of the nanotip is reduced below 1400 K to avoid static emission. Electrons are generated with the laser focused at the tip apex by one-photon photoemission from the W[100]/ZrO tip. Emission is restricted to the front facet of the emitter by chemical selectivity, the front facet having a lower work function compared to its neighbors. The tip is located in the electrostatic suppressor–extractor electrodes assembly usually encountered in SFEG sources. Experimental and numerical investigations of the electron beam properties as a function of the extractor, suppressor, and tip voltages (respectively U_{ext}, U_{sup}, U_{tip}) have shown that either a high electron yield of a high beam coherence can be selected depending on the value of a dimensionless parameter $\Gamma = (U_{tip} - U_{sup})/(U_{ext} - U_{sup})$ (Bormann et al., 2015). The linear dependence of the photoemitted current with the laser power confirmed the single-photon photoemission process (Feist et al., 2017). A peak brightness better than photocathode-based UTEMs was reported. Switching from the ultrafast to the conventional DC mode is achieved by increasing the tip temperature until the thermal emission regime of the Schottky emitter. This switching requires less than 1 h. The optical excitation of the sample can be done in two configurations: either at an angle of incidence of 55° relative to the electron beam (through the EDX port) or nearly parallel to the electron beam (through the line tube of the objective lens pole piece). The optical focal spot sizes achieved on the sample in the two configurations are about 50 μm (fwhm using 800 nm laser wavelength). Pump-probe delay, optical fluence, polarization state, and spot position are computer-controlled. Using linear photoemission from a Schottky emitter, a 9 Å focused beam diameter, 200 fs pulse duration, and 0.6 eV energy width could be demonstrated and the potential of the SFEG–UTEM for imaging, diffraction, and spectroscopy has been explored. The Göttingen group started in 2014 different time-resolved studies among which are the coherent interaction of free electrons with optical near-fields, optically induced magnetization dynamics, and the dynamics of structural and electronic phase transitions. These applications are discussed in Section 4 of this chapter.

3.6 UTEM with High-Brightness Electron Sources: Development of a UTEM Based on a Laser-Driven Cold-Field Emission Gun in Toulouse

As described in Section 2.3, cold field emission guns (CFEG) use metallic nanotips having an apex radius of the order of 100 nm and provide the highest brightness and the smallest energy spread of all sources implemented in conventional TEMs. Therefore, an ultrafast electron source based on a CFEG triggered by femtosecond laser pulses is expected to yield the ultrashort coherent electron pulses required for future ultrafast electron holography (Ehberger et al., 2015). Started in 2010 in CEMES-CNRS (Toulouse, France), the FemtoTEM project (A. Arbouet, F. Houdellier) aims at developing a UTEM based on a cold-field emission gun (Caruso, Houdellier, Abeilhou, & Arbouet, 2017; Houdellier, Caruso, Weber, Kociak, & Arbouet, 2018).

In the case of flat photocathodes, the size of the electron emission region is given by the laser focal spot size on the cathode ($\sim$20–30 µm). The poor brightness of these electron sources is a direct consequence of the large size of the emission region. Contrary to Schottky field emission guns, conventional cold field emission guns do not need a suppressor anode. Therefore the control of the electron beam properties through the Γ voltage ratio as in the laser-driven SFEG cannot be used in a CFEG. As a consequence, the optimization of the brightness of a laser-driven CFEG requires minimizing the region of the cathode excited by the laser beam using tight focusing with short focal distance optics. However, the addition of optical and optomechanical components such as mirrors and mirror holder (MH) close to the nanotip must carefully address issues related to vacuum and high voltage. In particular, the choice of the materials inserted close to the tip and their surface quality are critical. Finally, it is highly desirable that the modified electron source can be easily switched between the conventional DC mode for microscope alignment and routine sample characterization, and the laser-driven ultrafast mode for time-resolved experiments. Finally,

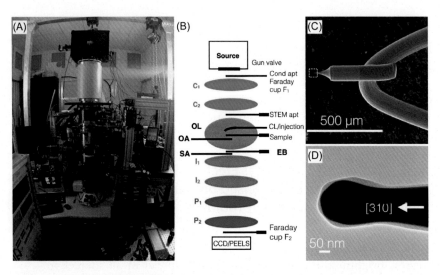

Figure 9 (A) Picture of the modified HF2000. The black box on top is the optical head designed to steer a femtosecond laser beam onto the apex of the field emission tip inside the electron gun. (B) Schematic of the TEM column. C_1 and C_2: first and second condenser lenses. OL: objective lens. OA: objective aperture. EB: electron biprism. SA: selected-area aperture. I_1 and I_2: intermediate lenses. P_1 and P_2: projector lenses. (C) Scanning Electron Micrograph (SEM) of a commercial HHT [310] oriented tungsten nanotip. (D) TEM image showing the (310) crystal facet of the tungsten tip (reprinted with permission from Houdellier et al., 2018).

these modifications must not deteriorate the main properties of the CFEG such as virtual source size, angular current density, etc.

The CFEG-UTEM developed in CEMES is based on a *Hitachi High Technologies* (HHT) HF2000 TEM fitted with a 200 kV cold field emission gun (Isakozawa et al., 1989). Several modifications are visible in Fig. 9A. The black box added on top of the TEM is called the *optical head*. It gathers the optical and opto–mechanical components required to (i) align a femtosecond laser beam on the nanotip inside the electron gun and (ii) control the laser power and polarization. The optical head is rigidly fixed onto a metallic housing called *gun-housing*. The latter is filled with SF_6 to electrically insulate the electron gun from the earth potential. Part of the optical set–up (mechanical delay stage) is visible on the optical table in the bottom right of Fig. 9A. Fig. 9B gives the details of the TEM column. The TEM column is basically unchanged below the extraction anode (located before the accelerating tube) compared to the original HF2000 except in the objective lens. The latter has been modified to include a light injec-

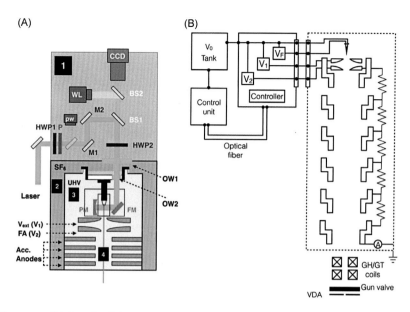

Figure 10 (A) Details of the optical head. HWP: half-wave plate. BS: beam-splitter. WL: white-light source. OW: optical window. FM: flat mirror. PM: parabolic mirror. (B) High voltage configuration of the HHT HF2000 TEM electron gun. V_0 is the acceleration voltage, V_1 the extraction voltage, V_2 the Focusing Anode (FA) voltage, and V_F the voltage applied to the tip during the flash-cleaning. GH: gun horizontal. GT: gun tilt. VDA = vacuum differential aperture. The anodes below V_2 are located inside the accelerating tube and the dotted rectangle stands for the gun-housing grounded at the earth potential. Reprinted with permission from Houdellier et al. (2018).

tion/cathodoluminescence system based on a parabolic mirror mounted on a XYZ stage (Kociak & Zagonel, 2017). The mirror is placed above the sample inside the objective lens pole piece gap and has a hole to let the electron beam go through. An inspection CCD camera allows monitoring the insertion of the parabolic mirror inside the objective lens and avoid mechanical interferences. The UTEM is equipped with one electrostatic Möellensted biprism above the intermediate 1 lens for electron holography, a Gatan 4kx4k USC1000 camera, and an electron energy loss spectrometer Gatan PEELS666. The ultrafast TEM developed in Toulouse used a standard monocrystalline [310] oriented tungsten cathode. A SEM and a TEM image of these nanoemitters are given respectively in Fig. 9C and D.

The laser system is based on a compact ultrafast fiber laser, delivering ultrashort (350 fs), high energy (up to 20 μJ) pulses at 1030 nm with a repetition rate tunable between single–shot and 2 MHz. Electron emission is triggered by the femtosecond pulses generated at 515 nm by Second Harmonic Generation (SHG) of the laser output in a beta barium borate (BBO) crystal. The 515 nm laser beam used to trigger electron emission first goes through a telescope used to optimize its focusing on the nanotip apex and then enters the optical head. As shown in Fig. 10A, the laser beam goes first through an attenuator composed of a half-wave plate and a polarizer. It is then redirected and precisely aligned on the nanotip using two mirrors mounted on piezo-driven positioning mirror mounts. The polarization of the laser is then controlled by a second half-wave plate. The laser beam enters first the SF_6 region and then the UHV region through two consecutive optical windows. The laser is then reflected first on a planar mirror and finally on a parabolic one. The focal distance of the latter is $f = 8$ mm which yields a focal spot diameter ($1/e^2$ diameter) of ~ 3 μm on the tip apex. With typical incident laser powers of 5 mW at a 1 MHz repetition rate, a probe current in the 1–3 pA range is obtained. Further details on the short-term and long term stability in ultrafast mode as well as measurements of the brightness and angular current density can be found in Houdellier et al. (2018).

The high voltage configuration of the HF2000 electron source is described in Fig. 10B. The tip and the mirror holder system are set at the acceleration voltage V_0 (usually −200 kV) and face the extracting anode set at a voltage V_1 relatively to the tip. The tip–anode distance is adjusted in a range of about 7–10 mm. The first anode of the accelerating tube, called focusing anode or gun lens, is set at a potential V_2 relatively to the tip. The other anodes located in the accelerating tube are used to equally distribute the acceleration voltage between the focusing anode and the earth potential (around 33 kV per stage for 200 kV total acceleration voltage). The ratio $R = V_2/V_1$ between the focusing anode and the extracting anode can be adjusted between 2 and 13. The ratio is used to set the position of the full gun cross-over which may be either real (condition selected for TEM applications to maximize the beam intensity) or virtual (to maximize the first illumination lens demagnification for Scanning TEM (STEM) applications) (Mamishin, Kubo, Cours, Monthioux, & Houdellier, 2017). It is worth noting that, when the TEM is operated in ultrafast mode, the electron emission is triggered by the laser pulses and V_1 becomes an additional parameter adjusted to optimize the electron beam properties.

To assess the influence of the additional components in the vicinity of the field-emission tip, we have computed the axial geometric aberrations of the ultrafast electron source using the EOD (*Electron Optics Design*) software.

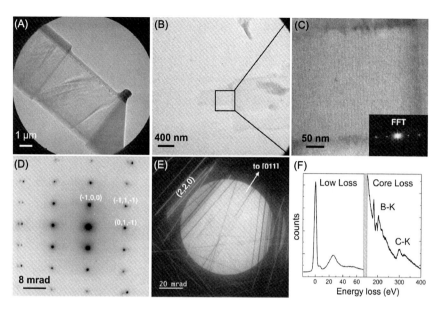

Figure 11 (A) Ultrafast conventional TEM image of a Si lamella (repetition rate $f_{rep} =$ 1 MHz, average number of electrons per pulse $N_e = 12.5$, integration time 100 s, Binning 1). (B) and (C) Ultrafast HREM image of a biological Catalase crystal ($f_{rep} = 2$ MHz, $N_e = 11$, 150 s, Binning 4). (D) Ultrafast SAED pattern taken along the [110] direction of a TiAl γ-phase crystal ($f_{rep} = 1$ MHz, $N_e = 12.5$, 150 s, Binning 2). (E) Ultrafast CBED pattern obtained near [110] direction of a Si crystal ($f_{rep} = 1$ MHz, $N_e = 12.5$, 100 s, Binning 1). (F) Ultrafast EELS spectrum of a boron nitride crystal ($f_{rep} = 2$ MHz, $N_e = 6.25$, low loss: 1 s/core loss: 1 min). Reprinted with permission from Houdellier et al. (2018).

Indeed, geometric aberrations affect the spot size formed by the gun and are one of the major brightness limiting factors of CFE sources. In real cross-over condition the spherical and chromatic aberration coefficients of the ultrafast CFEG are respectively $C_s \approx 60$ mm and $C_c \approx 12$ mm while in virtual cross-over condition they are respectively $C_s \approx 32$ mm and $C_c \approx 10$ mm which are in the same range as the ones of the original gun. To further confirm that the modifications of the electron gun have a negligible impact on the TEM performances, HREM images have been acquired in conventional DC mode on gold nanoparticles yielding an image lattice resolution of 0.2 nm at 200 kV of acceleration voltage.

The potential of the CFEG-UTEM for conventional and high-resolution imaging as well as electron diffraction in parallel (SAED, Selected Area Electron Diffraction) and convergent beam configuration (CBED), Electron Energy Loss Spectrometry (EELS) is illustrated in Fig. 11. The

UTEM was operated in ultrafast mode at an acceleration voltage of $V_0 = -150$ kV. An ultrafast conventional TEM image of a Si lamella and an ultrafast high-resolution image of a biological Catalase crystal are respectively shown in Fig. 11A and 11B. An ultrafast SAED pattern taken along the [110] direction of a TiAl γ-phase crystal and the ultrafast CBED pattern obtained near the [110] direction of a Si crystal are given in Fig. 11D and 11E. The ultrafast images and diffraction patterns obtained with ultrashort electron pulses (duration smaller than 400 fs at the sample) are of comparable quality as the ones obtained with the HF2000 in conventional mode. Importantly, the excitation of all lenses was the same in ultrafast and conventional mode as no modification was brought to the TEM electron optics. In ultrafast mode, a lattice image resolution of 0.9 nm was measured on a Crocidolite crystal at 150 kV under parallel illumination. The atomic structure of gold could not be resolved in ultrafast mode, probably because of mechanical instabilities on the high exposure time (150 s in this case) coming from the limited probe current. In convergent illumination mode, we achieved an ultimate spot size of 1.5 nm with 20 fA of probe current obtained for a $V_1 = 2$ kV and $R = 8$. The major difference between conventional and ultrafast modes lies in the level of emission and probe current. In standard DC mode the emission current is usually set around 10 µA, which can generate a probe current (in analysis mode using 30 µm STEM aperture) of approximately 100 pA. In ultrafast mode, the same electron optical condition ($V_1 = 4$ kV and $R = 5$) will yield an emission current around 2.5 pA for an incident average laser power of 6 mW and a repetition rate of 1 MHz. This is equivalent to approximately 15 electrons per pulse. The corresponding probe current is 80 fA. These values are mainly limited by the tip damage threshold and the laser repetition rate. Therefore, the exposure time has to be increased and the image resolution remains mainly limited by the microscope stability during the exposure time and the detector properties. Finally, ultrafast EELS has been performed on boron nitride sample. The features expected in the low-loss (bulk plasmon) and core-loss (boron and carbon K edge) regions could be observed. However, the spectra resolution remained limited by the electron energy spread. Indeed, depending on the number of electrons per pulse (from 1 to 20), the energy resolution lies between 1 and 1.5 eV.

With ultrafast TEMs based on nanoemitters, a completely new field of applications demanding electron beams with the highest brightness is now at reach. In Section 4, we will discuss the applications of the ultrafast TEMs based either on flat photocathodes or nanoemitters. Applications of high brightness UTEMs to ultrafast electron holography will be discussed in Section 5.

4. APPLICATIONS OF ULTRAFAST TRANSMISSION ELECTRON MICROSCOPY

Because of its ability to provide information with nanometer–femtosecond spatio-temporal resolution and to identify transient structures having lifetimes of the order of femtoseconds, ultrafast TEM has soon become a major tool in materials science, chemistry, and biology. Whereas a part of its applications are transpositions of conventional TEM studies in the time domain, ultrafast TEM has also come with a set of unique applications. Because it involves single electron pulses of femtosecond duration and very intense optical fields, UTEM has enabled the exploration of optical near-fields and the exploitation of electron–photon interactions to tailor electron pulses down to the attosecond timescale.

In this section, we review recent studies performed using ultrafast Transmission Electron Microscopes. We did not discriminate them by the technique used (imaging, diffraction, spectroscopy) but rather by their field of application. We will discuss UTEM studies for chemistry and biology first, followed by nanomechanics, nanomagnetism, and nano-optics.

4.1 Chemistry and Biology

Chemical and biological structures transform at the atomic scale on timescales ranging from a femtosecond to picoseconds. A vast body of literature exist on the study of ultrafast chemical and biological reactions using either ultrafast electron diffraction (UED) or ultrafast TEM (Hu, Vanacore, Yang, Miao, & Zewail, 2015; Park, Flannigan, & Zewail, 2012; Schäfer, Liang, & Zewail, 2011; Vanacore, van der Veen, & Zewail, 2015). Since the observation of the metal–insulator phase transition in vanadium dioxide (VO_2) which was the first example of an ultrafast structural phase transition observed in a UTEM (Grinolds et al., 2006), many studies have been reported and several reviews are now available (Zewail, 2006; Aseyev, Weber, & Ischenko, 2013; King et al., 2005). Recently the unique spatio-temporal resolution of UTEM operated in the single electron regime was exploited to investigate the ultrafast dynamics of single nano-objects having specific chemical functions or biological interest.

For instance in 2013 van der Veen et al. visualized the spin crossover dynamics of single isolated nanoparticles of the metal–organic framework

Fe(pyrazine)Pt(CN)$_4$ (van der Veen, Kwon, Tissot, Hauser, & Zewail, 2013). Using a small selected area (SA) aperture, they performed SA electron diffraction (SAED) of a nitrogen cooled isolated particle and monitored its transition from a diamagnetic low-spin to paramagnetic high-spin state exploiting the change of the (110) diffraction peak position combined with the change in particle dimensions using bright field imaging. Achieved on UEM2, this first observation of a phase transition on a single nano-object has been performed in stroboscopic pump-probe experiments at a repetition rate of 600 Hz using 10 ns electrons pulse. The technique has been pushed further in time resolution to study the dynamical nature of a catalytic active site in a single pristine JDF-L1 crystal (Yoo, Su, Thomas, & Zewail, 2016). Using 100 kHz stroboscopic pump-probe electron diffraction, the femtosecond atomic movement at the titanium active center could be evidenced based on the time-resolved intensities of diffracted SAED spots. As reported in Fig. 12, the evolution on a picosecond timescale of the (200) and (201) diffracted peak intensities was used to probe the time-dependence of the corresponding structure factors. The latter was ascribed to the displacement of titanium and oxygen atoms along their bonding direction. The Caltech team could observe the transient structure of the active site of a photoexcited catalytic process with a unique combination of high spatial and temporal resolution.

Ultrafast TEM has also been used to explore systems of biological interest. For instance time-resolved electron diffraction was exploited to detect movements in hydrated protein at picometer and nanosecond scales (Fitzpatrick, Lorenz, Vanacore, & Zewail, 2013). Biomechanical properties have been studied on samples such as amyloid or DNA (e.g., Lorenz & Zewail, 2013; Fitzpatrick, Park, & Zewail, 2013). Several other examples can be found in the literature (Zewail, 2006; Fitzpatrick, Park, et al., 2013).

Conical illumination has also been used to visualize the dynamics of a single particle inside a polycrystalline structure undergoing a phase transition (Liu, Kwon, Tang, & Zewail, 2014). Using such an innovative scanning illumination condition, the Caltech team performed time-resolved single nanoparticle dark-field imaging in the stroboscopic pump-probe configuration at a repetition rate of 2 kHz. The selectivity given by this specific illumination condition revealed that each particle embedded in the polycrystalline assembly has a characteristic structural dynamics associated with its size. More recently, the lattice thermalization and structural dynamics of layered transition-metal dichalcogenides has been explored using ultrafast TEM (Sun et al., 2015; Wei et al., 2017).

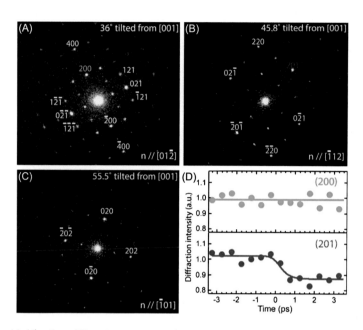

Figure 12 Ultrafast diffraction patterns observed along different zones axis and time dependent intensity changes of the (200) and (201) diffracted spots (reprinted with permission from Yoo et al., 2016).

4.2 Nanomechanics and Biomechanics

One major benefit of the Transmission Electron Microscope is the possibility to get information either in the real or reciprocal space. In the following we will split the discussion in two distinct parts first dealing with information gathered in reciprocal space using ultrafast diffraction and secondly regarding information obtained in real space using ultrafast TEM imaging.

Electron diffraction performed in a TEM is a very versatile technique due to the wide range of illumination conditions that can be implemented in the microscope (Zuo & Spence, 1992). For instance, parallel illumination electron diffraction can be obtained in combination with a selected area (SA) aperture located in the image plane of the objective lens. Selected area electron diffraction (SAED) patterns, constituted by sharp Bragg diffraction spots, can give precise information on the lattice parameters and the Bravais lattice symmetry of the crystal, but remain limited in spatial resolution by the size of the SA aperture (Morniroli, 2002). To improve the spatial resolution, electron microdiffraction can be implemented by focusing the electron beam with a small convergence angle α on the crystal. Small diffraction disks pattern can then be observed in the back focal plane of the objective lens and analyzed to retrieve lattice information from a smaller crystal area. Sharp spots pattern can also be obtained using nanobeam electron diffraction but needs dedicated illumination lenses to form a nanometric parallel electron beam (Zuo et al., 2004). Whether in SAED, nanobeam diffraction, or electron microdiffraction, the diffracted intensities cannot be directly related to the structural information (such as complex structure factor), like in X-ray diffraction, due to the strong dynamical contribution coming from multiple diffraction events (Reimer, 1993). Electron precession performed in SAED, nanobeam or electron microdiffraction conditions is used to minimize the contribution of dynamical effects in the diffracted intensities (Vincent & Midgley, 1994). Electron precession intensities can then be used to have more precise information on the crystal symmetry such as the Laue group (Morniroli, Stadelmann, Ji, & Nicolopoulos, 2010).

By increasing the condenser aperture size, it is possible to focus on the sample a sub-nanometric electron probe with a large convergence angle. If the convergence angle remains smaller than the smallest Bragg angle, a disk pattern known as Convergent Beam Electron Diffraction (CBED) pattern can be observed (Tanaka, 1997). Instead of diffraction spots, or small disks observed in SAED and electron microdiffraction, CBED yields transmitted and diffracted disks filled with useful rocking curve intensities. Zero order (ZOLZ) and high order Laue zone (HOLZ) diffracted intensities in a given zone axis pattern (ZAP) are acquired simultaneously in the CBED transmitted disk providing precise information on crystal space group symmetry, lattice parameters, complex structure factors, and charge densities (Schapink, Forghany, & Buxton, 1983; Zuo & Spence, 1991; Krämer & Mayer, 1999; Zuo, Kim, O'Keeffe, & Spence, 1999). When the convergence angle becomes larger than the smallest Bragg angle in the crystal, overlapping CBED patterns known as Kossel–Möellenstedt patterns are obtained, usually in the Large Angle CBED configuration (LACBED), from which precise information on crystalline symmetry and crystal defects can be obtained (Morniroli, 2006). Finally, Kikuchi diffraction originating from Bragg diffraction inside the cone of diffuse scattering, also provide rocking curve information in the inelastic background, and is visible around the elastic diffraction spots/disks (Williams & Carter, 1996).

The modulation of the atomic positions inside an optically excited sample modifies the diffraction conditions. Therefore, the ultrafast lattice dynamics induces transient modifications of the intensity and position of the characteristic features of a diffraction pattern. In the last decade, parallel beam selected area electron diffraction (SAED), Kikuchi diffraction and convergent-beam electron diffraction (CBED) have been used in combination with real-space imaging and electron spectroscopy to investigate the lattice dynamics in a wide range of materials. A large number of studies have been reported and discussed in several reviews (Zewail, 2006; Spencer Baskin & Zewail, 2014). In the following, we focus on recent stroboscopic electron diffraction experiments performed in UTEM that push further the space and time resolution.

Ultrafast SAED remains the most used configuration in these studies. For instance, the Caltech group performed ultrafast SAED on gold crystal films. Following optical excitation, atomic motions showed up as a modulation of the separation of the Bragg diffraction spots on the 100 ps timescale (Park, Kwon, Baskin, Barwick, & Zewail, 2009). A similar study on a graphite crystal was reported in which the dynamics of the Bragg spots positions and intensities was monitored. The Caltech team observed oscillations of spot positions and intensities with periods of 20 ps arising from acoustic resonances and used them to determine the dynamic elastic properties, which are anisotropic in graphite. Similar ultrafast SAED observations have been performed on multi-wall carbon nanotubes (MWCNT) revealing anisotropic atomic expansion dynamics of MWCNT following ultrafast heating. These phenomena were observed on the picosecond to microsecond timescale (Park, Flannigan, et al., 2012).

A strong improvement in space and time resolution has been demonstrated by Yurtsever et al. who implemented the first ultrafast CBED experiment on a Si wedge sample (Yurtsever & Zewail, 2009). From the changes of the HOLZ lines intensities on a picosecond timescale they could extract the time dependent Debye–Waller factor of Si (see Fig. 13).

Thanks to the CBED configuration they could study the dynamics from a region of the specimen having a diameter between 10 and 300 nm and showed that the structural changes within this small area occurred on a timescale of a few picoseconds. From the time-dependent Debye–Waller factor the root mean square atomic displacement could be retrieved and the temperature estimated.

Kikuchi lines, acquired in convergent beam illumination, have also been observed dynamically on the same sample (Yurtsever, Schaefer, & Zewail, 2012). As for ZOLZ dynamical lines in CBED/LACBED patterns, the position of Kikuchi lines are sensitive to crystal strain and curvature. The Caltech group observed oscillations of the Kikuchi lines positions with a period of 30 ps that were induced by propagating strain waves following the ultrafast heating of the Si slab. The convergent beam illumination allowed

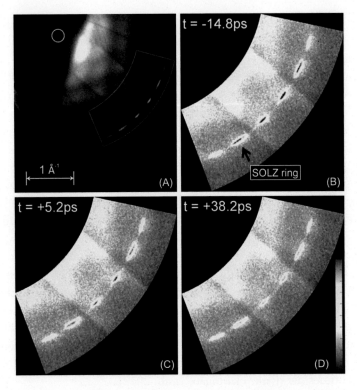

Figure 13 Temporal evolution of Si HOLZ line intensities observed close to the [011] ZAP (reproduced with permission from Yurtsever & Zewail, 2009).

them to observe both the longitudinal and transverse eigenmodes of the wedge structure with a spatial resolution of ≈100 nm.

Recently, using the ultrafast Schottky FE–TEM developed in the University of Göttingen, ultrafast CBED has also been performed close to the edge of a graphite crystal to study mechanical mode similar to those previously observed in SAED (Feist, da Silva, Liang, Ropers, & Schäfer, 2018). Clear oscillations of the HOLZ lines positions ascribed to dominant acoustic mechanical modes were detected. As reported in Fig. 14, the dynamics of the (452) HOLZ line intensity was also detected revealing the time evolution of the atomic mean square displacement as in the previous studies performed on the Si sample.

Thanks to the tridimensional information contained in a CBED pattern, the time dependence of the full deformation gradient tensor due to the membrane mechanical vibrations could be extracted from the dynamics of different HOLZ lines positions. This study expands the possibilities of

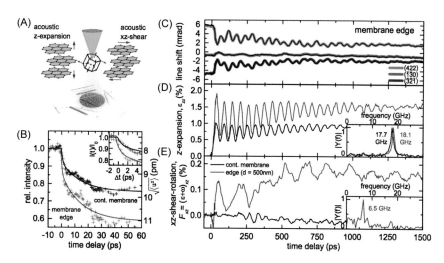

Figure 14 Time-dependent HOLZ-line evolution and dynamics of selected components of the deformation gradient tensor. (A) Two dominating mechanical modes have been observed: an out-of-plane z-axis expansion and an acoustic shear-rotation in the xz-plane. (B) (452) HOLZ-line intensity evolution and square root of atomic mean square displacement in the in-plane direction after optical excitation (red: membrane edge, black: continuous membrane). (C) Delay-dependent center-of-mass shift (black line) and reconstructed mean line position (colored line, background) of the (422), (130), and (321) HOLZ lines, probed at the graphite edge. (D) and (E) Reconstructed z-axis expansion (D) and in-plane xz-shear-rotation (E) components (red: membrane edge, black: continuous membrane) with respective Fourier analysis. Reprinted with permission from Feist et al. (2018).

ultrafast diffraction and exploits the sensitivity of HOLZ lines positions to the crystal strain state. It shows that the imaging of phonon modes up to the terahertz regime is possible with ultrafast CBED making it a promising tool for fields such as nanophononics for instance (Laude et al., 2017).

Nevertheless, the ultrafast CBED experiments reported to date have not reached the spatial resolution of CBED experiments with nanometric electron probes (Kim & Zuo, 2013; Chen et al., 2016). The best spatial resolution reported in Yurtsever et al. and Feist et al. was 10–300 nm and 28 nm respectively (Yurtsever & Zewail, 2009; Feist et al., 2018). Recent developments of high brightness electron sources (Feist et al., 2017; Houdellier et al., 2018) combined with optimized illumination optics should allow acquiring coherent ultrafast CBED patterns from nanometric or sub–nanometric areas. Combining ultrafast CBED with STEM acquisition (LeBeau, Findlay, Allen, & Stemmer, 2010), aberration correctors (Haider, Hartel, Muller, Uhlemann, & Zach, 2009) and modern fast pixelated detectors (Mac Raighne, Fernandez, Maneuski, McGrouther, & O'Shea, 2011; Yang et al., 2015) should considerably expand the range of applications of this technique and enable new ways of dynamically mapping fields (Krajnak, McGrouther, Maneuski, Shea, & McVitie, 2016). Moreover, no ultrafast quantitative CBED has yet been realized. Fitting time-dependent rocking curves obtained in ultrafast CBED should give access to the changes of the complex structure factor (Zuo & Spence, 1991; Zuo, Spence, & Hoier, 1989) and from the latter the atomic positions and charge density in complex material with unrivalled time and space resolution (Theissmann, Fuess, & Tsuda, 2012). Finally, the use of brighter ultrafast electron sources should enable the formation of coherent overlapping ultrafast CBED patterns, from which atomic scale information can be extracted (Terauchi et al., 1994; Zheng & Etheridge, 2013).

Ultrafast Transmission Electron Microscopes have also been used to provide information on the ultrafast dynamics directly in real space. For instance, the first generation Caltech UEM1 allowed to observe a light-induced reversible expansion–contraction of the crystalline quasi-one-dimensional semiconductor [Cu(TCNQ)] (Flannigan et al., 2007; Flannigan, Samartzis, Yurtsever, & Zewail, 2009). A nice illustration was given in a study by Baskin et al. in which the authors investigated the mechanical motion of arrays of cantilevers of different shapes. These so-called "nanomusical" systems, two examples of which are shown in Fig. 15A, were fabricated by focused ion beam milling of a layered $Ni/Ti/Si_3N_4$ thin film. Following optical excitation, the bending stress caused by differential thermal expansion in the multilayered film impulsively launched free vibrations inside the cantilevers. The motion following laser excitation was investigated using the second generation Caltech UEM2. The vibrations of the cantilevers are visible on the difference images displayed in Fig. 15A. Tilting the specimen allowed to identify the in-plane and out-of-plane cantilever bending and torsional motions in the measurements. The measured frequencies for the dominant mode were in good agreement with the natural frequencies of the transverse out-of-plane vibrations of an ideal clamped-free cantilever of uniform rectangular cross section. In a second set of experiments, the authors used high frequency, low fluence pulsed laser excitation. By varying the repetition rate, they could selectively and resonantly excite one cantilever out of the whole array (Baskin, Park, & Zewail, 2011).

Nanostructures supporting high frequency vibration modes with complex spatial structures clearly benefit from the ability of UTEM to provide

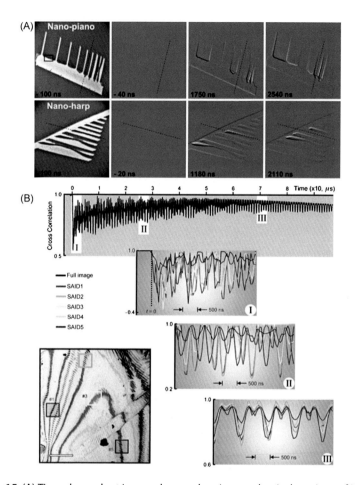

Figure 15 (A) Time-dependent image change showing mechanical motions of individual nanocantilevers following excitation by a single nanosecond laser pulse at a fluence of 2 mJ/cm². At left are images of the unperturbed states, 100 ns before arrival of the laser pulse. The six right-hand images are difference images, formed by subtracting the leftmost image from one recorded at the indicated time delay relative to the laser pulse, after normalization of the full image intensity. The sample was tilted to enhance the observability of out-of-plane motion. A dashed line is drawn as a reference, indicating the location of the fifth cantilever in each of the images. A uniform contrast range is used for display of the images in each difference image sequence. Reprinted with permission from Baskin et al. (2011). © 2011 American Chemical Society. (B) Time dependence of image cross correlation. The whole scan for 100 μs is made of 2000 images taken at 50 ns steps. Also depicted are the zoomed-in image cross-correlations of three representative time regimes (I, II, and III). In each zoomed-in panel, the selected-area image dynamics of five different regions are included. Note the evolution from the "chaotic" to the global resonance (drumming) behavior at long times. Reprinted with permission from Kwon et al. (2008b). © 2008 American Chemical Society.

images in real space at high temporal resolution. Kwon et al. investigated the ultrafast dynamics of a single crystal graphite film using UEM2. They acquired stroboscopically at kHz repetition rate selected-area-images using nanosecond electron pulses. Their results clearly evidenced the evolution from a multimode oscillation to a coherent resonance (global) mode. As shown in Fig. 15B, the motion is "chaotic" with several mechanical modes present over the micron scale on a timescale of a few microseconds. On a longer timescale, a single vibration with a well-defined fundamental frequency of 1.08 MHz survives before complete damping in about 200 μs. From this resonance frequency, the authors could determine an in-plane Young's modulus of 1.0 terapascal (Kwon, Barwick, Park, Baskin, & Zewail, 2008b).

4.3 Nanomagnetism

The response of magnetic materials to static or slowly varying external stimuli (magnetic field, pressure, temperature, etc.) has been extensively explored both experimentally and theoretically. For instance, the evolution of the magnetic configuration (including domain wall displacement) inside a ferromagnetic material placed in a slowly varying external magnetic field can be observed using magneto-optical microscopy and accurately computed using the Landau–Lifshitz and Gilbert equation (Stohr, 2016; Landau, Lifšic, Pitaevskij, & Landau, 2008). However, the dynamics of magnetic materials on ultrafast timescales is not fully understood and motivates an intense research (Koopmans, Kicken, van Kampen, & de Jonge, 2005; Kimel et al., 2005; Hansteen, Kimel, Kirilyuk, & Rasing, 2005; Battiato, Carva, & Oppeneer, 2010; Houssameddine et al., 2007). The development of ultrafast lasers has provided physicists with new tools to investigate the ultrafast dynamics and manipulate the spin degree of freedom in ferromagnetic materials. This led to the emergence of a very active field of research called femtomagnetism (Bigot et al., 2009; Beaurepaire, Merle, Daunois, & Bigot, 1996; Hohlfeld, Matthias, Knorren, & Bennemann, 1997; Aeschlimann et al., 1997).

The ultrafast magnetization dynamics of a magnetic sample excited by a femtosecond laser pulse can be described using the *three temperature model* (Beaurepaire et al., 1996; Kazantseva, Nowak, Chantrell, Hohlfeld, & Rebei, 2008; Hillebrands & Ounadjela, 2002). This phenomenological model describes the relaxation of a ferromagnet through the energy exchange between three thermodynamic reservoirs (electrons, phonons, and spins). The internal relaxation of each subsystem is assumed to be faster than its coupling with the other baths so that a temperature can be ascribed to each subsystem and the energy exchanges can be described by simple rate equations. However, this model does not account for the dynamics of strongly out-of-equilibrium magnetic systems nor the coherent coupling between the optical excitation and the spins. Beside fundamental issues, the dynamics of magnetic systems is also interesting for applications. For instance, the small size and high stability of magnetic vortex cores make them an interesting candidate for future magnetic data storage technology. Theoretical studies have predicted a vortex core switching time of only a few tens of picosecond (Hertel, Gliga, Fähnle, & Schneider, 2007).

New experiments providing time-resolved images of magnetic systems with nanometer/femtosecond resolutions are therefore required to deepen our understanding of their dynamics and explore their potential for applications. X-ray Magnetic Circular Dichroism (XMCD) performed on synchrotrons or free electrons lasers (FEL) is a very powerful technique yielding quantitative information on both the orbital and spin angular momentum in the sample (Bouldi et al., 2017; Stamm et al., 2007; Boeglin et al., 2010). Furthermore, lensless magnetic imaging is also possible on the instruments having the brightest sources (Eisebitt et al., 2004). In this latter case, images of magnetic nanostructures can be acquired and their ultrafast dynamics probed quantitatively in pump-probe experiments. The spatial resolution of these instruments is however limited to ~100 nm.

In the late 70s and beginning of 80s O. Bostanjoglo and coworkers (Bostanjoglo & Rosin, 1977, 1980) investigated the magnetization dynamics in a TEM for the first time. They used a stroboscopic illumination of the sample based on an electrostatic chopper (Petrov, Spivak, & Pavlyuchenko, 1972) in combination with an ultrasound oscillator. The latter generated a mechanical stress on the magnetostrictive specimen synchronized with the illumination. Magnetic domains were observed using the defocused mode of Lorentz imaging aka Fresnel imaging. Magnetic reversal has been observed thanks to domain wall motion as well as forced and free oscillations of Bloch line in NiFe samples, but the spatio-temporal resolution was limited to 30 nm/50 ns. 30 years later, H.S. Park et al. reported Fresnel imaging on the second generation UEM in Caltech (Park, Baskin, & Zewail, 2010). In their experiments, the microscope was set in Low Magnification mode without the use of a Lorentz lens and the images were recorded in the nanosecond stroboscopic mode at a low repetition rate (1 kHz). Domain

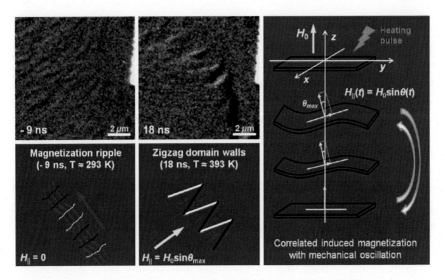

Figure 16 Domain wall nucleation and annihilation and schematic illustration of a possible mechanism of domain wall nucleation after laser excitation. Reprinted with permission from Park et al. (2010). © 2010 American Chemical Society.

wall nucleation, oscillation, and propagation inside Ni thin films deposited on various substrates could be observed (see Fig. 16).

Park and coworkers evidenced the nucleation of zig-zag domain walls induced by the deformation of the sample following its impulsive heating by the laser pump pulse. They could also observe the propagation of domain walls and quantify its velocity. The resolution of this first time-resolved investigation of a magnetic system in a UTEM was however limited to 10 ns and some 100 nm due to the absence of a Lorentz lens. The dynamics of skyrmions has also been investigated using the same type of photocathode-based UTEM equipped with a cryogenic cooling stage (between 77 and 300 K). They studied the dynamics associated with the light-induced writing and erasing of skyrmions in 60 nm thick nano-slab of FeGe combining usual continuous *in-situ* cryo-Lorentz TEM and nanosecond pump-probe cryo-Lorentz TEM (Rajeswari et al., 2015; Berruto et al., 2018). Recently Schliep et al. studied the dynamics of photo-induced magnetic domain walls in a FePt film using Fresnel imaging with picosecond resolution in a commercial TECNAI Femto UEM (Schliep, Quarterman, Wang, & Flannigan, 2017). As for Park et al. (2010), the microscope was not equipped with a Lorentz lens imaging system and was then limited in magnification and spatial resolution but a time resolution of 3 ps could nevertheless be obtained. As noticed in Schliep et al. (2017) an increase of the source brightness and the use of a Lorentz lens imaging system should significantly improve the temporal and spatial resolutions of UTEMs for the investigation of magnetic systems. The first step in this direction has been realized in Göttingen using the JEOL 2100 F equipped with a laser-driven Schottky field emitter described in Section 3.5. The light-induced demagnetization of a single magnetic vortex could be observed with a spatial resolution better than 100 nm and a temporal resolution of 700 fs (da Silva et al., 2017). From the time dependent changes in the Fresnel imaging contrast, a drop in saturation magnetization could be evidenced on an isolated Permalloy disc after 100 fs. This observation is in agreement with the three-temperature model. On a picosecond timescale, a partial recovery of the magnetization was observed. For the first time, the magnetization dynamics could be observed with a spatial resolution better than 100 nm and a temporal resolution of several hundreds of femtoseconds. This result demonstrates the interest of UTEMs equipped with high brightness ultrafast emitters for the investigation of the magnetization dynamics with high spatio-temporal resolution. The same instrument has also been used in conventional mode to study metastable magnetic structures in an optically excited iron layer deposited on a Si_3N_4 substrate but no time-resolved information has been extracted from this study (Eggebrecht et al., 2017).

The ultrafast TEM experiments reported so far have relied on the simplest magnetic imaging mode available in TEM that is Lorentz microscopy in Fresnel contrast imaging. However, the arsenal of TEM for the characterization of magnetic materials is much wider. For instance, with a Lorentz lens and an objective aperture in the diffraction plane, Foucault imaging and differential phase contrast can also provide additional information about the magnetization direction (Chapman & Kirk, 1997; Chapman, Batson, Waddell, & Ferrier, 1978; Chapman, Ploessl, & Donnet, 1992). Additionally, Fresnel contrast can also be analyzed using the so-called Transport Intensity Equation (TIE) to quantitatively reconstruct the product of the magnetic induction vector and thickness (Kohn, Habibi, & Mayo, 2016; Beleggia, Schofield, Volkov, & Zhu, 2004). By overlapping the electron beam transmitted by a magnetic sample with a reference beam traveling in vacuum, off-axis electron holography can quantitatively map the electron phase modifications due to electric or magnetic fields in the sample (see Section 5). The magnetic and electric potential can then be quantitatively extracted from the phase change using the well-known Aharonov–Bohm formula (Völkl, Allard, & Joy, 1999). As will be detailed in Section 5, off-axis electron holography requires the use of high quality electron beams with the highest spatial resolution possible and, as a consequence, the highest brightness (Harscher & Lichte, 1996). The development of ultrafast high brightness FE-TEMs based on laser-driven nanoemitters (SFEG in Göttingen and CFEG in Toulouse) enabling electron holography with ultrashort electron pulses is a first step toward time-resolved electron holography of magnetic materials (Feist et al., 2017; Houdellier et al., 2018). Combining these high brightness ultrafast electron sources with modern electron optics including spherical aberration corrector (Haider et al., 2009), and new direct electron detectors (Milazzo et al., 2010), will undoubtedly open new avenues for the quantitative study of the magnetization dynamics at the nanoscale.

4.4 Nano-Optics

The applications described so far in this chapter can be considered as direct extensions of conventional TEM techniques (imaging, diffraction, spectroscopy) to the time domain. In this section, we address new applications that rely on the interaction between ultrashort electron pulses and intense ultrashort laser pulses. They extend the potential of these new instruments by (i) giving access to the optical response of nanostructures with nanometer spatial resolution and down to tens of meV spectral resolution and (ii) enabling new strategies for the generation of femtosecond or attosecond electron pulses in the electron microscope. We first discuss electron/photon interactions before giving a short account of the most recent results.

Energy and momentum conservation laws prohibit the linear interaction between free electrons and free space light (de Abajo & Kociak, 2008; Barwick, Flannigan, & Zewail, 2009; Garcia de Abajo, 2010; Park & Zewail, 2010). This interaction becomes possible for instance when relativistic electrons move faster than light: this is the inverse Cherenkov effect (Kimura et al., 1995). Alternatively, energy and momentum conservation laws can be satisfied if an additional body (particle, nanostructure) provides the required energy and momentum. Kapitza and Dirac predicted in 1933 that a beam of free electrons can interact elastically and diffract on a stationary light wave (Kapitza & Dirac, 1933). In the particle picture, the Kapitza–Dirac effect is described as the interaction of a free electron with two photons by stimulated Compton scattering. The absence of sufficiently intense light sources has long impeded its experimental verification but the Kapitza–Dirac effect has since been detected unambiguously (Bucksbaum, Schumacher, & Bashkansky, 1988; Freimund, Aflatooni, & Batelaan, 2001). A metallic grating can also be used to meet the energy and momentum conservation requirements. For instance, electrons traveling in a direction perpendicular to the rulings of an illuminated grating can gain energy by the inverse Smith–Purcell effect (Mizuno, Pae, Nozokido, & Furuya, 1987). As will be seen in the following, the confinement of optical fields in the vicinity of optically excited nano-objects can also yield efficient interactions between electrons and light.

When an ultrashort electron pulse and a laser pulse arrive at the same time on a nanostructure, the interaction of the electrons with the optical near-field confined around the nanostructure can strongly modify the electron energy spectrum. Regularly spaced energy sidebands appear at $E_0 \pm n\hbar\omega$ in which ω is the angular frequency of the laser pulse. For low values of the incident optical intensity, first order perturbation theory can be used to compute the probability of energy gain/loss from the interaction between the moving charges and the electromagnetic field (de Abajo & Kociak, 2008):

$$P_{\text{gain/loss}}(\hbar\omega) = \left(\frac{e}{\hbar\omega}\right)^2 \left| \int dz \, E_z(x, y, z) e^{-i\omega z/v} \right|^2$$

In the latter, the electron is propagating along (OZ) at a velocity v. In the case of high power optical excitation, higher order interactions develop between the optical field and the free electrons yielding a more complex dynamics characterized by multiple photon sidebands. The occupation probability $P_{\text{gain/loss}}(N\hbar\omega)$ of the Nth photon sideband is then given by a Bessel function of the first kind of order N (García de Abajo, Asenjo-Garcia, & Kociak, 2010; Park & Zewail, 2010). It shows an oscillatory behavior with excitation intensity arising from interferences between the different quantum mechanical paths contributing to the transition probability (García de Abajo et al., 2010; Park & Zewail, 2010; Feist et al., 2015).

Electron energy gains originating from the interaction of fast electrons with optical near-fields were first observed in Caltech on the second generation UEM2. The associated technique was called photon-induced near-field electron microscopy. The energy spectrum of the electron pulses passing in the vicinity of a bundle of carbon nanotubes is shown in Fig. 17B, C. The spectrum acquired when the electron pulse precedes the optical excitation is given for reference. When the electron and optical pulses are

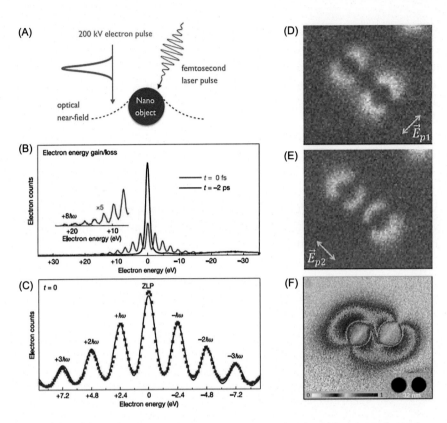

Figure 17 (A) Close-up of the interaction between an ultrashort electron pulse and an optically excited nano-object. (B) Comparison between the electron energy spectrum before (black) and during (red) optical excitation of a bundle of carbon nanotubes. The electron energies are referenced to the zero loss energy. (C) Magnified view of the electron energy gain spectrum at $t = 0$. The regularly spaced sidebands separated by the photon energy $\hbar\omega = 2.4$ eV correspond to emission or absorption of n photons by the fast electron (reprinted with permission from Barwick et al., 2009). (D) and (E) PINEM images taken on two silver nanoparticles separated by 70 nm for two different incident polarizations (reprinted with permission from Yurtsever & Zewail, 2012). © 2012 American Chemical Society. (F) PINEM image of two silver nanoparticles separated by an edge-to-edge distance of 32 nm. A narrow channel induced by the electromagnetic coupling between the dipolar near-fields of the two particles is visible (reprinted with permission from Yurtsever, Baskin, & Zewail, 2012). © 2012 American Chemical Society.

synchronized on the sample, their strong interaction induces a clear reduction of the zero loss peak intensity and the appearance of symmetric energy sidebands at $\pm n\hbar\omega$. These sidebands are associated with the absorption/emission of up to 8 photons by the fast particle.

The expression given above shows that the probability of energy gain/loss by the moving electron directly depends on the magnitude of the optical near-field. Three important points are worth mentioning here. First, the gain probability depends on the intensity of the optical excitation. Therefore, the gain/loss signal intensity can be boosted by increasing the laser power within the range set by the damage threshold of the sample. Second, the energy gain probability directly depends on the amplitude of the electric field along the electron trajectory. Therefore, by scanning the electron beam on the sample, it is possible to map the spatial distribution of the optical near-field with nanometer spatial resolution (de Abajo & Kociak, 2008; Barwick et al., 2009; García de Abajo et al., 2010; Park & Zewail, 2010). Third, contrary to Electron Energy Loss Spectroscopy (EELS) that probes all the optical excitations supported by the sample at a given energy, the optical excitations contributing to the gain signal can be selected through the polarization of the optical excitation. This influence of the incident polarization is visible in Fig. 17D, E in which the electric near-field intensity distribution characteristic of a dipole excitation is mapped on silver nanoparticles.

Following the pioneering work performed at Caltech in 2009, several systems of interest in nano-optics have been investigated. Spectrum images have been obtained on silver nanoparticles and the metallic copper–vacuum interface for different optical pump/electron probe delays (Yurtsever, van der Veen, & Zewail, 2012). As illustrated Fig. 17F, the optical near-field confined around dimers of silver nanoparticles could be mapped with a spatial resolution better than 10 nm (Yurtsever, Baskin, et al., 2012; Yurtsever & Zewail, 2012). The dual particle/wave character of the Fabry–Perot standing plasmon waves of silver nanowires has also been discussed in the context of PINEM experiments (Piazza et al., 2015).

Before limited to ~1 eV by the electron beam energy spread and spectrometer energy resolution, the spectral resolution of PINEM has been improved using the tunable light pulses delivered by an optical parametric amplifier to excite the sample and measure the evolution of the electron energy gain signal as a function of the energy of the optical excitation. Spectrally resolved photon-induced near-field electron microscopy (SRPINEM) allowed to obtain nm–fs-resolved maps of nanoparticle plasmons with an energy resolution determined by the laser line width (20 meV) and no longer limited by the electron beam energy spread or spectrometer resolution (Pomarico et al., 2017).

The faster electrons inside an electron pulse obviously reach the sample before electrons having a smaller kinetic energy. This correlation between energy and time called the chirp of an electron pulse is an important quantity that needs to be precisely characterized in UTEM. Its knowledge is in particular required before implementing pulse compression techniques. PINEM not only gives access to the electron pulse duration and energy spread but it can also be used to characterize in details the chirp of the electron pulses (Park, Kwon, & Zewail, 2012). The process of electron energy gain can also be used to decrease the duration of the electron pulses and therefore improve the temporal resolution of UTEM. Using an optical gating pulse shorter than the initial electron pulse and selecting only the electrons that have gained energy during their interaction with the gating pulse, it is possible to generate *in-situ* electron pulses with durations only limited by the gating pulse duration (Park, Kwon, et al., 2012; Hassan, Liu, Baskin, & Zewail, 2015; Hassan, Baskin, Liao, & Zewail, 2017). This photon gating technique, depicted in Fig. 18, may be used in the near future to generate attosecond electron pulses in a UTEM.

Coherent control schemes offer promising alternatives. The first demonstration of the coherent manipulation of the quantum state of free electrons in an electron microscope was achieved in Göttingen (Feist et al., 2015). In this study, the interaction between ultrashort electron pulses with optical near-fields was exploited to induce Rabi oscillations in the populations of electron momentum states. It was theoretically shown that after the interaction with the optical near-field, the electron wavepacket evolves into a train of attosecond electron pulses. Other coherent control schemes were later investigated by the Göttingen group. It was shown that the sequential interaction of free electron pulses with two spatially separated, phase controlled optical near-fields could be used to measure electronic dephasing with sub-cycle precision (Echternkamp, Feist, Schafer, & Ropers, 2016). As shown in Fig. 18, the interaction of an ultrashort electron packet with a two-color optical field can also be used to coherently control the electron wave function and yield attosecond electron pulses (Priebe et al., 2017). Similar two-color coherent control experiments have also been applied to control the electron emission from nanoemitters (Förster et al., 2016; Paschen et al., 2017). Recently, the modulation of a beam of 70 keV electrons by the electric field at the surface of a dielectric membrane was

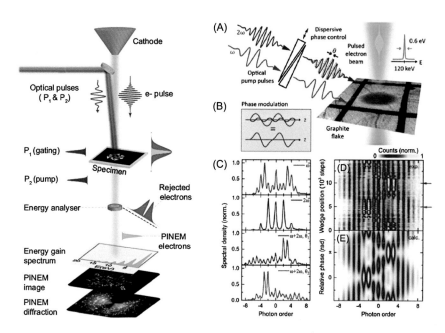

Figure 18 Left: Concept of photon gating in 4D electron microscopy. (A) The micro-scope column with one electron (dark blue) and two optical (red) pulses focused onto the specimen. One optical pulse is coincident with the electron pulse at the specimen to generate a PINEM signal. The resulting light blue PINEM pulse is filtered out from other electrons using an energy filter. The second optical pulse initiates the dynamics to be probed (reprinted with permission from Hassan et al., 2015). Right: Coherent control of the electron wave function using a phase-locked two-color optical field. (A) Optical pump pulses at frequencies ω ($\lambda = 800$ nm) and 2ω ($\lambda = 400$ nm) spatially and tem-porally overlapped with a pulsed electron beam on a single-crystalline graphite flake. The relative phase between the laser pulses is controlled using fused silica wedges. The reference electron energy spectrum (green) has a central energy of 120 keV and an energy width of 0.6 eV. (B) The electron-light-interaction modulates the phase of the electron wavefunction. Two-color laser fields yield a non-sinusoidal modulation (pur-ple curve). (C) Experimental electron energy spectra recorded for single-color (red and blue curves) and two-color excitation (purple and magenta curves). In the latter case, the spectra are strongly asymmetric and depend on the relative phase of the two colors ($\Theta_1 = \pi$, $\Theta_2 = 0$). (D) Electron energy spectrum as a function of the wedge position and therefore relative phase between the pulses. The spectra in (C) are taken from the positions marked by the purple and magenta arrows. (E) Corresponding simulated spec-tra (reprinted with permission from Priebe et al., 2017).

used to generate electron pulses with sub-optical cycle duration. The latter were used for diffraction and microscopy experiments (Morimoto & Baum, 2018).

The development of ultrafast TEM has opened a completely new field of research in which the strong interaction between ultrashort electron pulses and optical pulses can be exploited to explore the optical response of nanostructures with unprecedented spatial and spectral resolutions. Importantly, more and more sophisticated electron/photon interaction schemes are being developed to further improve the temporal resolution of UTEM. Whereas PINEM is related to the longitudinal component of the electron momentum, the transverse components may also be modified upon interaction with light. The Kapitza–Dirac effect offers such a possibility: the periodic motion of an electron in an oscillating field leads to a nonzero average transverse Lorentz force due to the magnetic field. This induces an elastic deflection of the electron. It has been theoretically predicted that an electron beam crossing the optical near-field associated with a plasmon resonance of a nanowire can undergo diffraction of its elastic and inelastic outgoing components (García de Abajo, Barwick, & Carbone, 2016). Contrary to the Kapitza–Dirac effect with free space light waves, this electron diffraction is due to the direct action of the electric field of the plasmon, without any contribution from the magnetic field. In a generalization of the Kapitza–Dirac effect, Peter Hommelhoff and coworkers have recently shown that the energy of sub-relativistic electrons can be strongly modulated on the few-femtosecond timescale via the interaction with a traveling wave created in vacuum by two colliding laser pulses at different frequencies (Kozák, Eckstein, Schönenberger, & Hommelhoff, 2018). This effect has recently enabled the all-optical generation of trains of free-electron pulses with a pulse duration of less than 300 as at an energy of 23.5 keV (Kozák, Schönenberger, & Hommelhoff, 2018).

In the last section of the chapter, we discuss the development of time-resolved electron holography experiments. We first highlight the important factors that need to be optimized in order to acquire time-resolved electron holograms. This important step would indeed open new horizons for the exploration of electric, magnetic, strain fields with unprecedented temporal resolution.

5. TOWARD TIME-RESOLVED ELECTRON HOLOGRAPHY WITH FEMTOSECOND ELECTRON PULSES

5.1 Introduction to Off-Axis Electron Holography

Off-axis electron holography in a Transmission Electron Microscope is a well-established technique yielding quantitative maps of electric, magnetic,

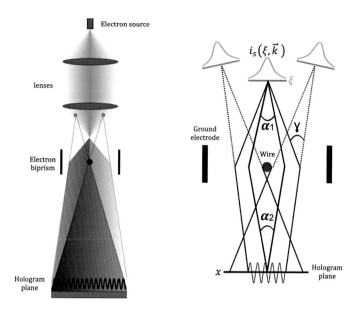

Figure 19 Geometry for acquiring electron holograms with the electron biprism located inside a transmission electron microscope.

and strain fields at the nanoscale (Völkl et al., 1999). The technique extracts the phase shift from the interference pattern originating from the overlap of two coherent electronic waves thanks to the action of a Möllenstedt electron biprism (Gabor, 1948). The biprism is a metal-coated quartz wire, with a radius of some hundreds of nanometers, brought to a potential relatively to an earth grounded electrode (Joy et al., 1993). As represented in Fig. 19, when the incoming electron beam crosses the biprism electric field, it is deflected and overlaps in a plane below the biprism. This deflection induces a phase shift between the two wavefronts one propagating freely in the TEM column, called the reference beam, and the other crossing the sample, called the sample beam (Tonomura, 1999).

After superposition in the image plane, an interferogram is obtained, with a carrier frequency $\mathbf{q_h}$ related to the phase difference between the two tilted wavefronts generated by the biprism. The intensity of the hologram for an incident plane wave $\Psi_0(\mathbf{r}) = A_0(\mathbf{r})e^{2\pi i \mathbf{k}\cdot\mathbf{r}}$ can be written as

$$I(\mathbf{r}) = \left| \Psi_0(\mathbf{r}) + \Psi_s(\mathbf{r}) \right|^2$$

$$I(\mathbf{r}) = A_0^2(\mathbf{r}) + A_s^2(\mathbf{r}) + 2CA_0(\mathbf{r})A_s(\mathbf{r})\cos\left(2\pi\,\mathbf{q_h}\cdot\mathbf{r} + \varphi_s(\mathbf{r})\right)$$

in which $\mathbf{r} = (x, y)$ are the coordinates in the sample plane, $\Psi_s(\mathbf{r})$ the sample beam wave function, $C = \frac{I_{max} - I_{min}}{I_{max} + I_{min}}$ the hologram contrast, and φ_s the sample beam phase shift coming from the interaction of the electron beam with the sample. The phase shift recorded in the electron hologram is sensitive to both the electrostatic potential and the in-plane component of the magnetic induction in the specimen (Tonomura, 1999). Neglecting dynamical diffraction effects, the phase shift in the transmitted beam can be written as

$$\varphi = \varphi^{elec} + \varphi^{mag}$$

$$\varphi = \frac{\pi}{\lambda E} \int_{-\infty}^{+\infty} V(\mathbf{r}, z)\, dz - \frac{e}{\hbar} \int_{-\infty}^{+\infty} A_z(\mathbf{r}, z)\, dz$$

where z is the propagation direction of the electrons, λ is the electron relativistic wavelength (2.51 pm at 200 keV), E the total electron energy, V the electrostatic potential, and A_z the component of the magnetic vector potential $\mathbf{A}$ parallel to the electron beam.

Generally the mean inner potential contribution is separated from the global electrostatic contribution and the total phase shift is written as

$$\varphi = C_E V_0 t(\mathbf{r}) + C_E \int_{-\infty}^{+\infty} V(\mathbf{r}, z)\, dz - \frac{e}{\hbar} \int_{-\infty}^{+\infty} A_z(\mathbf{r}, z)\, dz$$

in which $C_E = \frac{\pi}{\lambda E} = \frac{2\pi e}{\lambda} \frac{E_c + E_0}{E_c(E_c + 2E_0)}$ is a constant that depends on the microscope accelerating voltage, with E_c and E_0 respectively the kinetic energy of the accelerated electrons and the rest mass energy of the electron (511 keV), V_0 the mean inner potential of the sample, and $t(r)$ its thickness. C_E takes values of $7.28 \cdot 10^{-3}$ rad $\cdot$ V^{-1} $\cdot$ nm^{-1} at 200 keV.

Bright field electron holography can then be used to map the sample thickness (McCartney & Gajdardziska-Josifovska, 1994), the mean inner potential of a sample (Winkler et al., 2017; McCartney et al., 2010), the electrostatic field (de Knoop et al., 2014; Lichte et al., 2007) as well as the magnetic field (McCartney, Dunin-Borkowski, & Smith, 2005; Snoeck et al., 2008) with sub nanometric resolution.

To extract phase and amplitude information from an electron hologram, a Fourier filtering method is implemented. The Fourier transform of the hologram intensity can be written as

$$TF\big[I(\mathbf{r})\big] = \delta(\mathbf{q}) + TF\big[A_0^2(\mathbf{r}) + A_s^2(\mathbf{r})\big]$$
$$+ \delta(\mathbf{q} + \mathbf{q_h}) \otimes TF\big[CA_0(\mathbf{r})A_s(\mathbf{r})e^{i\varphi_s(\mathbf{r})}\big]$$
$$+ \delta(\mathbf{q} - \mathbf{q_h}) \otimes TF\big[CA_0(\mathbf{r})A_s(\mathbf{r})e^{-i\varphi_s(\mathbf{r})}\big]$$

This equation describes a peak at the origin of reciprocal space corresponding to the Fourier transform of the reference beam, a second peak centered at the origin corresponding to the Fourier transform of a conventional bright–field TEM image of the sample, and two sidebands peaks centered respectively at $\mathbf{q} = -\mathbf{q_h}$ and $\mathbf{q} = \mathbf{q_h}$ corresponding to the Fourier transform of the complex image wavefunction and the associated complex conjugate. To extract the sample beam complex wavefunction, one sideband peak is selected using an appropriate mask and inverse Fourier transformed. The phase can then be easily extracted from this complex wavefunction.

5.2 Relation Between the Contrast of the Electron Hologram and the Brightness of the Electron Source

The information that can be extracted from a hologram is related to the phase resolution, for a given spatial resolution. The latter must be optimized by carefully choosing a set of experimental conditions that enhance the hologram figure of merit (Chang, Dwyer, Boothroyd, & Dunin-Borkowski, 2015). These conditions are determined by a number of competing parameters such as the source beam spatial coherence, the camera transfer function, the instrument instabilities, etc.

The hologram figure of merit is given by the phase detection limit (Lehmann, 2004), which corresponds to the smallest detectable phase difference between adjacent pixels at a given signal-to-noise ratio $(\text{snr})_\varphi$ (de Ruijter & Weiss, 1993). The phase detection limit can be estimated using the standard deviation of the measured phase. As already discussed by Harscher and Lichte, the noise contribution in the hologram comes from the shot noise and the noise of the detection system (Harscher & Lichte, 1996). They showed that the standard deviation of the phase detection could be written as

$$\sigma_\varphi = \sqrt{\frac{2}{C^2 N}}$$

where N is the number of electrons per pixel during the measurement time. The phase detection limit as a function of the $(\text{snr})_\varphi$ and the standard deviation σ_φ can be expressed as

$$\delta\varphi = \sigma_\varphi \times (\text{snr})_\varphi = \frac{(\text{snr})_\varphi}{C}\sqrt{\frac{2}{N}}$$

To improve the phase detection limit, one needs to optimize the contrast C of the hologram together and the total number of electrons. Each reduction of contrast must be counterbalanced by a quadratic increase of the total number of electrons per pixels. The contrast is therefore central in off-axis electron holography and the relationship with experimental parameters needs to be properly described. As we shall demonstrate in the following, the best obtainable contrast is a function of the degree of coherence of the electron beam.

Like any interference phenomena, electron holography requires the highest spatial coherence. In the following we will show that the most important parameter controlling the degree of spatial coherence is the brightness of the electron gun. As in the case of conventional electron sources, this parameter will be critical to implement ultrafast electron holography with ultrashort electron pulses.

An ideal point source would be able to generate an ideal hologram with a theoretical contrast $C = 1$. Unfortunately, such an ideal electron source does not exist. A real source is defined by an intensity distribution in space and wavenumber $i_s(\xi, \mathbf{k})$. For the sake of simplicity, we consider in the following a 1-D source. Considering the source distribution as the sum of incoherent point sources without phase relation, we can write the hologram intensity as the sum of individual holograms generated by each independent point source. Considering $A_0(\mathbf{r}) = A_s(\mathbf{r}) = 1$ and $\varphi_s(\mathbf{r}) = 0$ the hologram intensity along the x direction perpendicular to the biprism wire can be rewritten as (Born et al., 1999)

$$I(x) = 2 \iint i_s(\xi, \mathbf{k})\big(1 + \cos(2\pi q_0 \cdot x + \varphi(\xi, \mathbf{k}))\big)\, d\xi\, d\mathbf{k}$$

where $q_0 = \Delta k_0 = k_0\alpha_2$ and $\varphi(\xi, \mathbf{k}) = 2\pi(k_0\alpha_1\xi + \kappa\alpha_2 x)$ with $\kappa = k - k_0$ (k_0 is the mean wavenumber of the source) and α_1, α_2 defined in Fig. 19.

Assuming that each point source located in ξ emits the same spectrum of wavenumbers $s(\mathbf{k})$ defined by $\int i(\xi)\, d\xi = 1$ and $\int s(\mathbf{k})\, d\mathbf{k} = 1$, the hologram intensity can be rewritten as

$$I(x) = 2\left(1 + \int s(\mathbf{k})\int i(\xi) \cos(2\pi(q_0 \cdot x + k_0\alpha_1\xi + \kappa\alpha_2 x))\, d\xi\, d\mathbf{k}\right)$$
$$I(x) = 2\big(1 + |\gamma|\cos(2\pi q_0 \cdot x + \rho)\big)$$

with $\gamma = |\gamma|e^{i\rho}$ the so-called complex degree of coherence. The latter is the product of the spatial and temporal degree of coherence $\gamma = \gamma_{sp}\gamma_{temp} = |\gamma_{sp}|e^{i\rho_{sp}}|\gamma_{temp}|e^{i\rho_{temp}}$. The temporal and spatial complex degrees of coherence can be written as

$$\gamma_{temp}(x) = \int s(\kappa)e^{-2\pi i\kappa\alpha_2 x}\, d\kappa$$

$$\gamma_{\text{sp}}(\alpha_1) = \int i(\xi) e^{-2\pi i k_0 \alpha_1 \xi} \, d\xi$$

The latter equations are respectively the well-known Wiener–Khintchine and Van Cittert–Zernike theorems which relate the temporal (resp. spatial) degree of coherence to the Fourier transform of the spectral (resp. spatial) power density of the source (Born et al., 1999). In the following we will always consider $|\gamma_{\text{temp}}| = 1$. Indeed for a high voltage source (some hundreds of keV), which exhibits an energy spread of ~ 1 eV, the complex degree of temporal coherence will limit the number of hologram fringes to 10^5. Usually an electron hologram has no more than 1000 fringes and the contribution of temporal coherence can be neglected provided that the energy spread remains below 10 eV (Lichte & Freitag, 2000). The main limiting factor in the hologram contrast in usual TEM conditions is therefore the spatial coherence.

It is clear from the last equation that the smallest sources are naturally expected to yield the best hologram contrast. In principle, it could be possible to obtain a hologram with any electron source. Indeed, one can achieve a given contrast at will using an appropriate demagnification of the electron source. However, a strong demagnification will induce a strong reduction of the current density in the hologram since the electron source is characterized by a fixed brightness. The brightness is an invariant parameter in the optical system (Born et al., 1999) defined by

$$B = \frac{I}{S\Omega}$$

where I is the emission current, S the emission surface, and Ω the emission solid angle.

To extract the brightness contribution in the hologram contrast, we will assume a Gaussian intensity distribution of the source:

$$i(\xi) = \frac{1}{\pi r_0^2} e^{-\xi^2 / r_0^2}$$

The complex degree of coherence can then be rewritten as

$$\gamma(\alpha_1) = \int i(\xi) e^{-2\pi i k_0 \alpha_1 \xi} \, d\xi = \int \frac{1}{\pi r_0^2} e^{-\xi^2 / r_0^2} e^{-2\pi i k_0 \alpha_1 \xi} \, d\xi$$

As previously detailed, the hologram contrast C is directly related to the absolute value of the complex degree of coherence which can be written as

$$\left|\gamma(\alpha_1)\right| = \left|\int \frac{1}{\pi r_0^2} e^{-\xi^2/r_0^2} e^{-2\pi i k_0 \alpha_1 \xi} \, d\xi\right| = \frac{1}{\pi r_0^2}\left|\int e^{-\xi^2/r_0^2 - 2\pi i k_0 \alpha_1 \xi} \, d\xi\right|$$

$$= e^{-(\pi k_0 \alpha_1 r_0)^2}$$

Taking $S = \pi r_0^2$ as the emission surface and $\Omega = \pi \alpha_1^2$ the emission solid angle, the emission current can be written $I = B(\pi r_0 \alpha_1)^2$. Because the degree of coherence directly corresponds to the experimental contrast of the hologram fringes $|\gamma| = e^{-(\pi k_0 \alpha_1 r_0)^2} = C$, the coherent current available to capture a hologram of contrast C is given by

$$I(C) = -B \frac{\ln(C)}{k_0^2}$$

The coherent current is then fully determined by the source brightness. This equation further shows that an increase in the acceleration voltage to decrease the aberrations and therefore improve the brightness does not improve the coherent current due to the k_0^2 term in the denominator (Lichte, 1993).

5.3 Implications for Time-Resolved Electron Holography

With the advent of high brightness ultrafast electron sources based on laser triggered Schottky field emission or cold field emission sources, it becomes possible to perform time-resolved electron holography experiments in the stroboscopic pump-probe configuration (Feist et al., 2017; Caruso et al., 2017; Houdellier et al., 2018). Fig. 20 shows the first off-axis electron hologram obtained using femtosecond electron pulses from a laser triggered CFEG (Houdellier et al., 2018).

However, the total amount of current in laser-driven mode is reduced by a factor $\sim f_{rep} \Delta t_{pulse}$ compared to the continuous mode, f_{rep} being the laser repetition rate and Δt_{pulse} the electron pulse duration. Furthermore, the brightness is an invariant parameter only influenced by the optical system aberrations. Therefore, the optimization of the quality of an ultrafast electron hologram will require to carefully adjust all other contributions affecting the phase detection limit. For instance, a rotationally symmetric coherent illumination is not necessary in off-axis electron holography (Lehmann, 2004). Indeed, a high degree of coherence is needed only in the direction perpendicular to the biprism wire. An elliptical illumination can then be used to improve the complex spatial degree of coherence in

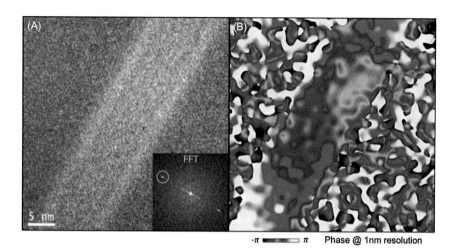

Figure 20 (A) Ultrafast off-axis electron hologram acquired in vacuum. (B) Corresponding phase map obtained at 1 nm resolution. Reprinted with permission from Houdellier et al. (2018).

the hologram direction by increasing the demagnification of the electron gun source size along this direction while increasing the current density in the wire direction and therefore decreasing the total exposure time.

In ultrafast electron holography the exposure time cannot be decreased below an optimum value determined by the current density in the hologram plane, the camera properties, and the microscope instabilities. Previously we assumed a Poisson distribution of the ultrafast electron beam, therefore summing as many electrons as possible in every pixel will reduce the noise contrast. But longer exposure times will also influence the hologram contrast through the microscope/biprism instabilities which can be accounted for by an additional factor C_{inst} in the contrast $C = |\gamma| C_{inst}$.

Due to the limited number of electrons provided by ultrafast high brightness electron sources, the detector quantum efficiency will have a crucial influence on the optimum exposure time, like in classical low dose electron microscopy (Yakovlev & Libera, 2008). Taking into account all the contributions discussed above, the phase detection limit can be written as

$$\delta\varphi = \frac{(snr)_\varphi}{|\gamma| C_{inst}} \sqrt{\frac{2}{N_{el} DQE}}$$

where N_{el} is the numbers of electrons per pixels, which can be expressed by $N_{el} = N_e \times f_{rep} \times \Delta t_{acqu}$ where N_e is the number of electrons per pulse,

f_{rep} is the repetition rate of the laser, Δt_{acqu} is the acquisition time, and DQE the quantum efficiency of the detector.

For instance, in the hologram reported in Fig. 20, the electron dose on the sample is ≈ 7 $e^-/\text{Å}^2$ per frame (150 s exposure time, $f_{rep} = 2$ MHz) and the pixel size is 0.5 Å. The contrast of the hologram is 20% and the DQE of the Gatan USC1000 camera is 0.057 at 1/2 of the Nyquist frequency. From these values, a standard deviation of the phase of $\sigma_\varphi = 1$ rad can be computed. It is generally admitted that a signal–noise ratio between 3 and 10 is necessary to properly determine the result of an experiment without ambiguity. As a consequence assuming an ideal situation given by a $(snr)_\varphi = 3$ and $C_{inst} = 1$ the phase detection limit between two points separated by 1 nm is $\delta\varphi_{1\text{ nm}} \approx 3$ rad. This value can be improved by decreasing the mask size in the Fourier space, *i.e.* increasing the "effective pixel" size and therefore the number of accumulated electrons.

It is clear that the number of electrons N_{el} contributes to the phase detection limit only through its square root, whereas the damping term C_{inst} has a linear contribution, which means that increasing the exposure time beyond a certain value will have a global detrimental effect on the phase detection limit that would not be compensated by an increase of the number of electrons per pixel.

To perform ultrafast electron holography it is then crucial to optimize the following competing parameters:

- $|\gamma|$ should be maximized using the highest brightness source and an optimized aspect ratio of the elliptical illumination to strongly demagnify the source size perpendicular to the biprism wire.
- N_{el} should be maximized using an optimized aspect ratio of the elliptical illumination to increase the current density on the detector without deteriorating $|\gamma|$ and by selecting an exposure time which minimizes the contribution of the damping term C_{inst}. To a lesser extent, this parameter can also be adjusted by increasing the number of electrons per pulse N_e or the repetition rate of the laser f_{rep}.

Finally, due to the limited brightness and limited exposure time available, the choice of the camera will be crucial. Integrating new direct electron-detection camera of high sensitivity should strongly enhance the possibility for ultrafast electron holography. Indeed, such detectors with an improved modulation transfer function (MTF) at high spatial frequencies allow for higher resolutions at lower magnifications, which leads to an increased effective field of view and current density, and then decrease of the exposure time (Chang, Dwyer, Barthel, Boothroyd, & Dunin-Borkowski, 2016).

Furthermore, such detectors with improved DQE will enhance the phase detection limit. At the Nyquist frequency the direct electron-detection detector has a DQE 50% higher than a standard CCD due to lower high frequency noise and will then contribute to improve the hologram noise without increasing the contribution from the instabilities of the microscope (Clough, Moldovan, & Kirkland, 2014).

Finding all the optimum experimental conditions between elliptical illumination aspect ratio, exposure time, and camera parameters for holography is a study under progress on the FE-UTEM in Toulouse.

6. CONCLUSION

Pioneered in the group of O. Bostanjoglo at the Technical University of Berlin in the 80s, the development of time-resolved Transmission Electron Microscopy has considerably benefited from evolutions in ultrafast laser and TEM technology. This enabled to reach, more than a decade ago, sub-nanometer/femtosecond spatio-temporal resolution. Indeed, by getting rid of the Coulomb repulsion inside the ultrashort electrons pulse, the concept of single electron imaging developed in Caltech improved by 6 orders of magnitude the temporal resolution of time-resolved TEMs. This first major breakthrough opened a new window on the ultrafast dynamics of nanoscale systems and enabled many applications in fields as diverse as nanomechanics, nanomagnetism, chemistry, biology, and nano-optics. Boosted by conceptual advances and far-reaching applications, a very dynamic field has emerged.

The vitality of the field is clear with an ever increasing number of laboratories now equipped with time-resolved TEMs in the U.S. (California Institute of Technology, Lawrence Livermore National Laboratory, University of Illinois, University of Minnesota, Purdue University), Germany (University of Göttingen), Sweden (KTH Royal Institute of Technology), Switzerland (EPFL), France (IPCMS Strasbourg, CEMES Toulouse), China (Beijing), Japan (Osaka University). Several other projects are under active development in Korea (Ulsan National Institute of Science & Technology), Israel (Technion), Russia, Sweden (Uppsala University), the U.S. (Pacific Northwest National Laboratory).

A lot of efforts are currently being made along several directions. First, the instrument itself is being improved. New high brightness ultrafast electron sources based on laser-driven nanoemitters have been developed for

UTEM investigations requiring coherent ultrashort electron pulses. Improvements of the electron optics capabilities are also under investigation to further improve the spatio-temporal resolution of DTEM and UTEM. Aberration correctors combined with new electrons detectors technologies will undoubtedly enhance the UTEM capabilities as they did for conventional TEMs. The latest improvements in ultrafast laser, vacuum technologies, and electronics will of course push further the resolutions. Exploiting electron/light interaction in the TEM column, new techniques are continuously flourishing to tailor the electron pulses for specific applications. Photon gating, optical or cavity-based compression schemes, coherent electron/photon interactions are developed to further decrease the temporal resolution. Some recent seminal studies show that attosecond timescale is now within the reach of ultrafast TEMs.

ACKNOWLEDGMENTS

The authors thank the Institut de Physique du CNRS, Agence Nationale de la Recherche for financial support (ANR grant ANR-14-CE26-0013), and Programme Investissements d'Avenir (Muse grant ANR-10-LABX-0037-NEXT). The authors acknowledge financial support from the European Union under the Seventh Framework Program under a contract for an Integrated Infrastructure Initiative (Reference 312483-ESTEEM2). We thank P. Abeilhou, S. Weber, and M. Pelloux for technical discussions and support. We are pleased to express here our gratitude for the continuous support and help from M. Kociak, M.J. Hÿtch, and E. Snoeck.

REFERENCES

Aeschlimann, M., Bauer, M., Pawlik, S., Weber, W., Burgermeister, R., Oberli, D., . . . Siegmann, H. C. (1997). Ultrafast spin-dependent electron dynamics in fcc Co. *Physical Review Letters*, *79*, 5158–5161. https://doi.org/10.1103/PhysRevLett.79.5158.

Aseyev, S. A., Weber, P. M., & Ischenko, A. A. (2013). Ultrafast electron microscopy for chemistry, biology and material science. *Journal of Analytical Sciences, Methods and Instrumentation*, *03*, 30. https://doi.org/10.4236/jasmi.2013.31005.

Bartels, A., Cerna, R., Kistner, C., Thoma, A., Hudert, F., Janke, C., . . . Dekorsy, T. (2007). Ultrafast time-domain spectroscopy based on high-speed asynchronous optical sampling. *Review of Scientific Instruments*, *78*, 035107. https://doi.org/10.1063/1.2714048.

Barwick, B., Flannigan, D. J., & Zewail, A. H. (2009). Photon-induced near-field electron microscopy. *Nature*, *462*, 902–906.

Barwick, B., Park, H. S., Kwon, O. H., Baskin, J. S., & Zewail, A. H. (2008). 4D imaging of transient structures and morphologies in ultrafast electron microscopy. *Science*, *322*, 1227–1231. https://doi.org/10.1126/science.1164000.

Baskin, J. S., Park, H. S., & Zewail, A. H. (2011). Nanomusical systems visualized and controlled in 4D electron microscopy. *Nano Letters*, *11*, 2183–2191. https://doi.org/10.1021/nl200930a.

Battiato, M., Carva, K., & Oppeneer, P. M. (2010). Superdiffusive spin transport as a mechanism of ultrafast demagnetization. *Physical Review Letters, 105*. https://doi.org/10.1103/PhysRevLett.105.027203.

Beaurepaire, E., Merle, J.-C., Daunois, A., & Bigot, J.-Y. (1996). Ultrafast spin dynamics in ferromagnetic nickel. *Physical Review Letters, 76*, 4250–4253. https://doi.org/10.1103/PhysRevLett.76.4250.

Beleggia, M., Schofield, M. A., Volkov, V. V., & Zhu, Y. (2004). On the transport of intensity technique for phase retrieval. *Ultramicroscopy, 102*, 37–49. https://doi.org/10.1016/j.ultramic.2004.08.004.

Berruto, G., Madan, I., Murooka, Y., Pomarico, E., Vanacore, G. M., Rajeswari, J., ... Carbone, F. (2018). Laser-induced skyrmion writing and erasing in an ultrafast cryo-TEM. arXiv preprint. arXiv:1709.00495.

Bigot, J.-Y., Vomir, M., & Beaurepaire, E. (2009). Coherent ultrafast magnetism induced by femtosecond laser pulses. *Nature Physics, 5*, 515–520. https://doi.org/10.1038/nphys1285.

Boeglin, C., Beaurepaire, E., Halté, V., López-Flores, V., Stamm, C., Pontius, N., ... Bigot, J.-Y. (2010). Distinguishing the ultrafast dynamics of spin and orbital moments in solids. *Nature, 465*, 458–461. https://doi.org/10.1038/nature09070.

Bormann, R., Gulde, M., Weismann, A., Yalunin, S. V., & Ropers, C. (2010). Tip-enhanced strong-field photoemission. *Physical Review Letters, 105*, 147601.

Bormann, R., Strauch, S., Schäfer, S., & Ropers, C. (2015). An ultrafast electron microscope gun driven by two-photon photoemission from a nanotip cathode. *Journal of Applied Physics, 118*. https://doi.org/10.1063/1.4934681.

Born, M., Wolf, E., Bhatia, A. B., Clemmow, P. C., Gabor, D., Stokes, A. R., ... Wilcock, W. L. (1999). *Principles of optics: Electromagnetic theory of propagation, interference and diffraction of light* (7th ed.). Cambridge: Cambridge University Press.

Bostanjoglo, O. (1999). High-speed electron microscopy. In P. W. Hawkes (Ed.), *Advances in imaging and electron physics* (pp. 21–61). Elsevier.

Bostanjoglo, O. (1989). Electron microscopy of fast processes. In P. W. Hawkes (Ed.), *Advances in electronics and electron physics* (pp. 209–279). Academic Press.

Bostanjoglo, O., Elschner, R., Mao, Z., Nink, T., & Weingartner, M. (2000). Nanosecond electron microscopes. *Ultramicroscopy, 81*, 141–147.

Bostanjoglo, O., & Heinricht, F. (1990). A reflection electron microscope for imaging of fast phase transitions on surfaces. *Review of Scientific Instruments, 61*, 1223–1229. https://doi.org/10.1063/1.1141952.

Bostanjoglo, O., Kornitzky, J., & Tornow, R. P. (1991). High-speed electron-microscopy of laser-induced vaporization of thin-films. *Journal of Applied Physics, 69*, 2581–2583.

Bostanjoglo, O., Niedrig, R., & Wedel, B. (1994). Ablation of metal films by picosecond laser pulses imaged with high-speed electron microscopy. *Journal of Applied Physics, 76*, 3045–3048. https://doi.org/10.1063/1.357484.

Bostanjoglo, O., & Nink, T. (1997). Liquid motion in laser pulsed Al, Co and Au films. *Applied Surface Science, 110*, 101–105.

Bostanjoglo, O., & Nink, T. (1996). Hydrodynamic instabilities in laser pulse-produced melts of metal films. *Journal of Applied Physics, 79*, 8725–8729. https://doi.org/10.1063/1.362499.

Bostanjoglo, O., & Rosin, T. (1980). Resonance oscillations of magnetic domain walls and Bloch lines observed by stroboscopic electron microscopy. *Physica Status Solidi A, 57*, 561–568. https://doi.org/10.1002/pssa.2210570212.

Bostanjoglo, O., & Rosin, T. (1977). Ultrasonically induced magnetic reversals observed by stroboscopic electron-microscopy. *Optica Acta*, *24*, 657–664.

Bostanjoglo, O., Tornow, R. P., & Tornow, W. (1987). Nanosecond transmission electron-microscopy and diffraction. *Journal of Physics E, Scientific Instruments*, *20*, 556–557.

Bostanjoglo, O., & Weingartner, M. (1997). Pulsed photoelectron microscope for imaging laser-induced nanosecond processes. *Review of Scientific Instruments*, *68*, 2456–2460.

Bouldi, N., Vollmers, N. J., Delpy-Laplanche, C. G., Joly, Y., Juhin, A., Sainctavit, P., ... Gerstmann, U. (2017). X-ray magnetic and natural circular dichroism from first principles: Calculation of K- and L_1-edge spectra. *Physical Review B*, *96*. https://doi.org/10.1103/PhysRevB.96.085123.

Bucksbaum, P. H., Schumacher, D. W., & Bashkansky, M. (1988). High-intensity Kapitza–Dirac effect. *Physical Review Letters*, *61*, 1182.

Carbone, F., Barwick, B., Kwon, O. H., Park, H. S., Baskin, J. S., & Zewail, A. H. (2009). EELS femtosecond resolved in 4D ultrafast electron microscopy. *Chemical Physics Letters*, *468*, 107–111. https://doi.org/10.1016/j.cplett.2008.12.027.

Carbone, F., Kwon, O.-H., & Zewail, A. H. (2009). Dynamics of chemical bonding mapped by energy-resolved 4D electron microscopy. *Science*, *325*, 181.

Caruso, G. M., Houdellier, F., Abeilhou, P., & Arbouet, A. (2017). Development of an ultrafast electron source based on a cold-field emission gun for ultrafast coherent TEM. *Applied Physics Letters*, *111*, 023101. https://doi.org/10.1063/1.4991681.

Chang, S. L. Y., Dwyer, C., Barthel, J., Boothroyd, C. B., & Dunin-Borkowski, R. E. (2016). Performance of a direct detection camera for off-axis electron holography. *Ultramicroscopy*, *161*, 90–97. https://doi.org/10.1016/j.ultramic.2015.09.004.

Chang, S. L. Y., Dwyer, C., Boothroyd, C. B., & Dunin-Borkowski, R. E. (2015). Optimising electron holography in the presence of partial coherence and instrument instabilities. *Ultramicroscopy*, *151*, 37–45. https://doi.org/10.1016/j.ultramic.2014.11.019.

Chapman, J. N., Batson, P. E., Waddell, E. M., & Ferrier, R. P. (1978). The direct determination of magnetic domain wall profiles by differential phase contrast electron microscopy. *Ultramicroscopy*, *3*, 203–214. https://doi.org/10.1016/S0304-3991(78)80027-8.

Chapman, J. N., & Kirk, K. J. (1997). Imaging magnetic structures in the transmission electron microscope. In G. C. Hadjipanayis (Ed.), *Magnetic hysteresis in novel magnetic materials* (pp. 195–206). Dordrecht: Springer Netherlands.

Chapman, J. N., Ploessl, R., & Donnet, D. M. (1992). Differential phase contrast microscopy of magnetic materials. *Ultramicroscopy*, *47*, 331–338. https://doi.org/10.1016/0304-3991(92)90162-D.

Chen, Z., Weyland, M., Ercius, P., Ciston, J., Zheng, C., Fuhrer, M. S., ... Findlay, S. D. (2016). Practical aspects of diffractive imaging using an atomic-scale coherent electron probe. *Ultramicroscopy*, *169*, 107–121. https://doi.org/10.1016/j.ultramic.2016.06.009.

Clough, R. N., Moldovan, G., & Kirkland, A. I. (2014). Direct detectors for electron microscopy. *Journal of Physics. Conference Series*, *522*, 012046. https://doi.org/10.1088/1742-6596/522/1/012046.

Crewe, A. V., Eggenberger, D. N., Wall, J., & Welter, L. M. (1968). Electron gun using a field emission source. *Review of Scientific Instruments*, *39*, 576–583. https://doi.org/10.1063/1.1683435.

da Silva, N. R., Möller, M., Feist, A., Ulrichs, H., Ropers, C., & Schäfer, S. (2017). Nanoscale mapping of ultrafast magnetization dynamics with femtosecond Lorentz microscopy. arXiv preprint. arXiv:1710.03307.

de Abajo, F. J. G., & Kociak, M. (2008). Electron energy-gain spectroscopy. *New Journal of Physics*, *10*, 073035.

de Knoop, L., Houdellier, F., Gatel, C., Masseboeuf, A., Monthioux, M., & Hÿtch, M. (2014). Determining the work function of a carbon-cone cold-field emitter by in situ electron holography. *Micron*, *63*, 2–8. https://doi.org/10.1016/j.micron.2014.03.005.

de Ruijter, W. J., & Weiss, J. K. (1993). Detection limits in quantitative off-axis electron holography. *Ultramicroscopy*, *50*, 269–283. https://doi.org/10.1016/0304-3991 (93)90196-5.

Dehm, G., Howe, J. M., & Zweck, J. (2012). Front matter. In *In-situ electron microscopy* (pp. I–XVIII). Wiley–VCH Verlag.

Domer, H., & Bostanjoglo, O. (2003). High-speed transmission electron microscope. *Review of Scientific Instruments*, *74*, 4369–4372. https://doi.org/10.1063/1.1611612.

Dunin-Borkowski, R. E., Kasama, T., & Harrison, R. J. (2015). Electron holography of nanostructured materials (Chapter 5). In *Nanocharacterisation: Vol. 2* (pp. 158–210). The Royal Society of Chemistry.

Echternkamp, K. E., Feist, A., Schafer, S., & Ropers, C. (2016). Ramsey-type phase control of free-electron beams. *Nature Physics*, *12*, 1000–1004.

Eggebrecht, T., Möller, M., Gatzmann, J. G., da Silva, N. R., Feist, A., Martens, U., ... Schäfer, S. (2017). Light-induced metastable magnetic texture uncovered by in situ Lorentz microscopy. *Physical Review Letters*, *118*. https://doi.org/10.1103/PhysRevLett.118.097203.

Ehberger, D., Hammer, J., Eisele, M., Krüger, M., Noe, J., Högele, A., ... Hommelhoff, P. (2015). Highly coherent electron beam from a laser-triggered tungsten needle tip. *Physical Review Letters*, *114*, 227601. https://doi.org/10.1103/PhysRevLett.114.227601.

Eisebitt, S., Lüning, J., Schlotter, W. F., Lörgen, M., Hellwig, O., Eberhardt, W., ... Stöhr, J. (2004). Lensless imaging of magnetic nanostructures by X-ray spectro-holography. *Nature*, *432*, 885–888. https://doi.org/10.1038/nature03139.

Elzinga, P. A., Kneisler, R. J., Lytle, F. E., Jiang, Y., King, G. B., & Laurendeau, N. M. (1987). Pump/probe method for fast analysis of visible spectral signatures utilizing asynchronous optical sampling. *Applied Optics*, *26*, 4303–4309. https://doi.org/10.1364/AO.26.004303.

Feist, A., Bach, N., da Silva, N. R., Danz, T., Möller, M., Priebe, K. E., ... Ropers, C. (2017). Ultrafast transmission electron microscopy using a laser-driven field emitter: Femtosecond resolution with a high coherence electron beam. 70th birthday of Robert Sinclair and 65th birthday of Nestor J. Zaluzec PICO 2017 – fourth conference on frontiers of aberration corrected electron microscopy. *Ultramicroscopy*, *176*, 63–73. https://doi.org/10.1016/j.ultramic.2016.12.005.

Feist, A., Echternkamp, K. E., Schauss, J., Yalunin, S. V., Schafer, S., & Ropers, C. (2015). Quantum coherent optical phase modulation in an ultrafast transmission electron microscope. *Nature*, *521*, 200–203.

Feist, A., da Silva, N. R., Liang, W., Ropers, C., & Schäfer, S. (2018). Nanoscale diffractive probing of strain dynamics in ultrafast transmission electron microscopy. *Structural Dynamics*, *5*, 014302. https://doi.org/10.1063/1.5009822.

Fitzpatrick, A. W. P., Lorenz, U. J., Vanacore, G. M., & Zewail, A. H. (2013). 4D cryo-electron microscopy of proteins. *Journal of the American Chemical Society*, *135*, 19123–19126. https://doi.org/10.1021/ja4115055.

Fitzpatrick, A. W. P., Park, S. T., & Zewail, A. H. (2013). Exceptional rigidity and biomechanics of amyloid revealed by 4D electron microscopy. *Proceedings of the National Academy of Sciences*, *110*, 10976–10981. https://doi.org/10.1073/pnas.1309690110.

Flannigan, D., Lobastov, V., & Zewail, A. (2007). Controlled nanoscale mechanical phenomena discovered with ultrafast electron microscopy. *Angewandte Chemie. International Edition in English, 46*, 9206–9210.

Flannigan, D. J., Samartzis, P. C., Yurtsever, A., & Zewail, A. H. (2009). Nanomechanical motions of cantilevers: Direct imaging in real space and time with 4D electron microscopy. *Nano Letters, 9*, 875–881. https://doi.org/10.1021/nl803770e.

Förster, M., Paschen, T., Krüger, M., Lemell, C., Wachter, G., Libisch, F., ... Hommelhoff, P. (2016). Two-color coherent control of femtosecond above-threshold photoemission from a tungsten nanotip. *Physical Review Letters, 117*, 217601. https://doi.org/10.1103/PhysRevLett.117.217601.

Freimund, D. L., Aflatooni, K., & Batelaan, H. (2001). Observation of the Kapitza–Dirac effect. *Nature, 413*, 142–143.

Gabor, D. (1948). A new microscopic principle. *Nature, 161*, 777–778. https://doi.org/10.1038/161777a0.

Gahlmann, A., Park, S. T., & Zewail, A. H. (2008). Ultrashort electron pulses for diffraction, crystallography and microscopy: Theoretical and experimental resolutions. *Physical Chemistry Chemical Physics, 10*, 2894–2909. https://doi.org/10.1039/b802136h.

Garcia de Abajo, F. J. (2010). Optical excitations in electron microscopy. *Reviews of Modern Physics, 82*, 209.

García de Abajo, F. J., Asenjo-Garcia, A., & Kociak, M. (2010). Multiphoton absorption and emission by interaction of swift electrons with evanescent light fields. *Nano Letters, 10*, 1859–1863. https://doi.org/10.1021/nl100613s.

García de Abajo, F. J., Barwick, B., & Carbone, F. (2016). Electron diffraction by plasmon waves. *Physical Review B, 94*, 041404. https://doi.org/10.1103/PhysRevB.94.041404.

Grinolds, M. S., Lobastov, V. A., Weissenrieder, J., & Zewail, A. H. (2006). Four-dimensional ultrafast electron microscopy of phase transitions. *Proceedings of the National Academy of Sciences, 103*, 18427–18431.

Haider, M., Hartel, P., Muller, H., Uhlemann, S., & Zach, J. (2009). Current and future aberration correctors for the improvement of resolution in electron microscopy. *Philosophical Transactions - Royal Society. Mathematical, Physical and Engineering Sciences, 367*, 3665–3682. https://doi.org/10.1098/rsta.2009.0121.

Haider, M., Uhlemann, S., Schwan, E., Rose, H., Kabius, B., & Urban, K. (1998). Electron microscopy image enhanced. *Nature, 392*, 768–769. https://doi.org/10.1038/33823.

Hansteen, F., Kimel, A., Kirilyuk, A., & Rasing, T. (2005). Femtosecond photomagnetic switching of spins in ferrimagnetic garnet films. *Physical Review Letters, 95*. https://doi.org/10.1103/PhysRevLett.95.047402.

Harscher, A., & Lichte, H. (1996). Experimental study of amplitude and phase detection limits in electron holography. *Ultramicroscopy, 64*, 57–66. https://doi.org/10.1016/0304-3991(96)00019-8.

Hassan, M. T., Baskin, J. S., Liao, B., & Zewail, A. H. (2017). High-temporal-resolution electron microscopy for imaging ultrafast electron dynamics. *Nature Photonics, 11*, 425. https://doi.org/10.1038/nphoton.2017.79.

Hassan, M. T., Liu, H., Baskin, J. S., & Zewail, A. H. (2015). Photon gating in four-dimensional ultrafast electron microscopy. *Proceedings of the National Academy of Sciences, 112*, 12944–12949. https://doi.org/10.1073/pnas.1517942112.

Hernandez Garcia, C., & Brau, C. A. (2002). Pulsed photoelectric field emission from needle cathodes. Proceedings of the 23rd international free electron laser conference and 8th FEL users workshop. *Nuclear Instruments & Methods in Physics Research. Section A,*

Accelerators, Spectrometers, Detectors and Associated Equipment, 483, 273–276. https://doi.org/10.1016/S0168-9002(02)00326-1.

Hertel, R., Gliga, S., Fähnle, M., & Schneider, C. M. (2007). Ultrafast nanomagnetic toggle switching of vortex cores. *Physical Review Letters, 98.* https://doi.org/10.1103/PhysRevLett.98.117201.

Hillebrands, B., & Ounadjela, K. (Eds.). (2002). *Spin dynamics in confined magnetic structures I. Topics in applied physics.* Berlin, Heidelberg: Springer.

Hohlfeld, J., Matthias, E., Knorren, R., & Bennemann, K. H. (1997). Nonequilibrium magnetization dynamics of nickel. *Physical Review Letters, 78*, 4861–4864. https://doi.org/10.1103/PhysRevLett.78.4861.

Hommelhoff, P., Kealhofer, C., & Kasevich, M. A. (2006). Ultrafast electron pulses from a tungsten tip triggered by low-power femtosecond laser pulses. *Physical Review Letters, 97*, 247402.

Hommelhoff, P., Sortais, Y., Aghajani-Talesh, A., & Kasevich, M. A. (2006). Field emission tip as a nanometer source of free electron femtosecond pulses. *Physical Review Letters, 96*, 077401.

Houdellier, F., Caruso, G. M., Weber, S., Kociak, M., & Arbouet, A. (2018). Development of a high brightness ultrafast transmission electron microscope based on a laser-driven cold field emission source. *Ultramicroscopy, 186*, 128–138. https://doi.org/10.1016/j.ultramic.2017.12.015.

Houssameddine, D., Ebels, U., Delaët, B., Rodmacq, B., Firastrau, I., Ponthenier, F., ... Dieny, B. (2007). Spin-torque oscillator using a perpendicular polarizer and a planar free layer. *Nature Materials, 6*, 447–453. https://doi.org/10.1038/nmat1905.

Hu, J., Vanacore, G. M., Yang, Z., Miao, X., & Zewail, A. H. (2015). Transient structures and possible limits of data recording in phase-change materials. *ACS Nano, 9*, 6728–6737. https://doi.org/10.1021/acsnano.5b01965.

Isakozawa, S., Kashikura, Y., Sato, Y., Takahashi, T., Ichihashi, M., & Murakoshi, H. (1989). *Proceedings of EMCA, 47*, 112–113.

Joy, D. C., Zhang, Y.-S., Zhang, X., Hashimoto, T., Bunn, R. D., Allard, L., ... Nolan, T. A. (1993). Practical aspects of electron holography. *Ultramicroscopy, 51*, 1–14. https://doi.org/10.1016/0304-3991(93)90130-P.

Kapitza, P. L., & Dirac, P. A. M. (1933). The reflection of electrons from standing light waves. *Mathematical Proceedings of the Cambridge Philosophical Society, 29*, 297–300.

Kazantseva, N., Nowak, U., Chantrell, R. W., Hohlfeld, J., & Rebei, A. (2008). Slow recovery of the magnetisation after a sub-picosecond heat pulse. *Europhysics Letters, 81*, 27004. https://doi.org/10.1209/0295-5075/81/27004.

Kim, K.-H., & Zuo, J.-M. (2013). Symmetry quantification and mapping using convergent beam electron diffraction. *Ultramicroscopy, 124*, 71–76. https://doi.org/10.1016/j.ultramic.2012.09.002.

Kimel, A. V., Kirilyuk, A., Usachev, P. A., Pisarev, R. V., Balbashov, A. M., & Rasing, T. (2005). Ultrafast non-thermal control of magnetization by instantaneous photomagnetic pulses. *Nature, 435*, 655–657. https://doi.org/10.1038/nature03564.

Kimura, W. D., Kim, G. H., Romea, R. D., Steinhauer, L. C., Pogorelsky, I. V., Kusche, K. P., ... Liu, Y. (1995). Laser acceleration of relativistic electrons using the inverse Cherenkov effect. *Physical Review Letters, 74*, 546.

King, W. E., Armstrong, M. R., Bostanjoglo, O., & Reed, B. W. (2007). High-speed electron microscopy. In *Science of microscopy* (pp. 406–444). New York: Springer.

King, W. E., Campbell, G. H., Frank, A., Reed, B., Schmerge, J. F., Siwick, B. J., ... Weber, P. M. (2005). Ultrafast electron microscopy in materials science, biology, and chemistry. *Journal of Applied Physics, 97*, 111101.

Kleinschmidt, H., & Bostanjoglo, O. (2001). Pulsed mirror electron microscope: A fast near-surface imaging probe. *Review of Scientific Instruments, 72*, 3898–3901.

Kociak, M., & Zagonel, L. F. (2017). Cathodoluminescence in the scanning transmission electron microscope. *Ultramicroscopy, 174*, 50–69.

Kohn, A., Habibi, A., & Mayo, M. (2016). Experimental evaluation of the 'transport-of-intensity' equation for magnetic phase reconstruction in Lorentz transmission electron microscopy. *Ultramicroscopy, 160*, 44–56. https://doi.org/10.1016/j.ultramic.2015.09.011.

Koopmans, B., Kicken, H. H. J. E., van Kampen, M., & de Jonge, W. J. M. (2005). Microscopic model for femtosecond magnetization dynamics. *Journal of Magnetism and Magnetic Materials, 286*, 271–275. https://doi.org/10.1016/j.jmmm.2004.09.079.

Kozák, M., Eckstein, T., Schönenberger, N., & Hommelhoff, P. (2018). Inelastic ponderomotive scattering of electrons at a high-intensity optical travelling wave in vacuum. *Nature Physics, 14*, 121–125. https://doi.org/10.1038/nphys4282.

Kozák, M., Schönenberger, N., & Hommelhoff, P. (2018). Ponderomotive generation and detection of attosecond free-electron pulse trains. *Physical Review Letters, 120*, 103203. https://doi.org/10.1103/PhysRevLett.120.103203.

Krajnak, M., McGrouther, D., Maneuski, D., Shea, V. O., & McVitie, S. (2016). Pixelated detectors and improved efficiency for magnetic imaging in STEM differential phase contrast. *Ultramicroscopy, 165*, 42–50. https://doi.org/10.1016/j.ultramic.2016.03.006.

Krämer, S., & Mayer, J. (1999). Using the Hough transform for HOLZ line identification in convergent beam electron diffraction. *Journal of Microscopy, 194*, 2–11. https://doi.org/10.1046/j.1365-2818.1999.00475.x.

Krivanek, O. L., Chisholm, M. F., Nicolosi, V., Pennycook, T. J., Corbin, G. J., Dellby, N., ... Pennycook, S. J. (2010). Atom-by-atom structural and chemical analysis by annular dark-field electron microscopy. *Nature, 464*, 571–574. https://doi.org/10.1038/nature08879.

Krivanek, O. L., Lovejoy, T. C., Dellby, N., Aoki, T., Carpenter, R. W., Rez, P., ... Crozier, P. A. (2014). Vibrational spectroscopy in the electron microscope. *Nature, 514*, 209–212. https://doi.org/10.1038/nature13870.

Kruger, M., Schenk, M., & Hommelhoff, P. (2011). Attosecond control of electrons emitted from a nanoscale metal tip. *Nature, 475*, 78–81.

Kwon, O. H., Barwick, B., Park, H. S., Baskin, J. S., & Zewail, A. H. (2008a). 4D visualization of embryonic, structural crystallization by single-pulse microscopy. *Proceedings of the National Academy of Sciences of the United States of America, 105*, 8519–8524. https://doi.org/10.1073/pnas.0803344105.

Kwon, O.-H., Barwick, B., Park, H. S., Baskin, J. S., & Zewail, A. H. (2008b). Nanoscale mechanical drumming visualized by 4D electron microscopy. *Nano Letters, 8*, 3557–3562. https://doi.org/10.1021/nl8029866.

LaGrange, T., Reed, B., DeHope, W., Shuttlesworth, R., & Huete, G. (2011). Movie mode dynamic transmission electron microscopy (DTEM): Multiple frame movies of transient states in materials with nanosecond time resolution. *Microscopy and Microanalysis, 17*, 458–459. https://doi.org/10.1017/S1431927611003163.

LaGrange, T., Reed, B. W., & Masiel, D. J. (2015). Movie-mode dynamic electron microscopy. *MRS Bulletin, 40*, 22–28. https://doi.org/10.1557/mrs.2014.282.

Landau, L. D., Lifšic, E. M., Pitaevskij, L. P., & Landau, L. D. (2008). *Electrodynamics of continuous media* (2nd ed., rev. and enl. ed.). *Course of theoretical physics: Vol. 26.* Amsterdam: Elsevier/Butterworth-Heinemann.

Laude, V., Belkhir, A., Alabiad, A. F., Addouche, M., Benchabane, S., Khelif, A., ... Baida, F. I. (2017). Extraordinary nonlinear transmission modulation in a doubly resonant acousto-optical structure. *Optica, 4,* 1245–1250. https://doi.org/10.1364/OPTICA.4.001245.

LeBeau, J. M., Findlay, S. D., Allen, L. J., & Stemmer, S. (2010). Position averaged convergent beam electron diffraction: Theory and applications. *Ultramicroscopy, 110,* 118–125. https://doi.org/10.1016/j.ultramic.2009.10.001.

Lehmann, M. (2004). Influence of the elliptical illumination on acquisition and correction of coherent aberrations in high-resolution electron holography. *Ultramicroscopy, 100,* 9–23. https://doi.org/10.1016/j.ultramic.2004.01.005.

Lichte, H. (1993). Parameters for high-resolution electron holography. *Ultramicroscopy, 51,* 15–20. https://doi.org/10.1016/0304-3991(93)90131-G.

Lichte, H., Formanek, P., Lenk, A., Linck, M., Matzeck, C., Lehmann, M., ... Simon, P. (2007). Electron holography: Applications to materials questions. *Annual Review of Materials Research, 37,* 539–588. https://doi.org/10.1146/annurev.matsci.37.052506.084232.

Lichte, H., & Freitag, B. (2000). Inelastic electron holography. *Ultramicroscopy, 81,* 177–186. https://doi.org/10.1016/S0304-3991(99)00188-6.

Liu, H., Kwon, O.-H., Tang, J., & Zewail, A. H. (2014). 4D imaging and diffraction dynamics of single-particle phase transition in heterogeneous ensembles. *Nano Letters, 14,* 946–954. https://doi.org/10.1021/nl404354g.

Lobastov, V. A., Srinivasan, R., Vigliotti, F., Ruan, C. Y., Feenstra, J. S., Chen, S. Y., ... Zewail, A. H. (2004). Ultrafast electron diffraction – from the gas phase to the condensed phase with picosecond and femtosecond resolution. *Ultrafast Optics IV, 95,* 419–435.

Lobastov, V. A., Srinivasan, R., & Zewail, A. H. (2005). Four-dimensional ultrafast electron microscopy. *Proceedings of the National Academy of Sciences of the United States of America, 102,* 7069–7073.

Lorenz, U. J., & Zewail, A. H. (2013). Biomechanics of DNA structures visualized by 4D electron microscopy. *Proceedings of the National Academy of Sciences, 110,* 2822–2827. https://doi.org/10.1073/pnas.1300630110.

Ma, Y.-Z., Hertel, T., Vardeny, Z., Fleming, G., & Valkunas, L. (2008). Ultrafast spectroscopy of carbon nanotubes. *Topics in Applied Physics, 111,* 321–352.

Mac Raighne, A., Fernandez, G. V., Maneuski, D., McGrouther, D., & O'Shea, V. (2011). Medipix2 as a highly flexible scanning/imaging detector for transmission electron microscopy. *Journal of Instrumentation, 6,* C01047. https://doi.org/10.1088/1748-0221/6/01/C01047.

Mamishin, S., Kubo, Y., Cours, R., Monthioux, M., & Houdellier, F. (2017). 200 keV cold field emission source using carbon cone nanotip: Application to scanning transmission electron microscopy. *Ultramicroscopy, 182,* 303–307.

McCartney, M. R., Agarwal, N., Chung, S., Cullen, D. A., Han, M.-G., He, K., ... Smith, D. J. (2010). Quantitative phase imaging of nanoscale electrostatic and magnetic fields using off-axis electron holography. *Ultramicroscopy, 110,* 375–382. https://doi.org/10.1016/j.ultramic.2010.01.001.

McCartney, M., Dunin-Borkowski, R., & Smith, D. (2005). Electron holography of magnetic nanostructures. In H. Hopster, & H. P. Oepen (Eds.), *Magnetic microscopy of nanostructures. Nanoscience and technology.* Berlin, Heidelberg: Springer.

McCartney, M. R., & Gajdardziska-Josifovska, M. (1994). Absolute measurement of normalized thickness, $t/\lambda i$, from off-axis electron holography. *Ultramicroscopy, 53,* 283–289. https://doi.org/10.1016/0304-3991(94)90040-X.

Milazzo, A.-C., Moldovan, G., Lanman, J., Jin, L., Bouwer, J. C., Klienfelder, S., ... Xuong, N.-H. (2010). Characterization of a direct detection device imaging camera for transmission electron microscopy. *Ultramicroscopy, 110,* 741–744. https://doi.org/10.1016/j.ultramic.2010.03.007.

Mizuno, K., Pae, J., Nozokido, T., & Furuya, K. (1987). Experimental evidence of the inverse Smith–Purcell effect. *Nature, 328,* 45–47.

Mohammed, O. F., Yang, D.-S., Pal, S. K., & Zewail, A. H. (2011). 4D scanning ultrafast electron microscopy: Visualization of materials surface dynamics. *Journal of the American Chemical Society, 133,* 7708 7711.

Morimoto, Y., & Baum, P. (2018). Diffraction and microscopy with attosecond electron pulse trains. *Nature Physics, 14,* 252–256. https://doi.org/10.1038/s41567-017-0007-6.

Morniroli, J. P. (2006). CBED and LACBED characterization of crystal defects. *Journal of Microscopy, 223,* 240–245. https://doi.org/10.1111/j.1365-2818.2006.01630.x.

Morniroli, J. P. (2002). *Large-angle convergent-beam electron diffraction applications to crystal defects.* Taylor & Francis Group, Société Française de Microscopie.

Morniroli, J.-P., Stadelmann, P., Ji, G., & Nicolopoulos, S. (2010). The symmetry of precession electron diffraction patterns. *Journal of Microscopy, 237,* 511–515. https://doi.org/10.1111/j.1365-2818.2009.03311.x.

Mukamel, S. (1999). *Principles of nonlinear optical spectroscopy. Oxford series on optical and imaging sciences.* USA: Oxford University Press.

Nink, T., Galbert, F., Mao, Z. L., & Bostanjoglo, O. (1999). Dynamics of laser pulse-induced melts in Ni–P visualized by high-speed transmission electron microscopy. *Applied Surface Science, 138,* 439–443.

Paarmann, A., Gulde, M., Muller, M., Schafer, S., Schweda, S., Maiti, M., ... Ernstorfer, R. (2012). Coherent femtosecond low-energy single-electron pulses for time-resolved diffraction and imaging: A numerical study. *Journal of Applied Physics, 112,* 113109.

Park, H. S., Baskin, J. S., Barwick, B., Kwon, O.-H., & Zewail, A. H. (2009). 4D ultrafast electron microscopy: Imaging of atomic motions, acoustic resonances, and moiré fringe dynamics. *Ultramicroscopy, 110,* 7–19. https://doi.org/10.1016/j.ultramic.2009.08.005.

Park, H. S., Baskin, J. S., Kwon, O. H., & Zewail, A. H. (2007). Atomic-scale imaging in real and energy space developed in ultrafast electron microscopy. *Nano Letters, 7,* 2545–2551. https://doi.org/10.1021/nl071369q.

Park, H. S., Baskin, J. S., & Zewail, A. H. (2010). 4D Lorentz electron microscopy imaging: Magnetic domain wall nucleation, reversal, and wave velocity. *Nano Letters, 10,* 3796–3803. https://doi.org/10.1021/nl102861e.

Park, H. S., Kwon, O. H., Baskin, J. S., Barwick, B., & Zewail, A. H. (2009). Direct observation of martensitic phase-transformation dynamics in iron by 4D single-pulse electron microscopy. *Nano Letters, 9,* 3954–3962. https://doi.org/10.1021/nl9032704.

Park, S. T., Flannigan, D. J., & Zewail, A. H. (2012). 4D electron microscopy visualization of anisotropic atomic motions in carbon nanotubes. *Journal of the American Chemical Society, 134,* 9146–9149. https://doi.org/10.1021/ja304042r.

Park, S. T., Kwon, O.-H., & Zewail, A. H. (2012). Chirped imaging pulses in four-dimensional electron microscopy: Femtosecond pulsed hole burning. *New Journal of Physics*, *14*, 053046. https://doi.org/10.1088/1367-2630/14/5/053046.

Park, S. T., & Zewail, A. H. (2010). Photon-induced near-field electron microscopy (PINEM): Theoretical and experimental. *New Journal of Physics*, *12*, 123028. https://doi.org/10.1088/1367-2630/12/12/123028.

Paschen, T., Förster, M., Krüger, M., Lemell, C., Wachter, G., Libisch, F., . . . Hommelhoff, P. (2017). High visibility in two-color above-threshold photoemission from tungsten nanotips in a coherent control scheme. *Journal of Modern Optics*, *64*, 1054–1060. https://doi.org/10.1080/09500340.2017.1281453.

Petrov, V. I., Spivak, G. V., & Pavlyuchenko, O. P. (1972). Electron microscopy of the magnetic structure of thin films. *Soviet Physics. Uspekhi*, *15*, 66–94. https://doi.org/10.1070/PU1972v015n01ABEH004946.

Piazza, L., Lummen, T. T. A., Quiñonez, E., Murooka, Y., Reed, B. W., Barwick, B., . . . Carbone, F. (2015). Simultaneous observation of the quantization and the interference pattern of a plasmonic near-field. *Nature Communications*, *6*, 6407. https://doi.org/10.1038/ncomms7407.

Pomarico, E., Madan, I., Berruto, G., Vanacore, G. M., Wang, K., Kaminer, I., . . . Carbone, F. (2017). meV resolution in laser-assisted energy-filtered transmission electron microscopy. *ACS Photonics*. https://doi.org/10.1021/acsphotonics.7b01393.

Priebe, K. E., Rathje, C., Yalunin, S. V., Hohage, T., Feist, A., Schäfer, S., & Ropers, C. (2017). Attosecond electron pulse trains and quantum state reconstruction in ultrafast transmission electron microscopy. *Nature Photonics*, *11*, 793–797. https://doi.org/10.1038/s41566-017-0045-8.

Rajeswari, J., Huang, P., Mancini, G. F., Murooka, Y., Latychevskaia, T., McGrouther, D., . . . Carbone, F. (2015). Filming the formation and fluctuation of skyrmion domains by cryo-Lorentz transmission electron microscopy. *Proceedings of the National Academy of Sciences*, *112*, 14212–14217. https://doi.org/10.1073/pnas.1513343112.

Reed, B. W., LaGrange, T., Shuttlesworth, R. M., Gibson, D. J., Campbell, G. H., & Browning, N. D. (2010). Solving the accelerator–condenser coupling problem in a nanosecond dynamic transmission electron microscope. *Review of Scientific Instruments*, *81*, 053706. https://doi.org/10.1063/1.3427234.

Reimer, L. (1993). Kinematical and dynamical theory of electron diffraction. In *Springer series in optical sciences. Transmission electron microscopy* (pp. 266–320). Berlin, Heidelberg: Springer.

Ropers, C., Solli, D. R., Schulz, C. P., Lienau, C., & Elsaesser, T. (2007). Localized multiphoton emission of femtosecond electron pulses from metal nanotips. *Physical Review Letters*, *98*, 043907.

Santala, M. K., Reed, B. W., Raoux, S., Topuria, T., LaGrange, T., & Campbell, G. H. (2013). Irreversible reactions studied with nanosecond transmission electron microscopy movies: Laser crystallization of phase change materials. *Applied Physics Letters*, *102*, 174105. https://doi.org/10.1063/1.4803921.

Schäfer, S., Liang, W., & Zewail, A. H. (2011). Structural dynamics of nanoscale gold by ultrafast electron crystallography. *Chemical Physics Letters*, *515*, 278–282. https://doi.org/10.1016/j.cplett.2011.09.042.

Schapink, F. W., Forghany, S. K. E., & Buxton, B. F. (1983). The symmetry of convergent-beam electron diffraction patterns from bicrystals. *Acta Crystallographica. Section A*, *39*, 805–813. https://doi.org/10.1107/S0108767383001580.

Schenk, M., Krüger, M., & Hommelhoff, P. (2010). Strong-field above-threshold photoemission from sharp metal tips. *Physical Review Letters, 105*, 257601.

Schliep, K. B., Quarterman, P., Wang, J.-P., & Flannigan, D. J. (2017). Picosecond Fresnel transmission electron microscopy. *Applied Physics Letters, 110*, 222404. https://doi.org/10.1063/1.4984586.

Shah, J. (1999). *Ultrafast spectroscopy of semiconductors and semiconductor nanostructures. Springer series in solid-state sciences*. Springer.

Sivis, M., Duwe, M., Abel, B., & Ropers, C. (2013). Extreme-ultraviolet light generation in plasmonic nanostructures. *Nature Physics, 9*, 304–309.

Snoeck, E., Gatel, C., Lacroix, L. M., Blon, T., Lachaize, S., Carrey, J., . . . Chaudret, B. (2008). Magnetic configurations of 30 nm iron nanocubes studied by electron holography. *Nano Letters, 8*, 4293–4298. https://doi.org/10.1021/nl801998x.

Spencer Baskin, J., & Zewail, A. H. (2014). Seeing in 4D with electrons: Development of ultrafast electron microscopy at Caltech. *Comptes Rendus Physique, 15*, 176–189. https://doi.org/10.1016/j.crhy.2013.11.002.

Stamm, C., Kachel, T., Pontius, N., Mitzner, R., Quast, T., Holldack, K., . . . Eberhardt, W. (2007). Femtosecond modification of electron localization and transfer of angular momentum in nickel. *Nature Materials, 6*, 740–743. https://doi.org/10.1038/nmat1985.

Stohr, J. (2016). *Magnetism: From fundamentals to nanoscale dynamics*. Springer.

Sun, S., Wei, L., Li, Z., Cao, G., Liu, Y., Lu, W. J., . . . Li, J. (2015). Direct observation of an optically induced charge density wave transition in 1T–TaSe$_2$. *Physical Review B, 92*, 224303. https://doi.org/10.1103/PhysRevB.92.224303.

Swanson, L. W., & Schwind, G. A. (2009). *Advances in imaging and electron physics*. Elsevier.

Tanaka, M. (1997). Convergent-beam electron diffraction. In *NATO ASI series. Electron crystallography* (pp. 77–113). Dordrecht: Springer.

Terauchi, M., Tsuda, K., Kamimura, O., Tanaka, M., Kaneyama, T., & Honda, T. (1994). Observation of lattice fringes in convergent-beam electron diffraction patterns. *Ultramicroscopy, 54*, 268–275. https://doi.org/10.1016/0304-3991(94)90126-0.

Theissmann, R., Fuess, H., & Tsuda, K. (2012). Experimental charge density of hematite in its magnetic low temperature and high temperature phases. *Ultramicroscopy, 120*, 1–9. https://doi.org/10.1016/j.ultramic.2012.04.006.

Tonomura, A. (1999). *Electron holography. Springer series in optical sciences*. Berlin, Heidelberg: Springer.

van der Veen, R. M., Kwon, O.-H., Tissot, A., Hauser, A., & Zewail, A. H. (2013). Single-nanoparticle phase transitions visualized by four-dimensional electron microscopy. *Nature Chemistry, 5*, 395–402. https://doi.org/10.1038/nchem.1622.

van der Veen, R. M., Penfold, T. J., & Zewail, A. H. (2015). Ultrafast core-loss spectroscopy in four-dimensional electron microscopy. *Structural Dynamics, 2*, 024302. https://doi.org/10.1063/1.4916897.

Vanacore, G. M., van der Veen, R. M., & Zewail, A. H. (2015). Origin of axial and radial expansions in carbon nanotubes revealed by ultrafast diffraction and spectroscopy. *ACS Nano, 9*, 1721–1729. https://doi.org/10.1021/nn506524c.

Vaz, C. A. F., Bland, J. A. C., & Lauhoff, G. (2008). Magnetism in ultrathin film structures. *Reports on Progress in Physics, 71*, 056501. https://doi.org/10.1088/0034-4885/71/5/056501.

Vincent, R., & Midgley, P. A. (1994). Double conical beam-rocking system for measurement of integrated electron diffraction intensities. *Ultramicroscopy, 53*, 271–282. https://doi.org/10.1016/0304-3991(94)90039-6.

Völkl, E., Allard, L. F., & Joy, D. C. (1999). *Introduction to electron holography*. Boston, MA: Springer US.

Wei, L., Sun, S., Guo, C., Li, Z., Sun, K., Liu, Y., ... Li, J. (2017). Dynamic diffraction effects and coherent breathing oscillations in ultrafast electron diffraction in layered 1T–TaSeTe. *Structural Dynamics*, *4*. https://doi.org/10.1063/1.4979643.

Weingartner, M., & Bostanjoglo, O. (1998). Pulsed photoelectron microscope for time-resolved surface investigations. *Surface & Coatings Technology*, *100*, 85–89.

Williams, D. B., & Carter, C. B. (1996). *Transmission electron microscopy: A textbook for materials science. Diffraction. II*. Springer.

Winkler, F., Tavabi, A. H., Barthel, J., Duchamp, M., Yucelen, E., Borghardt, S., ... Dunin-Borkowski, R. E. (2017). Quantitative measurement of mean inner potential and specimen thickness from high-resolution off-axis electron holograms of ultra-thin layered WSe$_2$. *Ultramicroscopy*, *178*, 38–47. https://doi.org/10.1016/j.ultramic.2016.07.016.

Yakovlev, S., & Libera, M. (2008). Dose-limited spectroscopic imaging of soft materials by low-loss EELS in the scanning transmission electron microscope. *Micron*, *39*, 734–740. https://doi.org/10.1016/j.micron.2007.10.019.

Yang, D.-S., Mohammed, O. F., & Zewail, A. H. (2010). Scanning ultrafast electron microscopy. *Proceedings of the National Academy of Sciences*, *107*, 14993–14998.

Yang, H., Jones, L., Ryll, H., Simson, M., Soltau, H., Kondo, Y., ... Nellist, P. D. (2015). 4D STEM: High efficiency phase contrast imaging using a fast pixelated detector. *Journal of Physics. Conference Series*, *644*, 012032.

Yoo, D.-K., Kwon, O.-H., Liu, H., Tang, J., & Zewail, A. H. (2015). Observing in space and time the ephemeral nucleation of liquid-to-crystal phase transitions. *Nature Communications*, *6*, 8639. https://doi.org/10.1038/ncomms9639.

Yoo, B.-K., Su, Z., Thomas, J. M., & Zewail, A. H. (2016). On the dynamical nature of the active center in a single-site photocatalyst visualized by 4D ultrafast electron microscopy. *Proceedings of the National Academy of Sciences*, *113*, 503–508. https://doi.org/10.1073/pnas.1522869113.

Yurtsever, A., Baskin, J. S., & Zewail, A. H. (2012). Entangled nanoparticles: Discovery by visualization in 4D electron microscopy. *Nano Letters*, *12*, 5027–5032. https://doi.org/10.1021/nl302824f.

Yurtsever, A., Schaefer, S., & Zewail, A. H. (2012). Ultrafast Kikuchi diffraction: Nanoscale stress–strain dynamics of wave-guiding structures. *Nano Letters*, *12*, 3772–3777. https://doi.org/10.1021/nl301644t.

Yurtsever, A., van der Veen, R. M., & Zewail, A. H. (2012). Subparticle ultrafast spectrum imaging in 4D electron microscopy. *Science*, *335*, 59–64.

Yurtsever, A., & Zewail, A. H. (2012). Direct visualization of near-fields in nanoplasmonics and nanophotonics. *Nano Letters*, *12*, 3334–3338. https://doi.org/10.1021/nl301643k.

Yurtsever, A., & Zewail, A. H. (2009). 4D nanoscale diffraction observed by convergent-beam ultrafast electron microscopy. *Science*, *326*, 708–712. https://doi.org/10.1126/science.1179314.

Zewail, A. H. (2006). 4D ultrafast electron diffraction, crystallography, and microscopy. *Annual Review of Physical Chemistry*, *57*, 65–103. https://doi.org/10.1146/annurev.physchem.57.032905.104748.

Zheng, C. L., & Etheridge, J. (2013). Measurement of chromatic aberration in STEM and SCEM by coherent convergent beam electron diffraction. *Ultramicroscopy*, *125*, 49–58. https://doi.org/10.1016/j.ultramic.2012.10.002.

Zuo, J. M., Gao, M., Tao, J., Li, B. Q., Twesten, R., & Petrov, I. (2004). Coherent nano-area electron diffraction. *Microscopy Research and Technique, 64*, 347–355. https://doi.org/10.1002/jemt.20096.

Zuo, J. M., Kim, M., O'Keeffe, M., & Spence, J. C. H. (1999). Direct observation of d-orbital holes and Cu–Cu bonding in Cu_2O. *Nature, 401*, 49–52. https://doi.org/10.1038/43403.

Zuo, J. M., & Spence, J. C. H. (1991). Automated structure factor refinement from convergent-beam patterns. *Ultramicroscopy, 35*, 185–196. https://doi.org/10.1016/0304-3991(91)90071-D.

Zuo, J. M., & Spence, J. C. H. (1992). *Electron microdiffraction*. Springer.

Zuo, J. M., Spence, J. C. H., & Hoier, R. (1989). Accurate structure-factor phase determination by electron diffraction in noncentrosymmetric crystals. *Physical Review Letters, 62*, 547–550. https://doi.org/10.1103/PhysRevLett.62.547.

CHAPTER TWO

Refinement of Generalized Mean Inequalities and Connections with Divergence Measures

Inder Jeet Taneja

Formerly Professor of Mathematics, Department of Mathematics, Federal University of Santa Catarina, Florianópolis, SC, Brazil
e-mail address: ijtaneja@gmail.com

Contents

Advances in Imaging and Electron Physics, Volume 207
ISSN 1076-5670
https://doi.org/10.1016/bs.aiep.2018.06.002

73

1. INTRODUCTION

In this section we shall give some divergence measures, distance measures, and mean divergences. For simplicity, let us consider

$$\Gamma_n = \left\{ P = (p_1, p_2, \ldots, p_n) \,\middle|\, p_i > 0, \sum_{i=1}^{n} p_i = 1 \right\}, \quad n \geq 2,$$

the set of all complete finite discrete probability distributions.

1.1 Divergence and Distance Measures

For all $P, Q \in \Gamma_n$, let us consider the following six measures well-known in the literature:

$$I(P\|Q) = \frac{1}{2}\left[\sum_{i=1}^{n} p_i \ln\left(\frac{2p_i}{p_i + q_i}\right) + \sum_{i=1}^{n} q_i \ln\left(\frac{2q_i}{p_i + q_i}\right) \right], \tag{1}$$

$$J(P\|Q) = \sum_{i=1}^{n} (p_i - q_i) \ln\left(\frac{p_i}{q_i}\right), \tag{2}$$

$$T(P\|Q) = \sum_{i=1}^{n} \left(\frac{p_i + q_i}{2}\right) \ln\left(\frac{p_i + q_i}{2\sqrt{p_i q_i}}\right), \tag{3}$$

$$\Delta(P\|Q) = \sum_{i=1}^{n} \frac{(p_i - q_i)^2}{p_i + q_i}, \tag{4}$$

$$h(P\|Q) = \frac{1}{2}\sum_{i=1}^{n} (\sqrt{p_i} - \sqrt{q_i})^2, \tag{5}$$

and

$$\Psi(P\|Q) = \sum_{i=1}^{n} \frac{(p_i - q_i)^2 (p_i + q_i)}{p_i q_i}. \tag{6}$$

The first three measures are classical divergence measures famous in the literature of information theory and statistics known as *Jensen–Shannon divergence* (Burbea & Rao, 1982), *J-divergence* (Jeffreys, 1946; Kullback & Leibler, 1951), and *Arithmetic–Geometric mean divergence* (Taneja, 1995) respectively. The last three measures are known as *triangular discrimination* (LeCam, 1986), *Hellingar's divergence* (Hellinger, 1909), and *symmetric chi-square divergence* (Dragomir, Sunde, & Buse, 2000; Taneja & Kumar, 2004;

Taneja, 2005a). The above six measures (1)–(6) lead us to the following inequality (Taneja, 2005b) given by

$$\frac{1}{4}\Delta(P\|Q) \le I(P\|Q) \le h(P\|Q) \le \frac{1}{8}J(P\|Q) \le T(P\|Q) \le \frac{1}{16}\Psi(P\|Q). \quad (7)$$

The measures I and T can be written in terms of *arithmetic* and *geometric means*:

$$I(P\|Q) = \sum_{i=1}^{n}\left[A(p_i \ln p_i, q_i \ln q_i) - A(p_i, q_i)\ln A(p_i, q_i)\right], \quad (8)$$

and

$$T(P\|Q) = \sum_{i=1}^{n} A(p_i, q_i)\ln\left(\frac{A(p_i, q_i)}{G(p_i, q_i)}\right), \quad (9)$$

where $A(a, b)$ and $G(a, b)$ are well-known *arithmetic* and *geometric means*. In view of $I(P\|Q) + T(P\|Q) = \frac{1}{4}J(P\|Q)$, the J *divergence* can also be written in terms of *arithmetic* and *geometric means*.

1.2 Mean Divergence Measures

Author (Taneja, 2006b) studied the following inequality

$$G(P\|Q) \le N_1(P\|Q) \le N_2(P\|Q) \le A(P\|Q), \quad (10)$$

where

$$G(P\|Q) = \sum_{i=1}^{n}\sqrt{p_i q_i}, \quad (11)$$

$$N_1(P\|Q) = \sum_{i=1}^{n}\left(\frac{p_i + q_i}{2}\right)^2, \quad (12)$$

$$N_2(P\|Q) = \sum_{i=1}^{n}\sqrt{\frac{p_i + q_i}{2}}\left(\frac{\sqrt{p_i} + \sqrt{q_i}}{2}\right), \quad (13)$$

and

$$A(P\|Q) = \sum_{i=1}^{n}\frac{p_i + q_i}{2} = 1. \quad (14)$$

The inequalities (10) admit non-negative differences given by

$$M_1(P\|Q) = D_{N_2 N_1}(P\|Q)$$
$$= \sum_{i=1}^{n} \left(\sqrt{\frac{p_i + q_i}{2}} \left(\frac{\sqrt{p_i} + \sqrt{q_i}}{2} \right) - \left(\frac{\sqrt{p_i} + \sqrt{q_i}}{2} \right)^2 \right), \tag{15}$$

$$M_2(P\|Q) = D_{N_2 G}(P\|Q) = \sum_{i=1}^{n} \left[\left(\frac{\sqrt{p_i} + \sqrt{q_i}}{2} \right) \sqrt{\frac{p_i + q_i}{2}} - \sqrt{p_i q_i} \right], \tag{16}$$

and

$$M_3(P\|Q) = D_{AN_2}(P\|Q)$$
$$= \sum_{i=1}^{n} \left[\left(\frac{p_i + q_i}{2} \right) - \left(\frac{\sqrt{p_i} + \sqrt{q_i}}{2} \right) \left(\sqrt{\frac{p_i + q_i}{2}} \right) \right]. \tag{17}$$

The following equalities hold

$$h(P\|Q) = D_{AN_1}(P\|Q) = D_{AG}(P\|Q) = D_{N_1 G}(P\|Q), \tag{18}$$

where, for example, $D_{AN_1} := A(P\|Q) - N_1(P\|Q)$, etc.

1.3 Parametric Generalizations

For all $P, Q \in \Gamma_n$, let us consider

$$L_t(P\|Q) = \sum_{i=1}^{n} \frac{(p_i - q_i)^2 (p_i + q_i)^2}{2^t (\sqrt{p_i q_i})^{t+1}}, \, t \in \mathbb{Z}. \tag{19}$$

In particular, $L_{-1}(a, b) = 2\Delta(a, b)$ and $L_1(a, b) = \frac{1}{2}\Psi(a, b)$, where Δ and Ψ are given by (4) and (6) respectively. Still, we have

$$L_0(P\|Q) = K(P\|Q) = \sum_{i=1}^{n} \frac{(p_i - q_i)^2}{\sqrt{p_i q_i}}, \tag{20}$$

$$L_2(P\|Q) = \frac{1}{2}F(P\|Q) = \sum_{i=1}^{n} \frac{(p_i^2 - q_i^2)^2}{4(p_i q_i)^{3/2}}, \tag{21}$$

and

$$L_3(P\|Q) = \frac{1}{8}L(P\|Q) = \sum_{i=1}^{n} \frac{(p_i - q_i)^2 (p_i + q_i)^3}{8(p_i q_i)^2}. \tag{22}$$

The measures K and F given by (20) and (21) are studied by Jain and Srivastava (2007), and Kumar and Johnson (2005) respectively. The measure L given by (22) appearing as a particular case of (19) is new. Some studies on it can be seen in Taneja (2006a, 2013c). For $t = -1, 0, 1, 2$, and 3, the measure (19) is convex in pair of probability distributions $(P, Q) \in \Delta_n \times \Delta_n$. Also for any $t \geq -1$, the measure (19) is monotonically increasing in t. This gives

$$\frac{1}{4}\Delta(P\|Q) \leq \frac{1}{8}K(P\|Q) \leq \frac{1}{16}\Psi(P\|Q) \leq \frac{1}{16}F(P\|Q) \leq \frac{1}{64}L(P\|Q). \quad (23)$$

Following author's work (Taneja, 2013a, 2013b), the 12 measures given by (1)–(6), (15)–(17), and (20)–(22) satisfy the following inequality

$$\frac{1}{4}\Delta(P\|Q) \leq I(P\|Q) \leq 4M_1(P\|Q) \leq \frac{4}{3}M_2(P\|Q)$$

$$\leq h(P\|Q) \leq 4M_3(P\|Q) \leq \frac{1}{8}J(P\|Q) \leq T(P\|Q)$$

$$\leq \frac{1}{8}K(P\|Q) \leq \frac{1}{16}\Psi(P\|Q) \leq \frac{1}{16}F(P\|Q) \leq \frac{1}{64}L(P\|Q). \quad (24)$$

The inequalities (24) include the measures given in Sections 1.1, 1.2, and 1.3.

2. COMBINED MEANS INEQUALITIES

This section brings inequalities based on seven means studied by Eve (2003) and Gini mean of order r and s (Gini, 1938). Instead, working with a pair of probability distributions as in Section 1, here we shall work with a pair of positive real numbers.

2.1 Seven Means

Let $a, b > 0$ be two positive numbers. Eves (2003) studied geometrical interpretations of the following *seven means*:

- **Arithmetic mean:** $A(a, b) = \frac{a+b}{2}$.
- **Geometric mean:** $G(a, b) = \sqrt{ab}$.
- **Harmonic mean:** $H(a, b) = \frac{2ab}{a+b}$.
- **Heronian mean:** $N(a, b) = \frac{a+\sqrt{ab}+b}{3}$.
- **Contra-harmonic mean:** $C(a, b) = \frac{a^2+b^2}{a+b}$.
- **Root-mean-square:** $S(a, b) = \sqrt{\frac{a^2+b^2}{2}}$.

- **Centroidal mean:** $R(a, b) = \frac{2(a^2 + ab + b^2)}{3(a+b)}$.

Along with geometric interpretations, Eves (2003) proved an inequality with these seven means:

$$H \leq G \leq N \leq A \leq R \leq S \leq C. \tag{25}$$

The following theorem gives the improvement over the inequalities (25) studied by author (Taneja, 2017):

Theorem 1 (Taneja, 2017). *The following inequalities hold:*

$$
\mathbf{H} \leq \begin{Bmatrix} \mathbf{G} \\ N + 5R - 5S \end{Bmatrix} \leq \begin{Bmatrix} \frac{3}{5}N + \frac{2}{5}H \\ S + H - A \leq \begin{Bmatrix} \frac{1}{3}S + \frac{2}{3}G \\ \frac{2}{9}C + \frac{7}{9}G \end{Bmatrix} \leq \mathbf{N} \end{Bmatrix}
$$

$$
\leq \begin{Bmatrix} A + \frac{1}{3}S - \frac{1}{3}C \\ A + \frac{1}{5}G - \frac{1}{5}R \end{Bmatrix} \leq \frac{5}{6}A + \frac{1}{6}H \leq \begin{Bmatrix} \frac{1}{4}S + \frac{3}{4}N \\ \frac{2}{3}S + \frac{1}{3}H \end{Bmatrix} \leq \mathbf{A}
$$

$$
\leq \begin{Bmatrix} \frac{1}{6}G + \frac{5}{6}R \leq \frac{4}{9}G + \frac{5}{9}C \\ \frac{1}{6}G + \frac{5}{6}R \\ \frac{6}{7}C + H - \frac{6}{7}N \\ S + H - G \end{Bmatrix} \leq \frac{1}{3}A + \frac{2}{3}S \end{Bmatrix} \leq \mathbf{R} \end{Bmatrix}
$$

$$
\leq \begin{Bmatrix} \mathbf{S} \leq \frac{1}{5}N + R - \frac{1}{5}H \\ 5A + G - 5N \end{Bmatrix} \leq \mathbf{C} \leq \frac{4}{7}C + \frac{3}{7}N \leq 3A + S - 3N. \tag{26}
$$

2.2 Gini Mean of Order r and s

The Gini mean (Gini, 1938) of order r and s is given by

$$
E_{r,s}(a, b) = \begin{cases} \left(\dfrac{a^r + b^r}{a^s + b^s} \right)^{\frac{1}{r-s}} & r \neq s \\[2ex] \exp\left(\dfrac{a^r \ln a + b^r \ln b}{a^r + b^r} \right) & r = s \neq 0 \\[2ex] \sqrt{ab} & r = s = 0. \end{cases} \tag{27}
$$

Due to monotonicity property of measure (27), we have the following inequalities:

$$P_1 \leq P_2 \leq P_3 \leq H \leq P_4 \leq G \leq N_1 \leq A \leq (P_5 \text{ or } S) \leq P_6, \tag{28}$$

where $P_1 = E_{-3,-2}$, $P_2 = E_{-2,-1}$, $P_3 = E_{-3/2,-1/2}$, $H = E_{-1,0}$, $P_4 = E_{-1/2,0}$, $G = E_{-1/2,1/2}$, $N_1 = E_{0,1/2}$, $A = E_{0,1}$, $P_5 = E_{1/2,1}$, $E_{0,2} = S$, and $P_6 = E_{1,2}$.

The means H, G, A, and S are *harmonic, geometric, arithmetic,* and the *square-root means* respectively (see Section 2.1). For more studies on measure (27) refer to Czinder and Pales (2005), Lehmer (1971), Chen (2008), Sánder (2004), Simic (2009a), Taneja (2012, 2014, 2017), etc.

2.3 Combined Inequalities

Taneja (2005a) studied the following inequalities:

$$H \leq G \leq N_1 \leq N \leq N_2 \leq A \leq S, \tag{29}$$

where

$$N_2(a, b) = \left(\frac{\sqrt{a} + \sqrt{b}}{2}\right)\left(\sqrt{\frac{a + b}{2}}\right)$$

and

$$N(a, b) - \frac{a + \sqrt{ab} + b}{3}.$$

The measure $N(a, b)$ is famous as *Heron's mean.* Some applications of the inequalities (29) in terms of probability distributions are done in Shi, Zhang, and Li (2010) and Simic (2009b).

Combining (25), (28), and (29), we have

$$\mathbf{P_1} \leq \mathbf{P_2} \leq \mathbf{P_3} \leq \mathbf{H} \leq \mathbf{P_4} \leq \mathbf{G} \leq \mathbf{N_1} \leq \mathbf{N} \leq \mathbf{N_2} \leq \mathbf{A} \leq \begin{Bmatrix} \mathbf{P_5} \\ \mathbf{R} \leq \mathbf{S} \end{Bmatrix} \leq \mathbf{C}, \tag{30}$$

where $P_6 = C$. We call the inequalities (30) as *combined means inequalities.* The work is based on the improvement of inequalities (30).

2.4 Nonnegative Differences

The inequalities appearing in (30) admit many nonnegative differences. Let's write them as:

$$D_{tp}(a, b) = b f_{tp}\left(\frac{a}{b}\right) = b\left[f_t\left(\frac{a}{b}\right) - f_p\left(\frac{a}{b}\right)\right], \tag{31}$$

where

$$f_{tp}(x) = f_t(x) - f_p(x), \quad f_t(x) \geq f_p(x), \quad \forall x > 0.$$

More precisely, the function $f : (0, \infty) \to \mathbb{R}$ appearing in (31) leads us to the following representations of inequalities (30):

$$f_{P_1}(x) \leq f_{P_2}(x) \leq f_{P_3}(x) \leq f_H(x) \leq f_{P_4}(x) \leq f_G(x) \leq f_{N_1}(x)$$

$$\leq f_N(x) \leq f_{N_2}(x) \leq f_A(x) \leq \begin{Bmatrix} f_{P_5}(x) \\ f_R(x) \leq f_S(x) \end{Bmatrix} \leq f_C(x). \qquad (32)$$

Equivalently,

$$\frac{x(x^2+1)}{x^3+1} \leq \frac{x(x+1)}{x^2+1} \leq \frac{x(\sqrt{x}+1)}{x^{3/2}+1} \leq \frac{2x}{1+x} \leq \frac{4x}{(\sqrt{x}+1)^2} \leq \sqrt{x}$$

$$\leq \left(\frac{\sqrt{x}+1}{2}\right)^2 \leq \frac{x+\sqrt{x}+1}{3} \leq \left(\frac{\sqrt{x}+1}{2}\right)\left(\sqrt{\frac{x+1}{2}}\right) \leq \frac{x+1}{2}$$

$$\leq \begin{Bmatrix} \left(\frac{x+1}{\sqrt{x}+1}\right)^2 \\ \frac{2(x^2+x+1)}{3(x+1)} \leq \sqrt{\frac{x^2+1}{2}} \end{Bmatrix} \leq \frac{x^2+1}{x+1}.$$

Remark 1. Based on the notations given in (31), we have the following equality relations:

(i) $3D_{CR} = 2D_{AH} = 2D_{CA} = D_{CH} = 6D_{RA} = \frac{3}{2}D_{RH} = \frac{(a-b)^2}{(a+b)} := \Delta$.

(ii) $3D_{AN} = D_{AG} = \frac{3}{2}D_{NG} = \frac{1}{2}(\sqrt{a}-\sqrt{b})^2 := h$.

The notation D_{AB} is understood as $D_{AB} = A - B$, where $A \geq B$. The measures Δ and h appearing in (i) and (ii) are the *triangular discrimination* and *Hellinger's distance* as given in (4) and (5) respectively for the probability distributions. For more details refer author's work (Taneja, 2017). Using the notations given in (31), the author (Taneja, 2017) proved a theorem giving inequalities among nonnegative differences arises due to inequalities (30).

Theorem 2 (Taneja, 2017). *Using the notations (31), the nonnegative difference arising due to (30) satisfies the following inequalities:*

$$\frac{1}{2}D^5_{HP_2}$$

$$\leq \begin{Bmatrix} \begin{Bmatrix} \frac{1}{3}D^{14}_{GP_2} \leq \frac{2}{7}D^{20}_{N_1P_2} \leq \frac{3}{11}D^{27}_{NP_2} \\ \frac{1}{8}D^{89}_{CP_1} \end{Bmatrix} \leq \begin{Bmatrix} \frac{1}{6}D^{88}_{CP_2} \\ \frac{2}{9}D^{54}_{P_5P_2} \end{Bmatrix} \\ \\ \begin{Bmatrix} D^4_{HP_3} \\ \frac{2}{5}D^9_{P_4P_2} \end{Bmatrix} \leq \frac{2}{3}D^8_{P_4P_3} \leq \begin{Bmatrix} \frac{2}{9}D^{54}_{P_5P_2} \\ 2D^7_{P_4H} \leq D^{12}_{GH} \leq \begin{Bmatrix} \frac{2}{3}D^{79}_{CP_5} \\ \frac{1}{5}D^{87}_{CP_3} \leq f^{67}_{SA} \end{Bmatrix} \\ \frac{1}{2}D^{13}_{GP_3} \leq 3D^{66}_{SR} \leq \frac{1}{6}D^{88}_{CP_2} \leq \frac{1}{5}D^{87}_{CP_3} \leq f^{67}_{SA} \end{Bmatrix} \end{Bmatrix}$$

$$\leq \begin{cases} \begin{rcases} \frac{1}{3}D_{SH}^{73} \leq \frac{1}{2}D_{AH}^{42} \\ \frac{4}{5}D_{SN_2}^{68} \\ \frac{3}{4}D_{SN}^{69} \end{rcases} \leq \begin{cases} \frac{2}{5}D_{SP_4}^{72} \\ \frac{3}{7}D_{CN}^{82} \end{cases} \leq \begin{cases} \frac{2}{5}D_{CN_1}^{83} \\ \frac{2}{7}D_{CP_4}^{85} \end{cases} \leq \frac{1}{3}D_{CG}^{84} \leq \frac{12}{11}D_{RN_2}^{57} \\[2em] \begin{rcases} \frac{4}{5}D_{SN_2}^{68} \\ \frac{3}{4}D_{SN}^{69} \end{rcases} \leq \frac{2}{3}D_{SN_1}^{70} \leq \begin{cases} \frac{1}{3}D_{CG}^{84} \leq \frac{12}{11}D_{RN_2}^{57} \\ \frac{1}{2}D_{SG}^{71} \end{cases} \end{cases}$$

$$\leq \begin{cases} \frac{2}{5}D_{P_5H}^{52} \leq \frac{3}{5}D_{RG}^{60} \\ \frac{6}{7}D_{RN_1}^{59} \end{cases} \leq 4D_{N_2N_1}^{30} \leq \frac{4}{3}D_{N_2G}^{31} \leq D_{AG}^{40} \leq 4D_{AN_2}^{37}$$

$$\leq \frac{2}{3}D_{P_5G}^{50} \leq D_{P_5N_1}^{49} \leq \frac{6}{5}D_{P_5N_3}^{48} \leq \frac{4}{3}D_{P_5N_2}^{47} \leq 2D_{P_5A}^{46}, \tag{33}$$

$$\frac{1}{2}D_{P_2P_1}^{1} \leq \frac{1}{3}D_{P_3P_1}^{3}$$

$$\leq \begin{cases} \frac{1}{4}D_{HP_1}^{6} \leq \frac{2}{9}D_{P_4P_1}^{10} \leq \frac{1}{5}D_{GP_1}^{15} \\[1em] \leq \begin{cases} \frac{2}{11}D_{N_1P_1}^{21} \leq \frac{3}{17}D_{N_3P_1}^{28} \leq \frac{4}{23}D_{N_2P_1}^{36} \leq \begin{cases} \frac{1}{6}D_{AP_1}^{45} \\ \frac{2}{7}D_{N_1P_2}^{20} \end{cases} \\ \frac{1}{3}D_{GP_2}^{14} \leq \frac{2}{7}D_{N_1P_2}^{20} \end{cases} \\[1.5em] \begin{rcases} \frac{1}{4}D_{HP_1}^{6} \\ D_{P_3P_2}^{2} \end{rcases} \leq \frac{1}{2}D_{HP_2}^{5} \end{cases}$$

$$\leq \frac{3}{11}D_{NP_2}^{27} \leq \frac{1}{15}D_{N_2P_2}^{35} \leq \begin{cases} \begin{rcases} \frac{3}{4}D_{SR}^{66} \\ \frac{2}{5}D_{N_1P_3}^{19} \end{rcases} \leq \frac{3}{8}D_{NP_3}^{26} \leq \frac{4}{11}D_{N_2P_3}^{34} \\[1em] \frac{1}{4}D_{AP_2}^{44} \leq \begin{cases} \frac{1}{5}D_{SP_2}^{75} \\ \frac{3}{14}D_{RP_2}^{64} \end{cases} \end{cases}$$

$$\leq \frac{1}{3}D_{AP_3}^{43} \leq \begin{cases} \frac{2}{3}D_{N_1H}^{18} \\ \frac{1}{4}D_{SP_3}^{74} \leq \frac{3}{11}D_{RP_3}^{63} \end{cases} \leq \frac{3}{5}D_{NH}^{25} \leq \frac{4}{7}D_{N_2H}^{33}$$

$$\leq \begin{cases} \begin{rcases} 12D_{N_2N}^{29} \\ \frac{1}{2}D_{AH}^{42} \end{rcases} \leq \begin{cases} D_{N_1P_4}^{17} \leq \frac{4}{5}D_{N_2P_4}^{32} \\ \frac{2}{7}D_{CP_4}^{85} \end{cases} \leq \frac{6}{7}D_{NP_4}^{24} \\[1.5em] \frac{1}{2}D_{AH}^{42} \leq \frac{4}{9}D_{CN_2}^{81} \leq \begin{cases} \frac{1}{3}D_{CG}^{84} \\ \frac{6}{13}D_{RP_4}^{61} \end{cases} \end{cases} \leq \frac{2}{3}D_{AP_4}^{41} \leq D_{AG}^{40}, \tag{34}$$

and

$$\begin{cases} \frac{1}{6}D_{AP_1}^{45} \leq \begin{cases} \frac{3}{20}D_{RP_1}^{65} \leq \frac{1}{7}D_{SP_1}^{76} \leq \frac{1}{4}D_{AP_2}^{44} \\ \frac{2}{13}D_{P_5P_1}^{55} \end{cases} \leq \frac{1}{3}D_{AP_3}^{43} \leq \frac{3}{11}D_{RP_3}^{63} \leq \frac{2}{7}D_{P_5P_3}^{53} \\[1.5em] \frac{1}{3}D_{GP_2}^{14} \leq \frac{2}{3}D_{CP_5}^{79} \leq 2D_{GP_4}^{11} \leq \frac{1}{2}D_{AH}^{42} \end{cases}$$

$$\leq \left\{ \begin{array}{l} \frac{3}{7}D_{CN}^{82} \leq \frac{2}{7}D_{CP_4}^{85} \\ \frac{4}{9}D_{CN_2}^{81} \end{array} \right\} \leq D_{CS}^{77} \leq D_{AG}^{40}. \tag{35}$$

According to inequalities (33), (34), and (35), the theorem below givens an improvement (Taneja, 2017) over the inequalities (24).

Theorem 3. *The following inequalities hold:*

$$\frac{1}{4}\Delta \leq \frac{1}{3}D_{CG}^{84} \leq \left\{ \begin{array}{l} \frac{12}{11}D_{RN_2}^{57} \leq \left\{ \begin{array}{l} \frac{2}{5}D_{P_5H}^{52} \leq \frac{3}{5}D_{RG}^{60} \\ \frac{6}{7}D_{RN_1}^{59} \end{array} \right\} \\ \frac{2}{3}D_{AP_4}^{41} \leq I \end{array} \right\} \leq 4M_1 \leq \frac{4}{3}M_2$$

$$\leq h \leq 4M_3 \leq \frac{2}{3}D_{P_5G}^{50} \leq \left\{ \begin{array}{l} D_{P_5N_1}^{49} \leq \frac{6}{5}D_{P_5N}^{48} \leq \frac{1}{3}D_{P_5N_2}^{47} \leq 2D_{P_5A}^{46} \\ \frac{1}{8}J \end{array} \right\} \leq T. \tag{36}$$

Remark 2. The following equalities hold:

$$D_{AH} = \frac{1}{2}\Delta, \quad D_{N_2N_1}^{30} = M_1, \quad D_{N_2G}^{31} = M_2, \quad D_{AG}^{40} = h, \quad \text{and} \quad D_{AN_2}^{37} = M_3.$$

As a consequence of Theorem 2, we have two groups of inequalities among the means given in (30).

2.5 Group I: Two Means Inequalities

This group gives inequalities having three means with *two means* in one side. See list below:

(i) Inequalities with P_1 and P_2

1. $P_1 \leq \frac{7P_2+3R}{10}$.
2. $P_2 \leq \frac{P_1+2P_3}{3}$.
3. $P_2 \leq \frac{H+P_1}{2}$.
4. $P_2 \leq \frac{3P_1+2G}{5}$.
5. $P_2 \leq \frac{3P_1+C}{4}$.
6. $P_2 \leq \frac{5P_1+4P_4}{9}$.
7. $P_2 \leq \frac{15P_1+8N_2}{23}$.
8. $P_2 \leq \frac{7P_1+4N_1}{11}$.
9. $P_2 \leq \frac{9P_1+4P_5}{13}$.
10. $P_2 \leq \frac{7P_3+2P_5}{9}$.
11. $P_2 \leq \frac{5P_1+2S}{7}$.
12. $P_2 \leq \frac{A+2P_1}{3}$.

(ii) Inequalities with P_3

13. $P_3 \leq \frac{P_1+3H}{4}$.
14. $P_3 \leq \frac{H+P_2}{2}$.
15. $P_3 \leq \frac{C+5P_2}{6}$.
16. $P_3 \leq \frac{2P_4+3P_2}{5}$.
17. $P_3 \leq \frac{A+3P_2}{4}$.
18. $P_3 \leq \frac{4P_2+S}{5}$.
19. $P_3 \leq \frac{11P_2+3R}{14}$.
20. $P_3 \leq \frac{5P_2+2N_1}{7}$.
21. $P_3 \leq \frac{2P_2+G}{3}$.
22. $P_3 \leq \frac{11P_2+4N_2}{15}$.

(iii) Inequalities with H

23. $H \leq \frac{4P_4+P_2}{5}$.

24. $H \leq \frac{2G+P_2}{3}$.

25. $H \leq \frac{2P_4+P_3}{3}$.

26. $H \leq \frac{8P_4+P_1}{9}$.

27. $H \leq \frac{5P_2+6N}{11}$.

28. $H \leq \frac{P_3+G}{2}$.

29. $H \leq \frac{A+2P_3}{3}$.

30. $H \leq \frac{4P_3+C}{5}$.

31. $H \leq \frac{3P_3+2N_1}{5}$.

32. $H \leq \frac{7P_3+4N_2}{11}$.

33. $H \leq \frac{5P_3+2P_5}{7}$.

34. $H \leq \frac{8P_3+3R}{11}$.

(iv) Inequalities with P_4

35. $P_4 \leq \frac{P_1+9G}{10}$.

36. $P_4 \leq \frac{G+H}{2}$.

37. $P_4 \leq \frac{3G+P_3}{4}$.

38. $P_4 \leq \frac{5H+S}{6}$.

39. $P_4 \leq \frac{4H+P_5}{5}$.

40. $P_4 \leq \frac{2H+N_1}{3}$.

41. $P_4 \leq \frac{A+3H}{4}$.

(v) Inequalities with G

42. $G \leq \frac{6N_1+P_2}{7}$.

43. $G \leq \frac{C+6P_4}{7}$.

44. $G \leq \frac{P_1+10N_1}{11}$.

45. $G \leq \frac{A+2P_4}{3}$.

46. $G \leq \frac{4P_4+S}{5}$.

47. $G \leq \frac{3P_4+P_5}{4}$.

48. $G \leq \frac{10P_4+3R}{13}$.

49. $\frac{S+3G}{4} \leq N_1$.

50. $\frac{C+5G}{6} \leq N_1$.

51. $\frac{G+2N_2}{3} \leq N_1$.

(vi) Inequalities with N_1

52. $N_1 \leq \frac{P_2+21N}{22}$.

53. $N_1 \leq \frac{P_1+33N}{34}$.

54. $N_1 \leq \frac{4N_2+P_4}{5}$.

55. $N_1 \leq \frac{9N+H}{10}$.

56. $N_1 \leq \frac{P_5+2G}{3}$.

57. $N_1 \leq \frac{P_3+15N}{16}$.

58. $\frac{7G+3R}{10} \leq N_1$.

59. $N_1 \leq \frac{H+S}{2}$.

(vii) Inequalities with N

60. $\frac{C+14N_1}{15} \leq N$.

61. $\frac{C+2P_4}{3} \leq N$.

62. $\frac{P_4+14N_2}{15} \leq N$.

63. $\frac{S+8N_1}{9} \leq N$.

64. $\frac{8P_4+7S}{15} \leq N$.

65. $N \leq \frac{11C+7P_2}{18}$.

66. $N \leq \frac{22P_5+5P_2}{27}$.

67. $N \leq \frac{P_1+68N_2}{69}$.

68. $N \leq \frac{P_2+44N_2}{45}$.

69. $N \leq \frac{P_5+5N_1}{6}$.

70. $N \leq \frac{H+20N_2}{21}$.

71. $N \leq \frac{P_3+32N_2}{33}$.

72. $N \leq \frac{7A+2P_4}{9}$.

(viii) Inequalities with N_2

73. $\frac{3R+11N_1}{14} \leq N_2$.

74. $\frac{C+3G}{4} \leq N_2$.

75. $\frac{S+5N_1}{6} \leq N_2$.

76. $N_2 \leq \frac{7A+H}{8}$.

77. $N_2 \leq \frac{P_1+23A}{24}$.

78. $N_2 \leq \frac{P_2+15A}{16}$.

79. $N_2 \leq \frac{P_5+9N}{10}$.

80. $N_2 \leq \frac{P_3+11A}{12}$.

81. $N_2 \leq \frac{A+3N}{4}$.

82. $N_2 \leq \frac{11N+R}{12}$.

83. $\frac{9N_1+C}{10} \leq N_2$.

84. $\frac{P_4+S}{2} \leq N_2$.

(ix) Inequalities with A

85. $\frac{9R+4P_4}{13} \le A.$

86. $\frac{2S+H}{3} \le A.$

87. $\frac{S+4N_2}{5} \le A.$

88. $\frac{S+3N}{4} \le A.$

89. $A \le \frac{3S+P_3}{4}.$

90. $A \le \frac{P_5+2N_2}{3}.$

91. $A \le \frac{4S+P_2}{5}.$

92. $A \le \frac{6R+P_2}{7}.$

93. $A \le \frac{P_1+12P_5}{13}.$

94. $A \le \frac{P_1+9R}{10}.$

95. $A \le \frac{P_4+3P_5}{4}.$

96. $A \le \frac{9R+2P_3}{11}.$

97. $\frac{2N+R}{3} \le A.$

(x) Inequalities with S

98. $S \le \frac{5C+2P_4}{7}.$

99. $S \le \frac{5C+4N_2}{9}.$

100. $S \le \frac{7C+P_1}{8}.$

(xi) Inequalities with P_5

101. $\frac{21R+P_3}{22} \le P_5.$

102. $\frac{3H+5C}{8} \le P_5.$

(xii) Inequalities with R

103. $\frac{11S+P_3}{12} \le R.$

104. $R \le \frac{P_1+20S}{21}.$

105. $\frac{A+2S}{3} \le R.$

106. $\frac{4N+3C}{7} \le R.$

107. $\frac{16N_2+11C}{27} \le R.$

2.6 Group II: Three Means Inequalities

This group gives inequalities having *four means* with *three means* on one side. See the list below:

(i) Inequalities with P_2

1. $P_2 \le P_5 + 3P_3 - 3P_4$

2. $P_2 \le C + 3P_3 - 3G$

3. $P_2 \le \frac{23N_1+14P_1-14N_2}{23}$

4. $P_2 \le \frac{18N+11P_1-11A}{18}$

5. $P_2 \le \frac{3P_5+2H-2S}{3}$

6. $P_2 \le \frac{16P_5+9P_1-9C}{16}$

(ii) Inequalities with P_3

7. $P_3 \le \frac{3N_1+2P_2-2N_2}{3}$

8. $P_3 \le N + 8R - 8S$

9. $P_3 \le \frac{13A+6P_1-6P_5}{13}$

(iii) Inequalities with H

10. $H \le A + 4P_4 - 4G$

11. $H \le \frac{2N_1+P_3-A}{2}$

12. $H \le \frac{4P_5+5G-5S}{4}$

13. $H \le \frac{11P_5+30N_2-30R}{11}$

14. $H \le \frac{11N+5P_3-5R}{11}$

15. $H \le P_3 + 3S - 3R$

16. $H \le S + 2P_5 - 2C$

17. $H \le \frac{4P_2+C-P_1}{4}$

(iv) Inequalities with P_4

18. $P_4 \leq N_1 + 12N - N_2$

19. $P_4 \leq \frac{2N_1 + H - A}{2}$

20. $P_4 \leq \frac{4C + 7H - 7A}{4}$

21. $P_4 \leq \frac{4S + 5H - 5A}{4}$

22. $P_4 \leq \frac{3H + C - P_5}{3}$

23. $P_4 \leq \frac{2A + G - C}{2}$

24. $P_4 \leq \frac{3G + P_5 - C}{3}$

25. $P_4 \leq P_5 + 2S - 2C$

26. $P_4 \leq \frac{27R + 26N_2 - 26C}{27}$

(v) Inequalities with G

27. $G \leq P_4 + 6N_2 - 6N$

28. $G \leq \frac{3H + 2C - 2P_5}{3}$

29. $G \leq \frac{3R + 2H - 2P_5}{3}$

30. $G \leq P_2 + 2C - 2P_5$

31. $G \leq P_5 + 6N_2 - 6A$

32. $G \leq \frac{4P_4 + 3R - 3A}{4}$

33. $G \leq C + 2N_1 - 2S$

34. $G \leq \frac{2N_1 + S - C}{2}$

35. $G \leq \frac{5H + C - P_3}{5}$

(vi) Inequalities with N_1

36. $N_1 \leq C + P_4 - S$

37. $N_1 \leq \frac{12R + 7G - 7S}{12}$

38. $N_1 \leq \frac{20N_2 + 3G - 3R}{20}$

(vii) Inequalities with N

39. $N \leq \frac{3N_1 + 2A - 2N_2}{3}$

40. $N \leq \frac{6C + 7H - 7A}{6}$

41. $N \leq \frac{27S + 8P_2 - 8P_5}{27}$

42. $N \leq \frac{4N_2 + R - S}{4}$

43. $N \leq \frac{9S + 8P_5 - 8C}{9}$

(viii) Inequalities with N_2

44. $N_2 \leq \frac{8C + 9H - 9A}{8}$

45. $N_2 \leq \frac{8A + P_4 - P_5}{8}$

46. $N_2 \leq \frac{14C + 9P_3 - 9P_5}{14}$

47. $N_2 \leq \frac{18S + 5P_2 - 5P_5}{18}$

48. $N_2 \leq \frac{4P_2 + 45S - 45R}{4}$

49. $N_2 \leq \frac{12N_3 + C - S}{12}$

50. $N_2 \leq \frac{6S + 5P_5 - 5C}{6}$

51. $N_2 \leq \frac{36R + 11G - 11C}{36}$

52. $N_2 \leq \frac{42N + C - P_4}{42}$

53. $N_2 \leq \frac{8G + 3P_5 - 3P_4}{8}$

(ix) Inequalities with A

54. $A \leq \frac{5S + P_3 - C}{5}$

(x) Inequalities with R

55. $R \leq \frac{3A + 2C - 2S}{3}$

56. $R \leq \frac{9P_2 + 14A - 14P_3}{9}$

(xi) Inequalities with P_5

57. $P_5 \leq C + P_3 - P_4$

(xii) Inequalities with S

58. $S \leq \frac{18R+C-P2}{18}$ **59.** $S \leq \frac{3P_2+5A-5P_3}{3}$

The aim of this work is to improve the inequalities (30) similar to Theorem 1. This we have done by using inequalities given in Groups I and II. Since these groups are divided according the number of means, the main results are also divided in two sections, i.e., one for Group I and another for Group II. Since there are many inequalities in each Group, the main results are divided in small propositions. In order to prove the results, we shall use frequently, the arguments given in the note below:

Note 1. *Let a and b be two positive numbers, i.e., $a > 0$ and $b > 0$. If $a^2 - b^2 \geq 0$, then we can conclude that $a \geq b$ because $a - b = (a^2 - b^2)/(a+b)$. If $b < 0$, then obviously, $a - (-b) = a + b > 0$ always holds. In some case this argument is applied twice.*

3. TWO MEANS INEQUALITIES

This section brings results based on inequalities given in Group I (Section 2.5). The results are divided in propositions.

Proposition 1. *The following inequalities hold:*

$$P_1 \leq P_2 \leq \frac{P_1 + 2P_3}{3} \leq \begin{cases} P_3 \\ \frac{H+P_1}{2} \leq \begin{cases} H \\ \frac{5P_1+4P_4}{9} \leq \frac{3P_1+2G}{5} \leq \text{(a)} \end{cases} \end{cases}$$

$$\text{(a)} \leq \begin{cases} G \\ \frac{7P_1+4N_1}{11} \leq \frac{15P_1+8N_2}{23} \leq \frac{A+2P_1}{3} \leq \begin{cases} \frac{5P_1+2S}{7} \leq \frac{3P_1+C}{4} \leq N_1 \\ \frac{9P_1+4P_5}{13} \leq N. \end{cases} \end{cases}$$

Proof. The proof is bases on the following results.

1. For $P_2 \leq \frac{P_1+2P_3}{3}$: We have to show that $\frac{1}{3}(P_1 + 2P_2 - 3P_3) \geq 0$. We can write $P_1 + 2P_2 - 3P_3 = bg_1(a/b)$, where

$$g_1(x) = \frac{u_1(x)}{(x^2+1)(x-\sqrt{x}+1)(x^3+1)}$$

with

$$u_1(x) = x\big(2x^{9/2} - 5x^4 + 3x^{7/2} + x^3 - 2x^{5/2} + x^2 + 3x^{3/2} - 5x + 2\sqrt{x}\big)$$
$$= x^{3/2}(\sqrt{x} - 1)^4\big(2x^2 + 3x^{3/2} + 3x + 3\sqrt{x} + 2\big).$$

Since $u_1(x) \geq 0$, $\forall x > 0$, hence proving the required result.

2. For $\frac{P_1 + 2P_3}{3} \leq \frac{H + P_1}{2}$: We have to show that $\frac{1}{6}(3H + P_1 - 4P_3) \geq 0$. We can write $3H + P_1 - 4P_3 = bg_2(a/b)$, where

$$g_2(x) = \frac{u_2(x)}{(x^3 + 1)(x - \sqrt{x} + 1)}$$

with

$$u_2(x) = x\big(3x^3 - 7x^{5/2} + x^2 + 6x^{3/2} + x - 7\sqrt{x} + 3\big)$$
$$= x(3x + 5\sqrt{x} + 3)(\sqrt{x} - 1)^4.$$

Since $u_2(x) \geq 0$, $\forall x > 0$, hence proving the required result.

3. For $\frac{H + P_1}{2} \leq \frac{5P_1 + 4P_4}{9}$: We have to show that $\frac{1}{18}(P_1 + 8P_4 - 9H) \geq 0$. We can write $P_1 + 8P_4 - 9H = bg_3(a/b)$, where

$$g_3(x) = \frac{u_3(x)}{(x^3 + 1)(\sqrt{x} + 1)^2}$$

with

$$u_3(x) = x\big(15x^3 + x^2 - 34x^{5/2} + x + 36x^{3/2} - 34\sqrt{x} + 15\big)$$
$$= x(15 + 26\sqrt{x} + 15)(\sqrt{x} - 1)^4.$$

Since $u_3(x) \geq 0$, $\forall x > 0$, hence proving the required result.

4. For $\frac{5P_1 + 4P_4}{9} \leq \frac{3P_1 + 2G}{5}$: We have to show that $\frac{2}{45}(P_1 + 9G - 10P_4) \geq 0$. We can write $P_1 + 9G - 10P_4 = bg_4(a/b)$, where

$$g_4(x) = \frac{u_4(x)}{(x^3 + 1)(\sqrt{x} + 1)^2}$$

with

$$u_4(x) = 9x^{9/2} - 21x^4 + 11x^{7/2} + 11x^{3/2} + x^2 + x^3 - 21x + 9\sqrt{x}$$
$$= \sqrt{x}(9x^2 + 15x^{3/2} + 17x + 15\sqrt{x} + 9)(\sqrt{x} - 1)^4.$$

Since $u_4(x) \geq 0$, $\forall x > 0$, hence proving the required result.

5. For $\frac{3P_1+2G}{5} \leq \frac{7P_1+4N_1}{11}$**:** We have to show that $\frac{2}{55}(P_1 + 10N_1 - 11G) \geq 0$. We can write $2P_1 + 20N_1 - 22G = bg_5(a/b)$, where $g_5(x) = \frac{1}{2}u_5(x)/(x^3 + 1)$ with

$$u_5(x) = 5x^4 - 12x^{7/2} + 7x^3 - 12\sqrt{x} + 7x + 5$$
$$= \left(5x^2 + 8x^{3/2} + 9x + 8\sqrt{x} + 5\right)(\sqrt{x} - 1)^4.$$

Since $u_5(x) \geq 0$, $\forall x > 0$, hence proving the required result.

6. For $\frac{7P_1+4N_1}{11} \leq \frac{15P_1+8N_2}{23}$**:** We have to show that $\frac{4}{253}(P_1 + 22N_2 - 23N_1) \geq 0$. We can write $P_1 + 22N_2 - 23N_1 = bg_6(a/b)$, where $g_6(x) = \frac{1}{4}u_6(x)/(x^3 + 1)$ with

$$u_6(x) = 22\sqrt{2x+2}(\sqrt{x} + 1)(x^3 + 1)$$
$$- \left(23x^4 + 46x^{7/2} + 19x^3 + 19x + 46\sqrt{x} + 23\right).$$

Let's consider

$$v_6(x) = \left[22\sqrt{2x+2}(\sqrt{x} + 1)(x^3 + 1)\right]^2$$
$$- \left(23x^4 + 46x^{7/2} + 19x^3 + 19x + 46\sqrt{x} + 23\right)^2.$$

After simplifications,

$$v_6(x) = (\sqrt{x} - 1)^4 \begin{pmatrix} 439x^6 + 1576x^{11/2} + 2616x^5 \\ +2952x^{9/2} + 2584x^4 + 1512x^{7/2} \\ +798x^3 + 1512x^{5/2} + 2584x^2 \\ +2952x^{3/2} + 2616x + 1576\sqrt{x} + 439 \end{pmatrix}.$$

In view of Note 1, $v_6(x) \geq 0$ implies $u_6(x) \geq 0$, $\forall x > 0$, hence proving the required result.

7. For $\frac{15P_1+8N_2}{23} \leq \frac{A+2P_1}{3}$**:** We have to show that $\frac{1}{69}(23A + P_1 - 24N_2) \geq 0$. We can write $23A + P_1 - 24N_2 = bg_7(a/b)$, where $g_7(x) = \frac{1}{2}u_7(x)/(x^3 + 1)$ with

$$u_7(x) = 23x^4 + 25x^3 + 25x + 23 - 12\sqrt{2x+2}(\sqrt{x} + 1)(x^3 + 1).$$

Let's consider

$$v_7(x) = \left(23x^4 + 25x^3 + 25x + 23\right)^2 - \left[12\sqrt{2x+2}(\sqrt{x} + 1)(x^3 + 1)\right]^2.$$

After simplifications,

$$v_7(x) = (\sqrt{x} - 1)^4 \begin{pmatrix} 241x^6 + 388x^{11/2} + 680x^5 + 780x^{9/2} \\ +688x^4 + 404x^{7/2} + 502x^3 + 404x^{5/2} \\ +688x^2 + 780x^{3/2} + 680x + 388\sqrt{x} + 241 \end{pmatrix}.$$

In view of Note 1, $v_7(x) \geq 0$ implies $u_7(x) \geq 0$, $\forall x > 0$, hence proving the required result.

8. For $\frac{A+2P_1}{3} \leq \frac{9P_1+4P_5}{13}$: We have to show that $\frac{1}{39}(P_1 + 12P_5 - 13A) \geq 0$. We can write $P_1 + 12P_5 - 13A = bg_8(a/b)$, where

$$g_8(x) = \frac{u_8(x)}{2(x^3 + 1)(\sqrt{x} + 1)^2}$$

with

$$u_8(x) = \begin{pmatrix} 11x^5 + 24x^4 - 26x^{9/2} - 22x^{7/2} + 13x^2 \\ +13x^3 + 24x - 22x^{3/2} - 26\sqrt{x} + 11 \end{pmatrix}.$$

After simplifications,

$$u_8(x) = (\sqrt{x} - 1)^4 \begin{pmatrix} 11x^3 + 18x^{5/2} + 30x^2 \\ +34x^{3/2} + 30x + 18\sqrt{x} + 11 \end{pmatrix}.$$

Since $u_8(x) \geq 0$, $\forall x > 0$, hence proving the required result.

9. For $\frac{A+2P_1}{3} \leq \frac{5P_1+2S}{7}$: We have to show that $\frac{1}{21}(P_1 + 6S - 7A) \geq 0$. We can write $P_1 + 6S - 7A = bg_9(a/b)$, where $g_9(x) = \frac{1}{2}u_9(x)/(x^3 + 1)$ with

$$u_9(x) = 6\sqrt{2x^2 + 2}(x^3 + 1) - (7x^4 + 5x^3 + 5x + 7).$$

Let's consider

$$v_9(x) = \left[6\sqrt{2x^2 + 2}(x^3 + 1)\right]^2 - (7x^4 + 5x^3 + 5x + 7)^2.$$

After simplifications,

$$v_9(x) = (x - 1)^4\left[23(x^4 + 1) + 20(x^3 + 1) + 2x(x - 1)^2 + x^2\right].$$

In view of Note 1, $v_9(x) \geq 0$ implies $u_9(x) \geq 0$, $\forall x > 0$, hence proving the required result.

10. For $\frac{5P_1+2S}{7} \leq \frac{3P_1+C}{4}$**:** We have to show that $\frac{1}{28}(P_1 + 7C - 8S) \geq 0$.
We can write $P_1 + 7C - 8S = bg_{10}(a/b)$, where $g_{10}(x) = u_{10}(x)/(x^3 + 1)$ with

$$u_{10}(x) = (x^2 + 1)(7x^2 - 6x + 7) - 4\sqrt{2x^2 + 2}(x^3 + 1)$$
$$= (x^2 + 1)[4(x^2 + 1) + 3(x - 1)^2] - 4\sqrt{2x^2 + 2}(x^3 + 1).$$

Let's consider

$$v_{10}(x) = \{(x^2 + 1)[4(x^2 + 1) + 3(x - 1)^2]\}^2$$
$$- [4\sqrt{2x^2 + 2}(x^3 + 1)]^2.$$

After simplifications,

$$v_{10}(x) = (x - 1)^4(x^2 + 1)(17x^2 - 16x + 17)$$
$$= (x - 1)^4(x^2 + 1)[9(x^2 + 1) + 8(x - 1)^2].$$

In view of Note 1, $v_{10}(x) \geq 0$ implies $u_{10}(x) \geq 0$, $\forall x > 0$, hence proving the required result.

11. For $\frac{3P_1+C}{4} \leq N_1$**:** We have to show that $\frac{1}{4}(4N_1 - 3P_1 - C) \geq 0$. We can write $4N_1 - 3P_1 - C = bg_{11}(a/b)$, where $g_{11}(x) = u_{11}(x)/(x^2 - x + 1)$ with

$$u_{11}(x) = 2x^{5/2} - x^2 - 2x^{3/2} - x + 2\sqrt{x}$$
$$= \sqrt{x}(2x + 3\sqrt{x} + 2)(\sqrt{x} - 1)^2.$$

Since $u_{11}(x) \geq 0$, $\forall x > 0$, hence proving the required result.

12. For $\frac{9P_1+4P_5}{13} \leq N$**:** We have to show that $\frac{1}{13}(13N - 9P_1 - 4P_5) \geq 0$.
We can write $13N - 9P_1 - 4P_5 = bg_{12}(a/b)$, where

$$g_{12}(x) = \frac{u_{12}(x)}{3(x^3 + 1)(\sqrt{x} + 1)^2}$$

with

$$u_{12}(x) = \begin{pmatrix} x^5 + 39x^{9/2} + x^4 - 15x^{7/2} - 26x^3 \\ -26x^2 - 15x^{3/2} + x + 39\sqrt{x} + 1 \end{pmatrix}.$$

After simplifications,

$$u_{12} = (\sqrt{x} - 1)^2 \begin{pmatrix} x^4 + 41x^{7/2} + 82x^3 + 108x^{5/2} \\ +108x^2 + 108x^{3/2} + 82x + 41\sqrt{x} + 1 \end{pmatrix}.$$

Since $u_{12}(x) \geq 0$, $\forall x > 0$, hence proving the required result.

13. For $\frac{P_1+2P_3}{3} \leq P_3$: We have to show that $\frac{1}{3}(P_3 - P_1) \geq 0$. It is true by definition. See inequalities (30).

14. For $\frac{H+P_1}{2} \leq H$: We have to show that $\frac{1}{2}(H - P_1) \geq 0$. It is true by definition. See inequalities (30).

15. For $\frac{5P_1+4P_4}{9} \leq P_4$: We have to show that $\frac{5}{9}(P_4 - P_1) \geq 0$. It is true by definition. See inequalities (30).

16. For $\frac{3P_1+2G}{5} \leq G$: We have to show that $\frac{3}{5}(G - P_1) \geq 0$. It is true by definition. See inequalities (30).

Combining Results 1 to 16 we get the proof of Proposition 1. $\square$

Proposition 2. *The following inequalities hold:*

$$P_3 \leq \frac{H+P_2}{2} \leq \begin{cases} \begin{cases} \frac{P_1+3H}{4} \leq \mathbf{H} \leq \frac{4P_3+C}{5} \leq \mathbf{N_1} \\ \mathbf{P_4} \\ \frac{2P_4+3P_2}{5} \leq \end{cases} \begin{cases} \frac{2P_2+G}{3} \leq \begin{cases} \mathbf{G} \\ \frac{5P_2+2N_1}{7} \leq (a) \end{cases} \end{cases} \end{cases}$$

$$(a) \leq \frac{11P_2+4N_2}{15} \leq \frac{A+3P_2}{4} \leq \begin{Bmatrix} \frac{4P_2+S}{5} \\ \frac{11P_2+3R}{14} \end{Bmatrix} \leq \frac{5P_2+C}{6} \leq \begin{Bmatrix} \frac{4P_3+C}{5} \\ \frac{7P_2+3R}{10} \\ \frac{7P_3+2P_5}{9} \end{Bmatrix}$$

$$\leq \mathbf{N_1}.$$

Proof. The proof is bases on the following results.

17. For $P_3 \leq \frac{H+P_2}{2}$: We have to show that $\frac{1}{2}(H + P_2 - 2P_3) \geq 0$. We can write $H + P_2 - 2P_3 = bg_{17}(a/b)$, where

$$g_{17}(x) = \frac{u_{17}(x)}{(x^2 + 1)(x+1)(x - \sqrt{x} + 1)}$$

with

$$u_{17}(x) = x\left(x^3 - 3x^{5/2} + 3x^2 - 2x^{3/2} - 3\sqrt{x} + 3x + 1\right)$$
$$= x(x + \sqrt{x} + 1)(\sqrt{x} - 1)^4.$$

Since $u_{17}(x) \geq 0$, $\forall x > 0$, hence proving the required result.

18. For $\frac{H+P_2}{2} \leq \frac{2P_4+3P_2}{5}$: We have to show that $\frac{1}{10}(4P_4 + P_2 - 5H) \geq 0$. We can write $4P_4 + P_2 - 5H = bg_{18}(a/b)$, where

$$g_{18}(x) = \frac{u_{18}(x)}{(x^2 + 1)(x+1)(\sqrt{x} + 1)^2}$$

with

$$u_{18}(x) = x\left(7x^3 + 9x^2 - 18x^{5/2} + 9x + 4x^{3/2} - 18\sqrt{x} + 7\right)$$
$$= x(\sqrt{x} - 1)^4(7x + 10\sqrt{x} + 7).$$

Since $u_{18}(x) \geq 0$, $\forall x > 0$, hence proving the required result.

19. For $\frac{2P_4 + 3P_2}{5} \leq \frac{2P_2 + G}{3}$: We have to show that $\frac{1}{15}(P_2 + 5G - 6P_4) \geq 0$. We can write $P_2 + 5G - 6P_4 = bg_{19}(a/b)$, where

$$g_{19}(x) = \frac{u_{19}(x)}{(x^2 + 1)(\sqrt{x} + 1)^2}$$

with

$$u_{19}(x) = 5x^{7/2} - 13x^3 + 7x^{3/2} + 2x^2 + 7x^{5/2} - 13x + 5\sqrt{x}$$
$$= \sqrt{x}(\sqrt{x} - 1)^4(5x + 7\sqrt{x} + 5).$$

Since $u_{19}(x) \geq 0$, $\forall x > 0$, hence proving the required result.

20. For $\frac{2P_2 + G}{3} \leq \frac{5P_2 + 2N_1}{7}$: We have to show that $\frac{1}{21}(P_2 + 6N_1 - 7G) \geq 0$. We can write $P_2 + 6N_1 - 7G = bg_{20}(a/b)$, where $g_{20}(x) = \frac{1}{2}u_{20}(x)/(x^2 + 1)$ with

$$u_{20}(x) = 3x^3 - 8x^{5/2} + 5x^2 - 8\sqrt{x} + 5x + 3$$
$$= (3x + 4\sqrt{x} + 3)(\sqrt{x} - 1)^4.$$

Since $u_{20}(x) \geq 0$, $\forall x > 0$, hence proving the required result.

21. For $\frac{5P_2 + 2N_1}{7} \leq \frac{11P_2 + 4N_2}{15}$: We have to show that $\frac{2}{105}(P_2 + 14N_2 - 15N_1) \geq 0$. We can write $P_2 + 14N_2 - 15N_1 = bg_{21}(a/b)$, where $g_{21}(x) = \frac{1}{4}u_{21}(x)/(x^2 + 1)$ with

$$u_{21}(x) = 14\sqrt{2x + 2}(\sqrt{x} + 1)(x^2 + 1)$$
$$- \left(15x^3 + 30x^{5/2} + 11x^2 + 11x + 30\sqrt{x} + 15\right).$$

Let's consider

$$v_{21}(x) = \left[14\sqrt{2x + 2}(\sqrt{x} + 1)(x^2 + 1)\right]^2$$
$$- \left(15x^3 + 30x^{5/2} + 11x^2 + 11x + 30\sqrt{x} + 15\right)^2.$$

After simplifications,

$$v_{21}(x) = (\sqrt{x} - 1)^4 \left(\begin{array}{l} 167x^4 + 552x^{7/2} + 760x^3 + 520x^{5/2} \\ +286x^2 + 520x^{3/2} + 760x + 552\sqrt{x} + 167 \end{array} \right).$$

In view of Note 1, $v_{21}(x) \geq 0$ implies $u_{21}(x) \geq 0$, $\forall x > 0$, hence proving the required result.

22. For $\frac{11P_2 + 4N_2}{15} \leq \frac{A + 3P_2}{4}$: We have to show that $\frac{1}{60}(15A + P_2 - 16N_2) \geq 0$. We can write $15A + P_2 - 16N_2 = bg_{22}(a/b)$, where $g_{22}(x) = \frac{1}{2}u_{22}(x)/(x^2 + 1)$ with

$$u_{22}(x) = (x + 1)(15x^2 + 2x + 15) - 8\sqrt{2x + 2}(\sqrt{x} + 1)(x^2 + 1).$$

Let's consider

$$v_{22}(x) = \left[(x + 1)(15x^2 + 2x + 15) \right]^2 - \left[8\sqrt{2x + 2}(\sqrt{x} + 1)(x^2 + 1) \right]^2.$$

After simplifications,

$$v_{22}(x) = (\sqrt{x} - 1)^4 (x + 1) \left(\begin{array}{l} 97x^3 + 132x^{5/2} + 103x^2 \\ +8x^{3/2} + 103x + 132\sqrt{x} + 97 \end{array} \right).$$

In view of Note 1, $v_{22}(x) \geq 0$ implies $u_{22}(x) \geq 0$, $\forall x > 0$, hence proving the required result.

23. For $\frac{A + 3P_2}{4} \leq \frac{4P_2 + S}{5}$: We have to show that $\frac{1}{20}(P_2 + 4S - 5A) \geq 0$. We can write $P_2 + 4S - 5A = bg_{23}(a/b)$, where $g_{23}(x) = \frac{1}{2}u_{23}(x)/(x^2 + 1)$ with

$$u_{23}(x) = 4\sqrt{2x^2 + 2}(x^2 + 1) - (5x^3 + 3x^2 + 3x + 5).$$

Let's consider

$$v_{23}(x) = \left[4\sqrt{2x^2 + 2}(x^2 + 1) \right]^2 - (5x^3 + 3x^2 + 3x + 5)^2.$$

After simplifications,

$$v_{23}(x) = (x - 1)^4 (7x^2 - 2x + 7) = (x - 1)^4 [6(x^2 + 1) + (x - 1)^2].$$

In view of Note 1, $v_{23}(x) \geq 0$ implies $u_{23}(x) \geq 0$, $\forall x > 0$, hence proving the required result.

24. For $\frac{A+3P_2}{4} \le \frac{11P_2+3R}{14}$: We have to show that $\frac{1}{28}(P_2 + 6R - 7A) \ge 0$. We can write $P_2 + 6R - 7A = bg_{24}(a/b)$, where

$$g_{24}(x) = \frac{(x-1)^4}{2(x^2+1)(x+1)}.$$

Since $g_{24}(x) \ge 0$, $\forall x > 0$, hence proving the required result.

25. For $\frac{4P_2+S}{5} \le \frac{5P_2+C}{6}$: We have to show that $\frac{1}{30}(P_2 + 5C - 6S) \ge 0$. We can write $P_2 + 5C - 6S = bg_{25}(a/b)$, where

$$g_{25}(x) = \frac{u_{25}(x)}{(x^2+1)(x+1)}$$

with

$$u_{25}(x) = 5x^4 + x^3 + 12x^2 + x + 5 - 3\sqrt{2x^2 + 2}(x^2 + 1)(x + 1).$$

Let's consider

$$v_{25}(x) = \left(5x^4 + x^3 + 12x^2 + x + 5\right)^2 - \left[3\sqrt{2x^2 + 2}(x^2 + 1)(x + 1)\right]^2.$$

After simplifications,

$$v_{25}(x) = (x-1)^4\left(7x^4 + 2x^3 + 15x^2 + 2x + 7\right).$$

In view of Note 1, $v_{25}(x) \ge 0$ implies $u_{25}(x) \ge 0$, $\forall x > 0$, hence proving the required result.

26. For $\frac{11P_2+3R}{14} \le \frac{5P_2+C}{6}$: We have to show that $\frac{1}{42}(2P_2 + 7C - 9R) \ge 0$. We can write $2P_2 + 7C - 9R = bg_{26}(a/b)$, where

$$g_{26}(x) = \frac{(x-1)^4}{(x^2+1)(x+1)}.$$

Since $g_{26}(x) \ge 0$, $\forall x > 0$, hence proving the required result.

27. For $\frac{5P_2+C}{6} \le \frac{4P_3+C}{5}$: We have to show that $\frac{1}{30}(24P_3 + C - 25P_2) \ge 0$. We can write $24P_3 + C - 25P = bg_{27}(a/b)$, where

$$g_{27}(x) = \frac{u_{27}(x)}{(x^2+1)(x+1)(x-\sqrt{x}+1)}$$

with

$$u_{27}(x) = \begin{pmatrix} x^5 - x^{9/2} + 25x^{7/2} - 49x^3 \\ +48x^{5/2} - 49x^2 + 25x^{3/2} - \sqrt{x} + 1 \end{pmatrix}.$$

After simplifications,

$$u_{27}(x) = (\sqrt{x} - 1)^2 \begin{pmatrix} x^4 + x^{7/2} + x^3 + 26x^{5/2} \\ +2x^2 + 26x^{3/2} + x + \sqrt{x} + 1 \end{pmatrix}.$$

Since $u_{27}(x) \geq 0$, $\forall x > 0$, hence proving the required result.

28. For $\frac{5P_2+C}{6} \leq \frac{7P_2+3R}{10}$: We have to show that $\frac{1}{30}(9R - 4P_2 - 5C) \geq 0$. We can write $9R - 4P_2 - 5C = bg_{28}(a/b)$, where

$$g_{28}(x) = \frac{(x^2 + 4x + 1)(x - 1)^2}{(x + 1)(x^2 + 1)}.$$

Since $g_{28}(x) \geq 0$, $\forall x > 0$, hence proving the required result.

29. For $\frac{5P_2+C}{6} \leq \frac{7P_3+2P_5}{9}$: We have to show that $\frac{1}{18}(14P_3 + 4P_5 - 3C - 15P_2) \geq 0$. We can write $14P_3 + 4P_5 - 3C - 15P_2 = bg_{29}(a/b)$, where

$$g_{29}(x) = \frac{u_{29}(x)}{(x + 1)(\sqrt{x} + 1)^2(x - \sqrt{x} + 1)}$$

with

$$u_{29}(x) = \begin{pmatrix} x^6 - 7x^{11/2} + 15x^5 - 2x^{9/2} + 17x^4 - 39x^{7/2} \\ +30x^3 - 39x^{5/2} + 17x^2 - 2x^{3/2} + 15x - 7\sqrt{x} + 1 \end{pmatrix}.$$

After simplifications,

$$u_{29}(x) = (\sqrt{x} - 1)^2 \begin{pmatrix} x^5 - 5x^{9/2} + 4x^4 + 11x^{7/2} + 35x^3 \\ +20x^{5/2} + 35x^2 + 11x^{3/2} + 4x - 5\sqrt{x} + 1 \end{pmatrix}.$$

Still, we need to show that the second expression inside the bracket of $u_{29}(x)$ is nonnegative? For simplicity, let's consider

$$h_{29}(x) = \begin{pmatrix} x^5 - 5x^{9/2} + 4x^4 + 11x^{7/2} + 35x^3 \\ +20x^{5/2} + 35x^2 + 11x^{3/2} + 4x - 5\sqrt{x} + 1 \end{pmatrix}.$$

Let's write

$$k_{29}(t) := h_{29}(t^2) = \begin{pmatrix} t^{10} - 5t^9 + 4t^8 + 11t^7 + 35t^6 \\ +20t^5 + 35t^4 + 11t^3 + 4t^2 - 5t + 1 \end{pmatrix}.$$

The polynomial equation $k_{29}(t) = 0$ is of 10th degree. It admits 10 solutions. All these 10 solutions are complex (not written here), i.e., it doesn't have any real solution. Since we are working with $t > 0$, then there are two possibilities, one when the graph of $k_{29}(t)$ is either in 1st quadrant or in 4th quadrant. Since $k_{29}(1) = 112$, this means that $k_{29}(t) > 0$, for all $t > 0$. Thus, $h_{29}(x) > 0$, for all $x > 0$, proving $u_{29}(x) \geq 0$, for all $x > 0$. This completes the proof of the result.

30. For $\frac{7P_3 + 3R}{10} \leq N_1$**:** We have to show that $\frac{1}{10}(10N_1 - 7P_3 - 3R) \geq 0$. We can write $10N_1 - 7P_3 - 3R = bg_{30}(a/b)$, where

$$g_{30}(x) = \frac{u_{30}(x)}{2(x+1)(x - \sqrt{x} + 1)}$$

with

$$u_{30}(x) = x^3 - 17x^2 + 9x^{5/2} + 14x^{3/2} - 17x + 9\sqrt{x} + 1$$
$$= (x^2 + 11x^{3/2} + 4x + 11\sqrt{x} + 1)(\sqrt{x} - 1)^2.$$

Since $u_{30}(x) \geq 0$, $\forall x > 0$, hence proving the required result.

31. For $\frac{7P_3 + 2P_5}{9} \leq N_1$**:** We have to show that $\frac{1}{9}(9N_1 - 7P_3 - 2P_5) \geq 0$. We can write $9N_1 - 7P_3 - 2P_5 = bg_{31}(a/b)$, where

$$g_{31}(x) = \frac{u_{31}(x)}{4(\sqrt{x} + 1)^2(x - \sqrt{x} + 1)}$$

with

$$u_{31}(x) = x^3 + 35x^{5/2} - 25x^2 - 22x^{3/2} - 25x + 35\sqrt{x} + 1$$
$$= (x^2 + 37x^{3/2} + 48x + 37\sqrt{x} + 1)(\sqrt{x} - 1)^2.$$

Since $u_{31}(x) \geq 0$, $\forall x > 0$, hence proving the required result.

32. For $\frac{H+P_2}{2} \leq \frac{P_1+3H}{4}$**:** We have to show that $\frac{1}{4}(P_1 + H - 2P_2) \geq 0$. We can write $P_1 + H - 2P_2 = bg_{32}(a/b)$, where

$$g_{32}(x) = \frac{x(x-1)^4}{(x^2+1)(x^3+1)}.$$

Since $g_{32}(x) \geq 0$, $\forall x > 0$, hence proving the required result.

33. For $\frac{P_1+3H}{4} \leq H$: We have to show that $\frac{1}{4}(H - P_1) \geq 0$. It is true by definition. See inequalities (30).

34. For $H \leq \frac{4P_3+C}{5}$: We have to show that $\frac{1}{5}(4P_3 + C - 5H) \geq 0$. We can write $4P_3 + C - 5H = bg_{34}(a/b)$, where

$$g_{34}(x) = \frac{u_{34}(x)}{(x+1)(x-\sqrt{x}+1)}$$

with

$$u_{34}(x) = x^3 - x^{5/2} - 5x^2 + 10x^{3/2} - 5x - \sqrt{x} + 1$$
$$= (x + 3\sqrt{x} + 1)(\sqrt{x} - 1)^4.$$

Since $u_{34}(x) \geq 0$, $\forall x > 0$, hence proving the required result.

35. For $\frac{4P_3+C}{5} \leq N_1$: We have to show that $\frac{1}{5}(5N_1 - 4P_3 - C) \geq 0$. We can write $5N_1 - 4P_3 - C = bg_{35}(a/b)$, where

$$g_{35}(x) = \frac{u_{35}(x)}{4(x+1)(x-\sqrt{x}+1)}$$

with

$$u_{35}(x) = x^3 - 15x^2 + 9x^{5/2} + 10x^{3/2} - 15x + 9\sqrt{x} + 1$$
$$= (x^2 + 11x^{3/2} + 6x + 11\sqrt{x} + 1)(\sqrt{x} - 1)^2.$$

Since $u_{35}(x) \geq 0$, $\forall x > 0$, hence proving the required result.

36. For $\frac{2P_4+3P_2}{5} \leq P_4$: We have to show that $\frac{3}{5}(P_4 - P_2) \geq 0$. It is true by definition. See inequalities (30).

37. For $\frac{2P_2+G}{3} \leq G$: We have to show that $\frac{2}{3}(G - P_2) \geq 0$. It is true by definition. See inequalities (30).

Combining Results 17 to 37 we get the proof of Proposition 2.   □

Proposition 3. *The following inequalities hold:*

$$H \leq \frac{2P_4 + P_3}{3}$$

$$\leq \begin{cases} \dfrac{4P_4+P_2}{5} \leq \begin{cases} \dfrac{2G+P_2}{3} \\ \dfrac{8P_4+P_1}{9} \\ \dfrac{8P_4+P_1}{9} \leq P_4 \end{cases} \leq \begin{cases} G \\ \dfrac{5P_2+6N}{11} \leq N_1 \end{cases} \\ \begin{matrix} \dfrac{4P_4+P_2}{5} \\ \dfrac{P_3+G}{2} \end{matrix} \leq \begin{cases} G \\ \dfrac{3P_3+2N_1}{5} \leq \dfrac{7P_3+4N_2}{11} \leq \dfrac{A+2P_3}{3} \leq \dfrac{8P_3+3R}{11} \leq \dfrac{5P_3+2P_5}{7} \leq N. \end{cases} \end{cases}$$

Proof. The proof is based on the following results.

38. For $H \leq \frac{2P_4+P_3}{3}$: We have to show that $\frac{1}{3}(2P_4 + P_3 - 3H) \geq 0$. We can write $2P_4 + P_3 - 3H = bg_{38}(a/b)$, where

$$g_{38}(x) = \frac{u_{38}(x)}{(\sqrt{x}+1)^2(x+1)(x-\sqrt{x}+1)}$$

with

$$u_{38}(x) = x(x^2 - 4x^{3/2} + 6x - 4\sqrt{x} + 1) = x(\sqrt{x}-1)^4.$$

Since $u_{38}(x) \geq 0$, $\forall x > 0$, hence proving the required result.

39. For $\frac{2P_4+P_3}{3} \leq \frac{4P_4+P_2}{5}$: We have to show that $\frac{1}{15}(2P_4+3P_2-5P_3) \geq 0$. We can write $2P_4 + 3P_2 - 5P_3 = bg_{39}(a/b)$, where

$$g_{39}(x) = \frac{3u_{39}(x)}{(\sqrt{x}+1)^2(x^2+1)(x-\sqrt{x}+1)}$$

with

$$u_{39}(x) = x(2x^3 - 5x^{5/2} + 2x^2 + 2x^{3/2} + 2x - 5\sqrt{x} + 2)$$
$$= x(2x + 3\sqrt{x} + 2)(\sqrt{x} - 1)^4.$$

Since $u_{39}(x) \geq 0$, $\forall x > 0$, hence proving the required result.

40. For $\frac{4P_4+P_2}{5} \leq \frac{8P_4+P_1}{9}$: We have to show that $\frac{1}{45}(4P_4+5P_1-9P_2) \geq 0$. We can write $4P_4 + 5P_1 - 9P_2 = bg_{40}(a/b)$, where

$$g_{40}(x) = \frac{u_{40}(x)}{(\sqrt{x}+1)^2(x^2+1)(x^3+1)}$$

with

$$u_{40}(x) = x\begin{pmatrix} 12x^5 - 8x^{9/2} - 13x^4 - 18x^{7/2} + 17x^3 \\ +20x^{5/2} + 17x^2 - 18x^{3/2} - 13x - 8\sqrt{x} + 12 \end{pmatrix}.$$

After simplifications,

$$u_{40}(x) = x(\sqrt{x}-1)^4\begin{pmatrix} 12x^3 + 40x^{5/2} + 75x^2 \\ +90x^{3/2} + 75x + 40\sqrt{x} + 12 \end{pmatrix}.$$

Since $u_{40}(x) \geq 0$, $\forall x > 0$, hence proving the required result.

41. For $\frac{4P_4+P_2}{5} \leq \frac{2G+P_2}{3}$**:** We have to show that $\frac{2}{15}(5G + P_2 - 6P_4) \geq 0$. It follows in view of Result 23 done before.

42. For $\frac{8P_4+P_1}{9} \leq \frac{5P_2+6N}{11}$**:** We have to show that $\frac{1}{99}(45P_2 + 54N - 88P_4 - 11P_1) \geq 0$. We can write $45P_2 + 54N - 88P_4 - 11P_1 = bg_{42}(a/b)$, where

$$g_{42}(x) = \frac{u_{42}(x)}{(\sqrt{x}+1)^2(x^2+1)(x^3+1)}$$

with

$$u_{42}(x) = \begin{pmatrix} 18 - 246x + 54\sqrt{x} + 122x^{3/2} + 122x^{11/2} + 54x^{13/2} \\ +18x^7 + 144x^{9/2} + 64x^{7/2} - 239x^4 - 246x^6 \\ +115x^5 - 239x^3 + 144x^{5/2} + 115x^2 \end{pmatrix}.$$

After simplifications,

$$u_{42}(x) = (\sqrt{x}-1)^4 \begin{pmatrix} 18x^5 + 126x^{9/2} + 150x^4 + 38x^{7/2} \\ -147x^3 - 198x^{5/2} - 147x^2 \\ +38x^{3/2} + 150x + 126\sqrt{x} + 18 \end{pmatrix}.$$

Still, we need to show that the second expression in the bracket of $u_{42}(x)$ is nonnegative. In order to show, let's consider

$$h_{42}(x) = \begin{pmatrix} 18x^5 + 126x^{9/2} + 150x^4 + 38x^{7/2} - 147x^3 \\ -198x^{5/2} - 147x^2 + 38x^{3/2} + 150x + 126\sqrt{x} + 18 \end{pmatrix}.$$

Then we can write

$$k_{42}(t) := h_{42}(t^2) = \begin{pmatrix} 18t^{10} + 126t^9 + 150t^8 + 38t^7 - 147t^6 \\ -198t^5 - 147t^4 + 38t^3 + 150t^2 + 126t + 18 \end{pmatrix}.$$

The polynomial equation $k_{42}(t) = 0$ is of 10th degree. It admits 10 solutions. Out of them 8 are complex (not written here) and two are reals -0.1779153623 and -5.620650106. Both these solutions are negative. Thus, we don't have any real positive solutions. Since we are working with $t > 0$, then there are two possibilities, one when either the graph of $k_{42}(t)$ is in 1st quadrant or in 4th quadrant. Since $k_{42}(t) = 172$, this means that $k_{42}(t) > 0$, for all $t > 0$. Thus we have $h_{42}(t) > 0$, for all $x > 0$. This proves the required result.

43. For $\frac{2G+P_2}{3} \leq \frac{5P_2+6N}{11}$**:** We have to show that $\frac{2}{33}(2P_2 + 9N - 11G) \geq 0$. We can write $2P_2 + 9N - 11G = bg_{43}(a/b)$, where $g_{43}(x) = u_{43}(x)/(x^2+1)$

with

$$u_{43}(x) = 3x^3 - 8x^{5/2} + 5x^2 + 5x - 8\sqrt{x} + 3$$
$$= (3x + 4\sqrt{x} + 3)(\sqrt{x} - 1)^4.$$

Since $u_{43}(x) \geq 0$, $\forall x > 0$, hence proving the required result.

44. For $\frac{2P_4+P_3}{3} \leq \frac{P_3+G}{2}$: We have to show that $\frac{1}{6}(P_3 + 3G - 4P_4) \geq 0$. We can write $P_3 + 3G - 4P_4 = bg_{44}(a/b)$, where

$$g_{44}(x) = \frac{u_{44}(x)}{(\sqrt{x}+1)^2(x-\sqrt{x}+1)}$$

with

$$u_{44}(x) = x^{5/2} - 4x^2 - 4x + 6x^{3/2} + \sqrt{x} = \sqrt{x}(\sqrt{x} - 1)^4.$$

Since $u_{44}(x) \geq 0$, $\forall x > 0$, hence proving the required result.

45. For $\frac{4P_4+P_2}{5} \leq \frac{3P_3+2N_1}{5}$: We have to show that $\frac{1}{5}(3P_3 + 2N_1 - 4P_4 - P_2) \geq 0$. We can write $3P_3 + 2N_1 - 4P_4 - P_2 = bg_{45}(a/b)$, where

$$g_{45}(x) = \frac{u_{45}(x)}{2(\sqrt{x}+1)^2(x^2+1)(x-\sqrt{x}+1)}$$

with

$$u_{45}(x) = \begin{pmatrix} x^5 + 3x^{9/2} - 25x^4 + 44x^{7/2} - 24x^3 \\ +2x^{5/2} - 24x^2 + 44x^{3/2} - 25x + 3\sqrt{x} + 1 \end{pmatrix}.$$

After simplifications,

$$u_{45}(x) = (\sqrt{x} - 1)^4 \begin{pmatrix} (x + 5\sqrt{x} + 1)(x-1)^2 \\ +2\sqrt{x}(x+1)(x-\sqrt{x}+1) \end{pmatrix}.$$

Since $u_{45}(x) \geq 0$, $\forall x > 0$, hence proving the required result.

46. For $\frac{P_3+G}{2} \leq \frac{3P_3+2N_1}{5}$: We have to show that $\frac{1}{10}(P_3 + 4N_1 - 5G) \geq 0$. We can write $P_3 + 4N_1 - 5G = bg_{46}(a/b)$, where $g_{46}(x) = u_{46}(x)/(x - \sqrt{x} + 1)$ with

$$u_{46}(x) = x^2 - 4x^{3/2} + 6x - 4\sqrt{x} + 1 = (\sqrt{x} - 1)^4.$$

Since $u_{46}(x) \geq 0$, $\forall x > 0$, hence proving the required result.

47. For $\frac{3P_3+2N_1}{5} \leq \frac{7P_3+4N_2}{11}$: We have to show that $\frac{2}{55}(P_3 + 10N_2 - 11N_1) \geq 0$. We can write $P_3 + 10N_2 - 11N_1 = bg_{47}(a/b)$, where $g_{47}(x) = \frac{1}{4}u_{47}(x)/(x - \sqrt{x} + 1)$ with

$$u_{47}(x) = 10\sqrt{2x+2}(\sqrt{x}+1)(x - \sqrt{x}+1)$$
$$- (11x^2 + 11x^{3/2} - 4x + 11\sqrt{x} + 11).$$

After simplifications,

$$u_{47}(x) = 10\sqrt{2x+2}(\sqrt{x}+1)(x - \sqrt{x}+1)$$
$$- \left[11(x^2+1) + 9\sqrt{x}(x+1) + 2\sqrt{x}(\sqrt{x}-1)^2\right].$$

Let's consider

$$v_{47}(x) = \left[10\sqrt{2x+2}(\sqrt{x}+1)(x - \sqrt{x}+1)\right]^2$$
$$- \left[11(x^2+1) + 9\sqrt{x}(x+1) + 2\sqrt{x}(\sqrt{x}-1)^2\right]^2.$$

This gives

$$v_{47}(x) = (\sqrt{x}-1)^4 \left(\begin{matrix} 6\sqrt{x}(\sqrt{x}-1)^2 + 79x^2 \\ +68x^{3/2} + x + 68\sqrt{x} + 79 \end{matrix}\right).$$

In view of Note 1, $v_{47}(x) \geq 0$ implies $u_{47}(x) \geq 0$, $\forall x > 0$, hence proving the required result.

48. For $\frac{7P_3+4N_2}{11} \leq \frac{A+2P_3}{3}$: We have to show that $\frac{1}{33}(11A + P_3 - 12N_2) \geq 0$. We can write $11A + P_3 - 12N_2 = bg_{48}(a/b)$, where $g_{48}(x) = \frac{1}{2}u_{48}(x)/(x - \sqrt{x} + 1)$ with

$$u_{48}(x) = 11x^2 - 11x^{3/2} + 24x - 11\sqrt{x} + 11 - 6\sqrt{2x+2}(x^{3/2}+1)$$
$$= 11(x+1)(x - \sqrt{x}+1) + 2x - 6\sqrt{2x+2}(x^{3/2}+1).$$

Let's consider

$$v_{48}(x) = \left[11(x+1)(x - \sqrt{x}+1) + 2x\right]^2 - \left[6\sqrt{2x+2}(x^{3/2}+1)\right]^2.$$

After simplifications,

$$v_{48}(x) = (\sqrt{x}-1)^4\left[26(x^2+1) + 23(x+1)(\sqrt{x}-1)^2 + 53x\right].$$

In view of Note 1, $v_{48}(x) \geq 0$ implies $u_{48}(x) \geq 0$, $\forall x > 0$, hence proving the required result.

49. For $\frac{A+2P_3}{3} \leq \frac{8P_3+3R}{11}$**:** We have to show that $\frac{1}{33}(2P_3+9R-11A) \geq 0$. We can write $2P_3 + 9R - 11A = bg_{49}(a/b)$, where

$$g_{49}(x) = \frac{u_{49}(x)}{2(x+1)(x-\sqrt{x}+1)}$$

with

$$u_{49}(x) = x^3 - x^{5/2} - 5x^2 + 10x^{3/2} - 5x - \sqrt{x} + 1$$
$$= (x + 3\sqrt{x} + 1)(\sqrt{x} - 1)^4.$$

Since $u_{49}(x) \geq 0$, $\forall x > 0$, hence proving the required result.

50. For $\frac{8P_3+3R}{11} \leq \frac{5P_2+6N}{11}$**:** We have to show that $\frac{1}{11}(5P_2 + 6N - 8P_3 - 3R) \geq 0$. We can write $5P_2 + 6N - 8P_3 - 3R = bg_{50}(a/b)$, where

$$g_{50}(x) = \frac{u_{50}(x)}{(x+1)(x^2+1)(x-\sqrt{x}+1)}$$

with

$$u_{50}(x) = \begin{pmatrix} 2x^{9/2} - 3x^4 - 3x^{7/2} + 7x^3 - 6x^{5/2} \\ +7x^2 - 3x^{3/2} - 3x + 2\sqrt{x} \end{pmatrix}$$
$$= \sqrt{x}(2x^2 + 5x^{3/2} + 5x + 5\sqrt{x} + 2)(\sqrt{x} - 1)^4.$$

Since $u_{50}(x) \geq 0$, $\forall x > 0$, hence proving the required result.

51. For $\frac{5P_3+2P_5}{7} \leq N$**:** We have to show that $\frac{1}{7}(7N - 5P_3 - 2P_5) \geq 0$. We can write $7N - 5P_3 - 2P_5 = bg_{51}(a/b)$, where

$$g_{51}(x) = \frac{u_{51}(x)}{3(\sqrt{x}+1)^2(x-\sqrt{x}+1)}$$

with

$$u_{51}(x) = x^3 + 20x^{5/2} - 19x^2 - 4x^{3/2} - 19x + 20\sqrt{x} + 1$$
$$= (x^2 + 22x^{3/2} + 24x + 22\sqrt{x} + 1)(\sqrt{x} - 1)^2.$$

Since $u_{51}(x) \geq 0$, $\forall x > 0$, hence proving the required result.

52. For $\frac{5P_2+6N}{11} \leq N_1$**:** We have to show that $\frac{1}{11}(11N_1 - 5P_2 - 6N) \geq 0$. We can write $11N_1 - 5P_2 - 6N = bg_{52}(a/b)$, where $g_{52}(x) = \frac{1}{4}u_{52}(x)/(x^2+1)$

with

$$u_{52}(x) = 3x^3 + 14x^{5/2} - 17x - 17x^2 + 14\sqrt{x} + 3$$
$$= \left(3x^2 + 20x^{3/2} + 20x + 20\sqrt{x} + 3\right)(\sqrt{x} - 1)^2.$$

Since $u_{52}(x) \geq 0$, $\forall x > 0$, hence proving the required result.

53. For $\frac{8P_4 + P_1}{9} \leq P_4$: We have to show that $\frac{1}{9}(P_4 - P_1) \geq 0$. It is true by definition. See inequalities (30).

54. For $\frac{P_3 + G}{2} \leq G$: We have to show that $\frac{1}{2}(G - P_3) \geq 0$. It is true by definition. See inequalities (30).

55. For $\frac{2G + P_2}{3} \leq G$: We have to show that $\frac{1}{3}(G - P_2) \geq 0$. It is true by definition. See inequalities (30).

Combining Results 38 to 55 we get the proof of Proposition 3.   □

Proposition 4. *The following inequalities hold:*

$$\mathbf{P_4} \leq \frac{G+H}{2} \leq \left\{ \begin{array}{l} \frac{3G+P_3}{4} \leq \frac{P_1+9G}{10} \leq \mathbf{G} \\ \frac{2H+N_1}{3} \leq \frac{5H+S}{6} \leq \frac{A+3H}{4} \leq \frac{4H+P_5}{5} \end{array} \right\} \leq \frac{3R+7G}{10} \leq \mathbf{N_1}.$$

Proof. The proof is based on the following results.

56. For $\mathbf{P_4} \leq \frac{G+H}{2}$: We have to show that $\frac{1}{2}(G + H - 2P_4) \geq 0$. We can write $G + H - 2P_4 = bg_{56}(a/b)$, where

$$g_{56}(x) = \frac{u_{48}(x)}{(x+1)(\sqrt{x}+1)^2}$$

with

$$u_{56}(x) = x^{5/2} + 6x^{3/2} - 4x^2 - 4x + \sqrt{x} = \sqrt{x}(\sqrt{x} - 1)^4.$$

Since $u_{56}(x) \geq 0$, $\forall x > 0$, hence proving the required result.

57. For $\frac{G+H}{2} \leq \frac{2H+N_1}{3}$: We have to show that $\frac{1}{6}(H + 2N_1 - 3G) \geq 0$. We can write $H + 2N_1 - 3G = bg_{57}(a/b)$, where $g_{57}(x) = \frac{1}{2}u_{57}(x)/(x+1)$ with

$$u_{57}(x) = x^2 - 4x^{3/2} + 6x - 4\sqrt{x} + 1 = (\sqrt{x} - 1)^4.$$

Since $u_{57}(x) \geq 0$, $\forall x > 0$, hence proving the required result.

58. For $\frac{2H+N_1}{3} \leq \frac{5H+S}{6}$**:** We have to show that $\frac{1}{6}(H+S-2N_1) \geq 0$. We can write $H+S-2N_1 = bg_{58}(a/b)$, where $g_{58}(x) = \frac{1}{2}u_{58}(x)/(x+1)$ with

$$u_{58}(x) = \sqrt{2x^2 + 2}(x+1) - [x^2 + x^{3/2} + \sqrt{x}(\sqrt{x}-1)^2 + \sqrt{x}+1].$$

Let's consider

$$v_{58}(x) = \left[\sqrt{2x^2 + 2}(x+1)\right]^2 - \left[x^2 + x^{3/2} + \sqrt{x}(\sqrt{x}-1)^2 + \sqrt{x}+1\right]^2.$$

After simplifications,

$$v_{58}(x) = (\sqrt{x}+1)^2(\sqrt{x}-1)^6.$$

In view of Note 1, $v_{58}(x) \geq 0$ implies $u_{58}(x) \geq 0$, $\forall x > 0$, hence proving the required result.

59. For $\frac{5H+S}{6} \leq \frac{A+3H}{4}$**:** We have to show that $\frac{1}{12}(3A - H - 2S) \geq 0$. We can write $3A - H - 2S = bg_{59}(a/b)$, where $g_{59}(x) = \frac{1}{2}u_{59}(x)/(x+1)$ with

$$u_{59}(x) = 3x^2 + 2x + 3 - 2(x+1)\sqrt{2x^2 + 2}.$$

Let's consider

$$v_{59}(x) = \left(3x^2 + 2x + 3\right)^2 - \left[2(x+1)\sqrt{2x^2 + 2}\right]^2.$$

After simplifications,

$$v_{59}(x) = (x-1)^4.$$

In view of Note 1, $v_{59}(x) \geq 0$ implies $u_{59}(x) \geq 0$, $\forall x > 0$, hence proving the required result.

60. For $\frac{5H+S}{6} \leq \frac{4H+P_5}{5}$**:** We have to show that $\frac{1}{30}(6P_5 - H - 5S) \geq 0$. We can write $6P_5 - H - 5S = bg_{60}(a/b)$, where

$$g_{60}(x) = \frac{u_{60}(x)}{2(x+1)(\sqrt{x}+1)^2}$$

with

$$u_{60}(x) = 4(x - \sqrt{x} + 1)\left(3x^2 + 3x^{3/2} + 8x + 3\sqrt{x} + 3\right) - 5\sqrt{2x^2 + 2}(x+1)(\sqrt{x}+1)^2.$$

Let's consider

$$v_{60}(x) = \left[4(x - \sqrt{x} + 1)(3x^2 + 3x^{3/2} + 8x + 3\sqrt{x} + 3)\right]^2$$
$$- \left[5\sqrt{2x^2 + 2}(x + 1)(\sqrt{x} + 1)^2\right]^2.$$

After simplifications,

$$v_{60}(x) = 2(\sqrt{x} - 1)^4 \begin{pmatrix} 47x^4 + 88x^{7/2} + 254x^3 + 280x^{5/2} \\ +422x^2 + 280x^{3/2} + 254x + 88\sqrt{x} + 47 \end{pmatrix}.$$

In view of Note 1, $v_{60}(x) \geq 0$ implies $u_{60}(x) \geq 0$, $\forall x > 0$, hence proving the required result.

61. For $\frac{4H+P_5}{5} \leq \frac{3R+7G}{10}$: We have to show that $\frac{1}{10}(3R + 7G - 8H - 2P_5) \geq 0$. We can write $3R + 7G - 8H - 2P_5 = bg_{61}(a/b)$, where $g_{61}(x) = u_{61}(x)/(x^3 + 1)$ with

$$u_{61}(x) = x^{7/2} - x^3 + \sqrt{x} - x$$
$$= \sqrt{x}(\sqrt{x} - 1)^2 (x^2 + x^{3/2} + x + \sqrt{x} + 1).$$

Since $u_{61}(x) \geq 0$, $\forall x > 0$, hence proving the required result.

62. For $\frac{G+H}{2} \leq \frac{3G+P_3}{4}$: We have to show that $\frac{1}{4}(G + P_3 - 2H) \geq 0$. We can write $G + P_3 - 2H = bg_{62}(a/b)$, where

$$g_{62}(x) = \frac{u_{62}(x)}{(x + 1)(x - \sqrt{x} + 1)}$$

with

$$u_{62}(x) = x^{5/2} - 4x^2 + 6x^{3/2} - 4x + \sqrt{x} = \sqrt{x}(\sqrt{x} - 1)^4.$$

Since $u_{62}(x) \geq 0$, $\forall x > 0$, hence proving the required result.

63. For $\frac{3G+P_3}{4} \leq \frac{P_1+9G}{10}$: We have to show that $\frac{1}{20}(2P_1 + 3G - 5P_3) \geq 0$. We can write $2P_1 + 3G - 5P_3 = bg_{63}(a/b)$, where

$$g_{63}(x) = \frac{u_{63}(x)}{(x^3 + 1)(x - \sqrt{x} + 1)}$$

with

$$u_{63}(x) = 3x^{9/2} - 6x^4 + x^{7/2} + 2x^3 + 2x^2 + x^{3/2} - 6x + 3\sqrt{x}$$
$$= \sqrt{x}(3x^2 + 6x^{3/2} + 7x + 6\sqrt{x} + 3)(\sqrt{x} - 1)^4.$$

Since $u_{63}(x) \geq 0$, $\forall x > 0$, hence proving the required result.

64. For $\frac{P_1+9G}{10} \le G$: We have to show that $\frac{1}{10}(G - P_1) \ge 0$. It is true by definition. See inequalities (30).

65. For $\frac{3R+7G}{10} \le G$: We have to show that $\frac{3}{10}(R - G) \ge 0$. It is true by definition. See inequalities (30).

66. For $\frac{3R+7G}{10} \le N_1$: We have to show that $\frac{1}{10}(10N_1 - 3R - 7G) \ge 0$. We can write $10N_1 - 3R - 7G = bg_{66}(a/b)$, where $g_{66}(x) = \frac{1}{2}u_{66}(x)/(x+1)$ with

$$u_{66}(x) = x^2 - 4x^{3/2} + 6x - 4\sqrt{x} + 1 = (\sqrt{x} - 1)^4.$$

Since $u_{66}(x) \ge 0$, $\forall x > 0$, hence proving the required result.

Combining Results 56 to 66 we get the proof of Proposition 4. □

Proposition 5. *The following inequalities hold:*

$$G \le \frac{4P_4 + S}{5} \le \frac{C + 6P_4}{7} \le \frac{10P_4 + 3R}{13} \le \frac{A + 2P_4}{3}$$

$$\le \left\{ \begin{array}{c} \frac{3P_4+P_5}{4} \\ \frac{6N_1+P_2}{7} \\ \frac{C+5G}{6} \\ \frac{S+3G}{4} \end{array} \right\} \le \left. \begin{array}{c} \frac{P_1+10N_1}{11} \\ \\ \frac{7G+3R}{10} \end{array} \right\} \le \frac{G+2N_2}{3} \right\} \le N_1.$$

Proof. The proof is based on the following results.

67. For $G \le \frac{4P_4+S}{5}$: We have to show that $\frac{1}{5}(4P_4 + S - 5G) \ge 0$. We can write $4P_4 + S - 5G = bg_{67}(a/b)$, where $g_{67}(x) = \frac{1}{2}u_{67}(x)/(\sqrt{x} + 1)^2$ with

$$u_{67}(x) = \sqrt{2x^2 + 2(\sqrt{x} + 1)^2} - 2\sqrt{x}[2(x + 1) + 3(\sqrt{x} - 1)^2].$$

Let's consider

$$v_{67}(x) = \left[\sqrt{2x^2 + 2}(\sqrt{x} + 1)^2\right]^2 - \left\{2\sqrt{x}[2(x + 1) + 3(\sqrt{x} - 1)^2]\right\}^2.$$

After simplifications,

$$v_{67}(x) = 2(x + 10\sqrt{x} + 1)(\sqrt{x} - 1)^6.$$

In view of Note 1, $v_{67}(x) \ge 0$ implies $u_{67}(x) \ge 0$, $\forall x > 0$, hence proving the required result.

68. For $\frac{4P_4+S}{5} \le \frac{C+6P_4}{7}$: We have to show that $\frac{1}{35}(5C+2P_4-7S) \ge 0$. We can write $5C+2P_4-7S = bg_{68}(a/b)$, where

$$g_{68}(x) = \frac{u_{68}(x)}{2(x+1)(\sqrt{x}+1)^2}$$

with

$$u_{68}(x) = 10x^3 + 20x^{5/2} + 26x^2 + 26x + 20\sqrt{x} + 10$$
$$- 7\sqrt{2x^2+2}(x+1)(\sqrt{x}+1)^2.$$

Let's consider

$$v_{68}(x) = \left(10x^3 + 20x^{5/2} + 26x^2 + 26x + 20\sqrt{x} + 10\right)^2$$
$$- \left[7\sqrt{2x^2+2}(x+1)(\sqrt{x}+1)^2\right]^2.$$

After simplifications,

$$v_{68}(x) = 2(\sqrt{x}-1)^4 \begin{pmatrix} x^4 + 8x^{7/2} + 94x^3 + 264x^{5/2} \\ +386x^2 + 264x^{3/2} + 94x + 8\sqrt{x} + 1 \end{pmatrix}.$$

In view of Note 1, $v_{68}(x) \ge 0$ implies $u_{68}(x) \ge 0$, $\forall x > 0$, hence proving the required result.

69. For $\frac{C+6P_4}{7} \le \frac{10P_4+3R}{13}$: We have to show that $\frac{1}{91}(21R-8P_4-13C) \ge 0$. We can write $21R - 8P_4 - 13C = bg_{69}(a/b)$, where

$$g_{69}(x) = \frac{u_{69}(x)}{(x+1)(\sqrt{x}+1)^2}$$

with

$$u_{69}(x) = x^3 - 17x^2 + 2x^{5/2} - 17x + 28x^{3/2} + 2\sqrt{x} + 1$$
$$= (x+6\sqrt{x}+1)(\sqrt{x}-1)^4.$$

Since $u_{69}(x) \ge 0$, $\forall x > 0$, hence proving the required result.

70. For $\frac{10P_4+3R}{13} \le \frac{A+2P_4}{3}$: We have to show that $\frac{1}{39}(13A-4P_4-9R) \ge 0$. We can write $13A - 4P_4 - 9R = bg_{70}(a/b)$, where

$$g_{70}(x) = \frac{u_{70}(x)}{2(x+1)(\sqrt{x}+1)^2}$$

with

$$u_{70}(x) = x^3 - 17x^2 + 2x^{5/2} - 17x + 28x^{3/2} + 2\sqrt{x} + 1$$
$$= (x + 6\sqrt{x} + 1)(\sqrt{x} - 1)^4.$$

Since $u_{70}(x) \geq 0$, $\forall x > 0$, hence proving the required result.

71. For $\frac{A+2P_4}{3} \leq \frac{3P_4+P_5}{4}$**:** We have to show that $\frac{1}{12}(P_4 + 3P_5 - 4A) \geq 0$. We can write $P_4 + 3P_5 - 4A = bg_{71}(a/b)$, where $g_{71}(x) = u_{71}(x)/(\sqrt{x} + 1)^2$ with

$$u_{71}(x) = x^2 + 6x - 4x^{3/2} - 4\sqrt{x} + 1 = (\sqrt{x} - 1)^4.$$

Since $u_{71}(x) \geq 0$, $\forall x > 0$, hence proving the required result.

72. For $\frac{A+2P_4}{3} \leq \frac{6N_1+P_2}{7}$**:** We have to show that $\frac{1}{21}(18N_1 + 3P_2 - 7A - 14P_4) \geq 0$. We can write $18N_1 + 3P_2 - 7A - 14P_4 = bg_{72}(a/b)$, where

$$g_{72}(x) = \frac{u_{72}(x)}{(x^2 + 1)(\sqrt{x} + 1)^2}$$

with

$$u_{72}(x) = \begin{pmatrix} x^4 + 11x^{7/2} - 33x^3 + 17x^{5/2} \\ +8x^2 + 17x^{3/2} - 33x + 11\sqrt{x} + 1 \end{pmatrix}$$
$$= (x^2 + 15x^{3/2} + 21x + 15\sqrt{x} + 1)(\sqrt{x} - 1)^4.$$

Since $u_{72}(x) \geq 0$, $\forall x > 0$, hence proving the required result.

73. For $\frac{A+2P_4}{3} \leq \frac{C+5G}{6}$**:** We have to show that $\frac{1}{6}(C + 5G - 2A - 4P_4) \geq 0$. We can write $C + 5G - 2A - 4P_4 = bg_{73}(a/b)$, where

$$g_{73}(x) = \frac{u_{73}(x)}{(x + 1)(\sqrt{x} + 1)^2}$$

with

$$u_{73}(x) = 5x^{5/2} + 6x^{3/2} - 8x^2 - 8x + 5\sqrt{x}$$
$$= \sqrt{x}(5x + 2\sqrt{x} + 5)(\sqrt{x} - 1)^2.$$

Since $u_{73}(x) \geq 0$, $\forall x > 0$, hence proving the required result.

74. For $\frac{A+2P_4}{3} \leq \frac{S+3G}{4}$**:** We have to show that $\frac{1}{12}(3S + 9G - 4A - 8P_4) \geq 0$. We can write $3S + 9G - 4A - 8P_4 = bg_{74}(a/b)$, where $g_{74}(x) = \frac{1}{2}u_{74}(x)/(\sqrt{x} + 1)^2$ with

$$u_{74}(x) = 3\sqrt{2x^2 + 2}(\sqrt{x} + 1)^2 - 2[2(\sqrt{x} - 1)^4 + 3\sqrt{x}(\sqrt{x} + 1)^2].$$

Let's consider

$$u_{74}(x) = \left[3\sqrt{2x^2 + 2}(\sqrt{x}+1)^2\right]^2$$
$$- 4\left[2(\sqrt{x}-1)^4 + 3\sqrt{x}(\sqrt{x}+1)^2\right]^2.$$

After simplifications,

$$v_{74}(x) = 2(\sqrt{x}-1)^2 \begin{pmatrix} x^3 + 78x^{5/2} + 15x^2 \\ +388x^{3/2} + 15x + 78\sqrt{x} + 1 \end{pmatrix}.$$

In view of Note 1, $v_{74}(x) \geq 0$ implies $u_{74}(x) \geq 0$, $\forall x > 0$, hence proving the required result.

75. For $\frac{6N_1+P_2}{7} \leq \frac{P_1+10N_1}{11}$: We have to show that $\frac{1}{77}(7P_1 + 4N_1 - 11P_2) \geq 0$. We can write $7P_1 + 4N_1 - 11P_2 = bg_{75}(a/b)$, where

$$g_{75}(x) = \frac{u_{75}(x)}{(x^2 + 1)(x^3 + 1)}$$

with

$$u_{75}(x) = \begin{pmatrix} x^6 + 2x^{11/2} - 3x^5 - 10x^4 + 2x^{7/2} \\ +16x^3 + 2x^{5/2} - 10x^2 - 3x + 2\sqrt{x} + 1 \end{pmatrix}.$$

After simplifications,

$$u_{75}(x) = (\sqrt{x}-1)^4 \begin{pmatrix} x^4 + 6x^{7/2} + 15x^3 + 28x^{5/2} \\ +35x^2 + 28x^{3/2} + 15x + 6\sqrt{x} + 1 \end{pmatrix}.$$

Since $u_{75}(x) \geq 0$, $\forall x > 0$, hence proving the required result.

76. For $\frac{C+5G}{6} \leq \frac{7G+3R}{10}$: We have to show that $\frac{1}{30}(9R - 4G - 5C) \geq 0$. We can write $9R - 4G - 5C = bg_{76}(a/b)$, where $g_{76}(x) = u_{76}(x)/(x+1)$ with

$$u_{76}(x) = \left(x^2 + 6x - 4x^{3/2} - 4\sqrt{x} + 1\right) = (\sqrt{x}-1)^4.$$

Since $u_{76}(x) \geq 0$, $\forall x > 0$, hence proving the required result.

77. For $\frac{S+3G}{4} \leq \frac{7G+3R}{10}$: We have to show that $\frac{1}{20}(6R - G - 5S) \geq 0$. We can write $6R - G - 5S = bg_{77}(a/b)$, where $g_{77}(x) = \frac{1}{2}u_{77}(x)/(x+1)$ with

$$u_{77}(x) = (x+1)(\sqrt{x}-1)^2 + 7x^2 + 6x + 7 - 5\sqrt{2x^2 + 2}(x+1).$$

Let's consider

$$v_{77}(x) = \left[(x+1)(\sqrt{x}-1)^2 + 7x^2 + 6x + 7\right]^2$$
$$- \left[5\sqrt{2x^2 + 2}(x+1)\right]^2.$$

After simplifications,

$$v_{77}(x) = 2(\sqrt{x}-1)^4\left(7x^2 + 12x^{3/2} + 22x + 12\sqrt{x} + 7\right).$$

In view of Note 1, $v_{77}(x) \geq 0$ implies $u_{77}(x) \geq 0$, $\forall x > 0$, hence proving the required result.

78. For $\frac{P_1+10N_1}{11} \leq \frac{G+2N_2}{3}$: We have to show that $\frac{1}{33}(11G + 22N_2 - 3P_1 - 30N_1) \geq 0$. We can write $11G + 22N_2 - 3P_1 - 30N_1 = bg_{78}(a/b)$, where $g_{78}(x) = \frac{1}{2}u_{78}(x)/(x^3 + 1)$ with

$$u_{78}(x) = 11\sqrt{2x^2 + 2}(\sqrt{x}+1)(x^3 + 1)$$
$$- \left(15x^4 + 8x^{7/2} + 21x^3 + 21x + 8\sqrt{x} + 15\right).$$

Let's consider

$$v_{78}(x) = \left[11\sqrt{2x^2 + 2}(\sqrt{x}+1)(x^3 + 1)\right]^2$$
$$- \left(15x^4 + 8x^{7/2} + 21x^3 + 21x + 8\sqrt{x} + 15\right)^2.$$

After simplifications,

$$v_{78}(x) = (\sqrt{x}-1)^2 \begin{pmatrix} 17x^7 + 278x^{13/2} + 329x^6 + 528x^{11/2} \\ +528x^5 + 528x^{9/2} + 382x^4 + 628x^{7/2} \\ +382x^3 + 528x^{5/2} + 528x^2 + 528x^{3/2} \\ +329x + 278\sqrt{x} + 17 \end{pmatrix}.$$

In view of Note 1, $v_{78}(x) \geq 0$ implies $u_{78}(x) \geq 0$, $\forall x > 0$, hence proving the required result.

79. For $\frac{7G+3R}{10} \leq \frac{G+2N_2}{3}$: We have to show that $\frac{1}{30}(20N_2 - 11G - 9R) \geq 0$. We can write $20N_2 - 11G - 9R = bg_{79}(a/b)$, where $g_{79}(x) = u_{79}(x)/(x+1)$ with

$$u_{79}(x) = 5\sqrt{2x+2}(\sqrt{x}+1)(x+1) - (6x^2 + 11x^{3/2} + 6x + 11\sqrt{x} + 6).$$

Let's consider

$$v_{79}(x) = \left[5\sqrt{2x+2}(\sqrt{x}+1)(x+1)\right]^2$$
$$- \left(6x^2 + 11x^{3/2} + 6x + 11\sqrt{x} + 6\right)^2.$$

After simplifications,

$$v_{79}(x) = \left(14x^2 + 24x^{3/2} + 19x + 24\sqrt{x} + 14\right)(\sqrt{x} - 1)^4.$$

In view of Note 1, $v_{79}(x) \geq 0$ implies $u_{79}(x) \geq 0$, $\forall x > 0$, hence proving the required result.

80. For $\frac{G+2N_2}{3} \leq N_1$: We have to show that $\frac{1}{3}(3N_1 - G - 2N_2) \geq 0$. We can write $3N_1 - G - 2N_2 = bg_{80}(a/b)$, where $g_{80}(x) = \frac{1}{4}u_{80}(x)$ with

$$u_{80}(x) = 3x + 2\sqrt{x} + 3 - 2(\sqrt{x} + 1)\sqrt{2x + 2}.$$

Let's consider

$$v_{80}(x) = (3x + 2\sqrt{x} + 3)^2 - \left[2(\sqrt{x} + 1)\sqrt{2x + 2}\right]^2.$$

After simplifications,

$$v_{80}(x) = (\sqrt{x} - 1)^4.$$

In view of Note 1, $v_{80}(x) \geq 0$ implies $u_{80}(x) \geq 0$, $\forall x > 0$, hence proving the required result.

81. For $\frac{3P_4+P_5}{4} \leq N_1$: We have to show that $\frac{1}{4}(4N_1 - 3P_4 - P_5) \geq 0$. We can write $4N_1 - 3P_4 - P_5 = bg_{81}(a/b)$, where $g_{81}(x) = u_{81}(x)/(\sqrt{x} + 1)^2$ with

$$u_{81}(x) = x^{3/2} + \sqrt{x} - 2x = \sqrt{x}(\sqrt{x} - 1)^2.$$

Since $u_{81}(x) \geq 0$, $\forall x > 0$, hence proving the required result.

Combining Results 67 to 81 we complete the proof of Proposition 5.

□

Proposition 6. *The following inequalities hold:*

$$\mathbf{N_1} \leq \frac{4N_2 + P_4}{5} \leq \begin{cases} \frac{9N+H}{10} \leq \begin{cases} \frac{P_3+15N}{16} \leq \frac{P_2+21N}{22} \leq \frac{P_1+33N}{34} \\ \frac{C+14N_1}{15} \\ \frac{S+8N_1}{9} \\ \frac{8P_4+7S}{15} \end{cases} \leq \frac{P_4+14N_2}{15} \\ \begin{cases} \frac{P_5+2G}{3} \\ \frac{8P_4+7S}{15} \end{cases} \leq \frac{C+2P_4}{3} \end{cases} \leq \mathbf{N}.$$

Proof. The proof is based on the following results.

82. For $N_1 \leq \frac{4N_2+P_4}{5}$: We have to show that $\frac{1}{5}(4N_2+P_4-5N_1) \geq 0$. We can write $4N_2 + P_4 - 5N_1 = bg_{82}(a/b)$, where $g_{82}(x) = \frac{1}{4}u_{82}(x)/(\sqrt{x}+1)^2$ with

$$u_{82}(x) = 4\sqrt{2x+2}(\sqrt{x}+1)^3 - (5x^2 + 20x^{3/2} + 14x + 20\sqrt{x} + 5).$$

Let's consider

$$v_{82}(x) = \left[4\sqrt{2x+2}(\sqrt{x}+1)^3\right]^2 - \left(5x^2 + 20x^{3/2} + 14x + 20\sqrt{x} + 5\right)^2.$$

After simplifications,

$$v_{82}(x) = \left(7x^2 + 20x^{3/2} + 10x + 20\sqrt{x} + 7\right)(\sqrt{x} - 1)^4.$$

In view of Note 1, $v_{82}(x) \geq 0$ implies $u_{82}(x) \geq 0$, $\forall x > 0$, hence proving the required result.

83. For $\frac{4N_2+P_4}{5} \leq \frac{9N+H}{10}$: We have to show that $\frac{1}{10}(9N + H - 8N_2 - 2P_4) \geq 0$. We can write $9N + H - 8N_2 - 2P_4 = bg_{83}(a/b)$, where

$$g_{83}(x) = \frac{u_{83}(x)}{(x+1)(\sqrt{x}+1)^2}$$

with

$$u_{83}(x) = 3x^3 + 9x^{5/2} + 9x^2 + 22x^{3/2} + 9x + 9\sqrt{x} + 3$$
$$- 2\sqrt{2x+2}(x+1)(\sqrt{x}+1)^3.$$

Let's consider

$$v_{83}(x) = \left(3x^3 + 9x^{5/2} + 9x^2 + 22x^{3/2} + 9x + 9\sqrt{x} + 3\right)^2$$
$$- \left[2\sqrt{2x+2}(x+1)(\sqrt{x}+1)^3\right]^2.$$

After simplifications,

$$v_{83}(x) = (\sqrt{x} - 1)^4 \left(\begin{array}{l} x^4 + 10x^{7/2} + 25x^3 + 34x^{5/2} \\ + 52x^2 + 34x^{3/2} + 25x + 10\sqrt{x} + 1 \end{array}\right).$$

In view of Note 1, $v_{83}(x) \geq 0$ implies $u_{83}(x) \geq 0$, $\forall x > 0$, hence proving the required result.

84. For $\frac{4N_2+P_4}{5} \leq \frac{P_5+2G}{3}$**:** We have to show that $\frac{1}{15}(5P_5 + 10G -$ $12N_2 - 3P_4) \geq 0$. We can write $5P_5 + 10G - 12N_2 - 3P_4 = bg_{84}(a/b)$, where $g_{84}(x) = u_{84}(x)/(\sqrt{x}+1)^2$ with

$$u_{84}(x) = 5x^2 + 10x^{3/2} + 18x + 10\sqrt{x} + 5 - 3\sqrt{2x+2}(\sqrt{x}+1)^3.$$

Let's consider

$$u_{84}(x) = \left(5x^2 + 10x^{3/2} + 18x + 10\sqrt{x} + 5\right)^2$$
$$- 3\left[\sqrt{2x+2}(\sqrt{x}+1)^3\right]^2.$$

After simplifications,

$$v_{84}(x) = \left(7x^2 + 20x^{3/2} + 30x + 20\sqrt{x} + 7\right)(\sqrt{x} - 1)^4.$$

In view of Note 1, $v_{84}(x) \geq 0$ implies $u_{84}(x) \geq 0$, $\forall x > 0$, hence proving the required result.

85. For $\frac{9N+H}{10} \leq \frac{P_3+15N}{16}$**:** We have to show that $\frac{1}{80}(5P_3 + 3N - 8H) \geq 0$. We can write $5P_3 + 3N - 8H = bg_{85}(a/b)$, where

$$g_{85}(x) = \frac{u_{85}(x)}{(x+1)(x-\sqrt{x}+1)}$$

with

$$u_{85}(x) = x^3 + 16x^{3/2} - 9x^2 - 9x + 1 = (x + 4\sqrt{x} + 1)(\sqrt{x} - 1)^4.$$

Since $u_{85}(x) \geq 0$, $\forall x > 0$, hence proving the required result.

86. For $\frac{9N+H}{10} \leq \frac{C+15P_1}{16}$**:** We have to show that $\frac{1}{30}(2C + 28N_1 - 27N - 3H) \geq 0$. We can write $2C + 28N_1 - 27N - 3H = bg_{86}(a/b)$, where $g_{86}(x) = 5u_{86}(x)/(x+1)$ with

$$u_{86}(x) = x^{3/2} - 2x + \sqrt{x} = \sqrt{x}(\sqrt{x} - 1)^2.$$

Since $u_{86}(x) \geq 0$, $\forall x > 0$, hence proving the required result.

87. For $\frac{9N+H}{10} \leq \frac{S+8N_1}{9}$**:** We have to show that $\frac{1}{90}(10S + 80N_1 - 81N - 9H) \geq 0$. We can write $10S + 80N_1 - 81N - 9H = bg_{87}(a/b)$, where $g_{87}(x) = u_{87}(x)/(x+1)$ with

$$u_{87}(x) = 5\sqrt{2x^2 + 2}(x+1) - (7x^2 - 13x^{3/2} + 32x - 13\sqrt{x} + 7).$$

After simplifications,

$$u_{87}(x) = 5\sqrt{2x^2 + 2}(x+1)$$
$$- \left[7(x+1)(\sqrt{x}-1)^2 + 18x + \sqrt{x}(x+1) \right].$$

Let's consider

$$v_{87}(x) = \left[5\sqrt{2x^2 + 2}(x+1) \right]^2$$
$$- \left[7(x+1)(\sqrt{x}-1)^2 + 18x + \sqrt{x}(x+1) \right]^2.$$

This gives

$$v_{87}(x) = (\sqrt{x}-1)^2 \left(\begin{array}{c} x^3 + 109\sqrt{x}(x^2+1) + 380x^{3/2} \\ +75\sqrt{x}(x+1)(\sqrt{x}-1)^2 + 1 \end{array} \right).$$

In view of Note 1, $v_{87}(x) \geq 0$ implies $u_{87}(x) \geq 0$, $\forall x > 0$, hence proving the required result.

88. For $\frac{9N+H}{10} \leq \frac{8P_4+7S}{15}$: We have to show that $\frac{1}{30}(16P_4 + 14S - 27N - 3H) \geq 0$. We can write $16P_4 + 14S - 27N - 3H = bg_{88}(a/b)$, where

$$g_{88}(x) = \frac{u_{88}(x)}{(x+1)(\sqrt{x}+1)^2}$$

with

$$u_{88}(x) = 7\sqrt{2x^2 + 2}(x+1)(\sqrt{x}+1)^2$$
$$- \left[2(x+1) + (\sqrt{x}-1)^2 \right](3x^2 + 11x^{3/2} + 11\sqrt{x} + 3).$$

Let's consider

$$v_{88}(x) = \left[7\sqrt{2x^2 + 2}(x+1)(\sqrt{x}+1)^2 \right]^2$$
$$- \left\{ \left[2(x+1) + (\sqrt{x}-1)^2 \right](3x^2 + 11x^{3/2} + 11\sqrt{x} + 3) \right\}^2.$$

After simplifications,

$$v_{88}(x) = (\sqrt{x}-1)^2 \left(\begin{array}{c} 17x^5 - 60x^{9/2} + 152x^4 + 1054x^{7/2} \\ -73x^3 + 2300x^{5/2} - 73x^2 \\ +1054x^{3/2} + 152x - 60\sqrt{x} + 17 \end{array} \right).$$

Still, we need to show that the second expression in the bracket of $v_{88}(x)$ is nonnegative. In order to show, let's consider

$$h_{88}(x) = \begin{pmatrix} 17x^5 - 60x^{9/2} + 152x^4 + 1054x^{7/2} - 73x^3 \\ +2300x^{5/2} - 73x^2 + 1054x^{3/2} + 152x - 60\sqrt{x} + 17 \end{pmatrix}.$$

Then we can write

$$k_{88}(t) := h_{88}(t^2) = \begin{pmatrix} 17t^{10} - 60t^9 + 152t^8 + 1054t^7 - 73t^6 \\ +2300t^5 - 73t^4 + 1054t^3 + 152t^2 - 60t + 17 \end{pmatrix}.$$

The polynomial equation $k_{88}(t) = 0$ is of 10th degree. It admits 10 solutions. Out of them 8 solutions are complex (not written here) and two are reals -0.3177553967 and -2.908434428. Both the solutions are negative. Thus, we don't have any real positive solution. Since we are working with $t > 0$, then there are two possibilities, one when the graph of $k_{88}(t)$ is either in 1st quadrant or in 4th quadrant. Since $k_{88}(t) = 4480$, this means that $k_{88}(t) > 0$, for all $t > 0$. Thus we have $h_{88}(t) > 0$, for all $x > 0$. In view of Note 1, we get the required result.

89. For $\frac{8P_4 + 7S}{15} \leq \frac{C + 2P_4}{3}$: We have to show that $\frac{1}{15}(5C + 2P_4 - 7S) \geq 0$. We can write $5C + 2P_4 - 7S = bg_{89}(a/b)$, where

$$g_{89}(x) = \frac{u_{89}(x)}{2(x+1)(\sqrt{x}+1)^2}$$

with

$$u_{89}(x) = \left(10x^3 + 20x^{5/2} + 26x^2 + 26x + 20\sqrt{x} + 10\right)$$
$$- 7\sqrt{2x^2 + 2}(x+1)(\sqrt{x}+1)^2.$$

Let's consider

$$v_{89}(x) = \left(10x^3 + 20x^{5/2} + 26x^2 + 26x + 20\sqrt{x} + 10\right)^2$$
$$- \left[7\sqrt{2x^2 + 2}(x+1)(\sqrt{x}+1)^2\right]^2.$$

After simplifications,

$$v_{89}(x) = 2(\sqrt{x} - 1)^2 \begin{pmatrix} x^4 + 8x^{7/2} + 94x^3 + 264x^{5/2} \\ +386x^2 + 264x^{3/2} + 94x + 8\sqrt{x} + 1 \end{pmatrix}.$$

In view of Note 1, $v_{89}(x) \geq 0$ implies $u_{89}(x) \geq 0$, $\forall x > 0$, hence proving the required result.

90. For $\frac{P_5+2G}{3} \leq \frac{C+2P_4}{3}$: We have to show that $\frac{1}{3}(C+2P_4 - P_5 - 2G) \geq 0$. We can write $C+2P_4 - P_5 - 2G = bg_{90}(a/b)$, where

$$g_{90}(x) = \frac{2u_{90}(x)}{(x+1)(\sqrt{x}+1)^2}$$

with

$$u_{90}(x) = x^2 - 2x^{3/2} + x = x(\sqrt{x}-1)^2.$$

Since $u_{90}(x) \geq 0$, $\forall x > 0$, hence proving the required result.

91. For $\frac{P_3+15N}{16} \leq \frac{P_2+21N}{22}$: We have to show that $\frac{1}{176}(8P_2 + 3N - 11P_3) \geq 0$. We can write $8P_2 + 3N - 11P_3 = bg_{91}(a/b)$, where

$$g_{91}(x) = \frac{u_{91}(x)}{(x^2+1)(x-\sqrt{x}+1)}$$

with

$$u_{91}(x) = x^4 - 2x^3 - 8x^{5/2} + 18x^2 - 8x^{3/2} - 2x + 1$$
$$= (x^2 + 4x^{3/2} + 8x + 4\sqrt{x} + 1)(\sqrt{x}-1)^4.$$

Since $u_{91}(x) \geq 0$, $\forall x > 0$, hence proving the required result.

92. For $\frac{P_2+21N}{22} \leq \frac{P_1+33N}{34}$: We have to show that $\frac{1}{374}(11P_1 + 6N - 17P_2) \geq 0$. We can write $11P_1 + 6N - 17P_2 = bg_{92}(a/b)$, where

$$g_{92}(x) = \frac{u_{92}(x)}{(x^2+1)(x^3+1)}$$

with

$$u_{92}(x) = \begin{pmatrix} 2x^6 + 2x^{11/2} - 4x^5 - 15x^4 + 2x^{7/2} \\ +26x^3 + 2x^{5/2} - 15x^2 - 4x + 2\sqrt{x} + 2 \end{pmatrix}.$$

After simplifications,

$$u_{92}(x) = (\sqrt{x}-1)^4 \begin{pmatrix} 2x^4 + 10x^{7/2} + 24x^3 + 44x^{5/2} \\ +55x^2 + 44x^{3/2} + 24x + 10\sqrt{x} + 2 \end{pmatrix}.$$

Since $u_{92}(x) \geq 0$, $\forall x > 0$, hence proving the required result.

93. For $\frac{P_1+33N}{34} \leq N$**:** We have to show that $\frac{1}{34}(N - P_1) \geq 0$. It is true by definition. See inequalities (30).

94. For $\frac{C+14N_1}{15} \leq \frac{P_4+14N_2}{15}$**:** We have to show that $\frac{1}{15}(P_4 + 14N_2 - C - 14N_1) \geq 0$. We can write $P_4 + 14N_2 - C - 14N_1 = bg_{94}(a/b)$, where

$$g_{94}(x) = \frac{u_{94}(x)}{2(x+1)(\sqrt{x}+1)^2}$$

with

$$u_{94}(x) = 7\sqrt{2x+2}(x+1)(\sqrt{x}+1)^3$$
$$- \left(9x^3 + 32x^{5/2} + 43x^2 + 56x^{3/2} + 43x + 32\sqrt{x} + 9\right).$$

Let's consider

$$v_{94}(x) = \left[7\sqrt{2x+2}(x+1)(\sqrt{x}+1)^3\right]^2$$
$$- \left(9x^3 + 32x^{5/2} + 43x^2 + 56x^{3/2} + 43x + 32\sqrt{x} + 9\right)^2.$$

After simplifications,

$$v_{94}(x) = (\sqrt{x} - 1)^4 \begin{pmatrix} 17x^4 + 80x^{7/2} + 184x^3 + 288x^{5/2} \\ +318x^2 + 288x^{3/2} + 184x + 80\sqrt{x} + 17 \end{pmatrix}.$$

In view of Note 1, $v_{94}(x) \geq 0$ implies $u_{94}(x) \geq 0$, $\forall x > 0$, hence proving the required result.

95. For $\frac{S+8N_1}{9} \leq \frac{P_4+14N_2}{15}$**:** We have to show that $\frac{1}{90}(3P_4 + 42N_2 - 5S - 40N_1) \geq 0$. We can write $3P_4 + 42N_2 - 5S - 40N_1 = bg_{95}(a/b)$, where

$$g_{95}(x) = \frac{u_{95}(x)}{2(x+1)(\sqrt{x}+1)^2}$$

with

$$u_{95}(x) = (\sqrt{x}+1)^2\left[21(\sqrt{x}+1)\sqrt{2x+2} - 5\sqrt{2x^2+2}\right]$$
$$- \left(20x^2 + 80x^{3/2} + 96x + 80\sqrt{x} + 20\right).$$

Before proceeding further, we shall show that the second expression in the first line of $u_{95}(x)$ is nonnegative. For simplicity, let's write

$$p_{95}(x) = 21(\sqrt{x}+1)\sqrt{2x+2} - 5\sqrt{2x^2+2}.$$

Let's consider

$$h_{95}(x) = \left[21(\sqrt{x}+1)\sqrt{2x+2}\right]^2 - \left[5\sqrt{2x^2+2}\right]^2.$$

After simplifications,

$$h_{95}(x) = 4\left(208x^2 + 441x^{3/2} + 441x + 441\sqrt{x} + 208\right).$$

In view of Note 1, $h_{95}(x) \geq 0$ implies $p_{95}(x) \geq 0$, $\forall x > 0$. Now we shall show that $u_{95}(x) \geq 0$, $\forall x > 0$. Let's consider

$$v_{95}(x) = \left\{(\sqrt{x}+1)^2\left[21(\sqrt{x}+1)\sqrt{2x+2} - 5\sqrt{2x^2+2}\right]\right\}^2$$
$$- \left(20x^2 + 80x^{3/2} + 96x + 80\sqrt{x} + 20\right)^2.$$

After simplifications,

$$v_{95}(x) = \begin{pmatrix} 532x^4 + 2292^{7/2} + 4172x^3 + 4572x^{5/2} \\ +3744x^2 + 4572x^{3/2} + 4172x + 2292\sqrt{x} + 532 \end{pmatrix}$$
$$- 210\sqrt{2x+2}\sqrt{2x^2+2}(\sqrt{x}+1)^5.$$

Still, we are unable to decide the nonnegativity of $v_{95}(x)$. Let's consider

$$w_{95}(x) = \begin{pmatrix} 532x^4 + 2292^{7/2} + 4172x^3 + 4572x^{5/2} \\ +3744x^2 + 4572x^{3/2} + 4172x + 2292\sqrt{x} + 532 \end{pmatrix}^2$$
$$- \left[210\sqrt{2x+2}\sqrt{2x^2+2}(\sqrt{x}+1)^5\right]^2.$$

After simplifications,

$$w_{95}(x) = 16(\sqrt{x}-1)^4 \begin{pmatrix} 6664x^6 + 68824x^{11/2} + 333929x^5 \\ +1015494x^{9/2} + 2151337x^4 \\ +3328718x^{7/2} + 3843508x^3 \\ +3328718x^{(5/2)} + 2151337x^2 \\ +1015494x^{3/2} + 333929x \\ +68824\sqrt{x} + 6664 \end{pmatrix}.$$

In view of Note 1, the nonnegativity of $w_{95}(x)$ proves the nonnegativity of $v_{95}(x)$, giving $u_{95}(x) \geq 0$, $\forall x > 0$, hence proving the required result.

96. For $\frac{8P_4+7S}{15} \leq \frac{P_4+14N_2}{15}$**:** We have to show that $\frac{7}{15}(2N_2 - P_4 - S) \geq 0$. We can write $2N_2 - P_4 - S = bg_{96}(a/b)$, where $g_{96}(x) = \frac{1}{2}u_{96}(x)/(\sqrt{x}+1)^2$

with

$$u_{96}(x) = (\sqrt{x}+1)^2[(\sqrt{x}+1)\sqrt{2x+2} - \sqrt{2x^2+2}] - 8x.$$

Before proceeding further, we shall show that the second expression in the first line of $u_{96}(x)$ is nonnegative. For simplicity, let's write

$$p_{96}(x) = (\sqrt{x}+1)\sqrt{2x+2} - \sqrt{2x^2+2}.$$

Let's consider

$$h_{96}(x) = \left[(\sqrt{x}+1)\sqrt{2x+2}\right]^2 - \left[\sqrt{2x^2+2}\right]^2.$$

After simplifications,

$$h_{96}(x) = 4\sqrt{x}(x + \sqrt{x} + 1).$$

In view of Note 1, $h_{96}(x) \geq 0$ implies $p_{96}(x) \geq 0$, $\forall x > 0$. Now we shall show that $u_{96}(x) \geq 0$, $\forall x > 0$. Let's consider

$$v_{96}(x) = \left\{(\sqrt{x}+1)^2[(\sqrt{x}+1)\sqrt{2x+2} - \sqrt{2x^2+2}]\right\}^2 - (8x)^2.$$

After simplifications,

$$v_{96}(x) = \begin{pmatrix} 4x^4 + 20x^{7/2} + 44x^3 + 60x^{5/2} \\ +60x^{3/2} + 44x + 20\sqrt{x} + 4 \end{pmatrix} - 2\sqrt{2x+2}\sqrt{2x^2+2}(\sqrt{x}+1)^5.$$

Still, we are unable to decide the nonnegativity of $v_{96}(x)$. Let's apply again Note 1. Let's consider

$$w_{96}(x) = \begin{pmatrix} 4x^4 + 20x^{7/2} + 44x^3 + 60x^{5/2} \\ +60x^{3/2} + 44x + 20\sqrt{x} + 4 \end{pmatrix}^2 - \left[2\sqrt{2x+2}\sqrt{2x^2+2}(\sqrt{x}+1)^5\right]^2.$$

After simplifications,

$$w_{96}(x) = 16x(\sqrt{x}-1)^4 \begin{pmatrix} x^4 + 14x^{7/2} + 65x^3 + 158x^{5/2} \\ +228x^2 + 158x^{3/2} + 65x + 14\sqrt{x} + 1 \end{pmatrix}.$$

In view of Note 1, the nonnegativity of $w_{96}(x)$ proves the nonnegativity of $v_{96}(x)$, giving $u_{96}(x) \geq 0$, $\forall x > 0$, hence proving the required result.

97. For $\frac{P_4+14N_2}{15} \leq$ **N:** We have to show that $\frac{1}{15}(15N - P_4 - 14N_2) \geq 0$. We can write $15N - P_4 - 14N_2 = bg_{97}(a/b)$, where $g_{97}(x) = \frac{1}{2}u_{97}(x)/(\sqrt{x}+1)^2$ with

$$u_{97}(x) = 10x^2 + 30x^{3/2} + 32x + 30\sqrt{x} + 10 - 7\sqrt{2x+2}(\sqrt{x}+1)^3.$$

Let's consider

$$v_{97}(x) = \left(10x^2 + 30x^{3/2} + 32x + 30\sqrt{x} + 10\right)^2 - \left[7\sqrt{2x+2}(\sqrt{x}+1)^3\right]^2.$$

After simplifications,

$$v_{97}(x) = 2(\sqrt{x}-1)^4\left(x^2 + 10x^{3/2} + 20x + 10\sqrt{x} + 1\right).$$

In view of Note 1, $v_{97}(x) \geq 0$ implies $u_{97}(x) \geq 0$, $\forall x > 0$, hence proving the required result.

98. For $\frac{C+2P_4}{3} \leq$ **N:** We have to show that $\frac{1}{3}(3N - C - 2P_4) \geq 0$. We can write $3N - C - 2P_4 = bg_{98}(a/b)$, where

$$g_{98}(x) = \frac{u_{98}(x)}{(x+1)(\sqrt{x}+1)^2}$$

with

$$u_{98}(x) = x^{5/2} + 6x^{3/2} - 4x^2 - 4x + \sqrt{x} = \sqrt{x}(\sqrt{x}-1)^4.$$

Since $u_{98}(x) \geq 0$, $\forall x > 0$, hence proving the required result.

Combining Results 82 to 98 we get the proof of Proposition 6. $\qquad\square$

Proposition 7. *The following inequalities hold:*

$$N \leq \frac{H+20N_2}{21} \leq \begin{cases} \frac{P_3+32N_2}{33} \leq \frac{P_2+44N_2}{45} \leq \frac{P_1+68N_2}{69} \leq \begin{cases} N_2 \\ \frac{11C+7P_2}{18} \leq \begin{cases} P_5 \\ R \end{cases} \end{cases} \\ \left.\begin{array}{c} \frac{P_3+32N_2}{33} \leq \frac{P_2+44N_2}{45} \\ \frac{P_5+5N_1}{6} \leq \frac{7A+2P_4}{9} \end{array}\right\} \leq \begin{cases} A \\ \frac{22P_5+5P_2}{27} \leq P_5. \end{cases} \end{cases}$$

Proof. The proof is based on the following results.

99. For N $\leq \frac{H+20N_2}{21}$**:** We have to show that $\frac{1}{21}(H + 20N_2 - 21N) \geq 0$. We can write $H + 20N_2 - 21N = bg_{99}(a/b)$, where $g_{99}(x) = u_{99}(x)/(x+1)$

with

$$u_{99}(x) = 5\sqrt{2x+2}(\sqrt{x}+1)(x+1)$$
$$- (7x^2 + 7x^{3/2} + 12x + 7\sqrt{x} + 7).$$

Let's consider

$$v_{99}(x) = \left[5\sqrt{2x+2}(\sqrt{x}+1)(x+1)\right]^2$$
$$- \left(7x^2 + 7x^{3/2} + 12x + 7\sqrt{x} + 7\right)^2.$$

After simplifications,

$$v_{99}(x) = (\sqrt{x}-1)^4\left(x^2 + 6x^{3/2} + x + 6\sqrt{x} + 1\right).$$

In view of Note 1, $v_{99}(x) \geq 0$ implies $u_{99}(x) \geq 0$, $\forall x > 0$, hence proving the required result.

100. For $\frac{H+20N_2}{21} \leq \frac{P_3+32N_2}{33}$: We have to show that $\frac{1}{231}(7P_3 + 4N_2 - 11H) \geq 0$. We can write $7P_3 + 4N_2 - 11H = bg_{100}(a/h)$, where

$$g_{100}(x) = \frac{u_{100}(x)}{(x+1)(x-\sqrt{x}+1)}$$

with

$$u_{100}(x) = \sqrt{2x+2}(\sqrt{x}+1)(x+1)(x-\sqrt{x}+1)$$
$$- x(15x - 22\sqrt{x} + 15).$$

After simplifications,

$$u_{100}(x) = \sqrt{2x+2}(\sqrt{x}+1)(x+1)(x-\sqrt{x}+1)$$
$$- x\left[4(x+1) + 11(\sqrt{x}-1)^2\right].$$

Let's consider

$$v_{100}(x) = \left[\sqrt{2x+2}(\sqrt{x}+1)(x+1)(x-\sqrt{x}+1)\right]^2$$
$$- \left\{x\left[4(x+1) + 11(\sqrt{x}-1)^2\right]\right\}^2.$$

After simplifications,

$$v_{100}(x) = (\sqrt{x}-1)^4\left(\begin{array}{l}2x^4 + 8x^{7/2} + 26x^3 + 31x^{5/2} + 31x^{3/2} \\ +37x^{3/2}(\sqrt{x}-1)^2 + x^2 + 26x + 8\sqrt{x} + 2\end{array}\right).$$

In view of Note 1, $v_{100}(x) \geq 0$ implies $u_{100}(x) \geq 0$, $\forall x > 0$, hence proving the required result.

101. For $\frac{P_3+32N_2}{33} \le \frac{P_2+44N_2}{45}$: We have to show that $\frac{1}{495}(11P_2 + 4N_2 - 15P_3) \ge 0$. We can write $11P_2 + 4N_2 - 15P_3 = bg_{101}(a/b)$, where

$$g_{101}(x) = \frac{u_{101}(x)}{(x^2 + 1)(x - \sqrt{x} + 1)}$$

with

$$u_{101}(x) = \sqrt{2x + 2}(\sqrt{x} + 1)(x - \sqrt{x} + 1)(x^2 + 1) - x(4x^2 + 11x^{3/2} - 22x + 11\sqrt{x} + 4).$$

After simplifications,

$$u_{101}(x) = \sqrt{2x + 2}(\sqrt{x} + 1)(x - \sqrt{x} + 1)(x^2 + 1) - x[4(x^2 + 1) + 11\sqrt{x}(\sqrt{x} - 1)^2].$$

Let's consider

$$v_{101}(x) = [\sqrt{2x + 2}(\sqrt{x} + 1)(x - \sqrt{x} + 1)(x^2 + 1)]^2 - \{x[4(x^2 + 1) + 11\sqrt{x}(\sqrt{x} - 1)^2]\}^2.$$

After simplifications,

$$v_{101}(x) = (\sqrt{x} - 1)^4 \begin{pmatrix} 2x^6 + 8x^{11/2} + 22x^5 + 52x^{9/2} + 94x^4 \\ +21x^{7/2} + x^3 + 39x^{5/2}(\sqrt{x} - 1)^2 \\ +21x^{5/2} + 94x^2 + 52x^{3/2} + 22x + 8\sqrt{x} + 2 \end{pmatrix}.$$

In view of Note 1, $v_{101}(x) \ge 0$ implies $u_{101}(x) \ge 0$, $\forall x > 0$, hence proving the required result.

102. For $\frac{P_2+44N_2}{45} \le \frac{P_1+68N_2}{69}$: We have to show that $\frac{1}{1035}(15P_1 + 8N_2 - 23P_2) \ge 0$. We can write $15P_1 + 8N_2 - 23P_2 = bg_{102}(a/b)$, where

$$g_{102}(x) = \frac{u_{102}(x)}{(x^3 + 1)(x^2 + 1)}$$

with

$$u_{102}(x) = 2\sqrt{2x + 2}(\sqrt{x} + 1)(x^3 + 1)(x^2 + 1) - x[8(x^4 + x^3 + x + 1) + 15x(x - 1)^2].$$

Let's consider

$$v_{102}(x) = [2\sqrt{2x + 2}(\sqrt{x} + 1)(x^3 + 1)(x^2 + 1)]^2 - \{x[8(x^4 + x^3 + x + 1) + 15x(x - 1)^2]\}^2.$$

After simplifications,

$$v_{102}(x) = (\sqrt{x} - 1)^4 \begin{pmatrix} 8x^{10} + 48x^{19/2} + 160x^9 + 400x^{17/2} + 784x^8 \\ +1360x^{15/2} + 1856x^7 + 2064x^{13/2} \\ +1783x^6 + 860x^{11/2} + 218x^5 + 860x^{9/2} \\ +1783x^4 + 2064x^{7/2} + 1856x^3 + 1360x^{5/2} \\ +784x^2 + 400x^{3/2} + 160x + 48\sqrt{x} + 8 \end{pmatrix}.$$

In view of Note 1, $v_{102}(x) \geq 0$ implies $u_{102}(x) \geq 0$, $\forall x > 0$, hence proving the required result.

103. For $\frac{P_1 + 68P_2}{69} \leq N_2$: We have to show that $\frac{1}{69}(N_2 - P_1) \geq 0$. It is true by definition. See inequalities (30).

104. For $\frac{P_1 + 68N_2}{69} \leq \frac{11C + 7P_2}{18}$: We have to show that

$$\frac{1}{414}(253C + 161P_2 - 6P_1 - 408N_2) \geq 0.$$

We can write $253C + 161P_2 - 6P_1 - 408N_2 = hg_{104}(a/b)$, where

$$g_{104}(x) = \frac{u_{104}(x)}{(x^3 + 1)(x^2 + 1)}$$

with

$$u_{104}(x) = 253x^6 - 98x^5 + 920x^4 - 518x^3 + 920x^2 - 98x + 253 \\ - 102\sqrt{2x + 2}(\sqrt{x} + 1)(x^3 + 1)(x^2 + 1).$$

After simplifications,

$$u_{104}(x) = \begin{pmatrix} 7(7x^4 + 27x^2 + 7)(\sqrt{x} - 1)^2(\sqrt{x} + 1)^2 \\ +204x^6 + 612x^4 + 612x^2 + 204 \end{pmatrix} \\ - 102\sqrt{2x + 2}(\sqrt{x} + 1)(x^3 + 1)(x^2 + 1).$$

Let's consider

$$v_{104}(x) = \begin{pmatrix} 7(7x^4 + 27x^2 + 7)(\sqrt{x} - 1)^2(\sqrt{x} + 1)^2 \\ +204x^6 + 612x^4 + 612x^2 + 204 \end{pmatrix}^2 \\ - \left[102\sqrt{2x + 2}(\sqrt{x} + 1)(x^3 + 1)(x^2 + 1) \right]^2.$$

After simplifications,

$$v_{104}(x) = (\sqrt{x} - 1)^4 \times w_{104}(x),$$

where

$$w_{104}(x) = \begin{pmatrix} 43201x^{10} + 131188x^{19/2} + 174342x^9 + 41428x^{17/2} \\ +13911x^8 + 290024x^{15/2} + 500724x^7 + 36328x^{13/2} \\ +x^6 + 257420x^{11/2} + 808290x^5 + 257420x^{9/2} + x^4 \\ +74176x^{7/2}(x^2+1)(\sqrt{x}-1)^2 + 36328x^{7/2} \\ +500724x^3 + 290024x^{5/2} + 13911x^2 + 41428x^{3/2} \\ +174342x + 131188\sqrt{x} + 43201 \end{pmatrix}.$$

In view of Note 1, $v_{104}(x) \geq 0$ implies $u_{104}(x) \geq 0$, $\forall x > 0$, hence proving the required result.

105. For $\frac{11C+7P_2}{18} \leq R$: We have to show that $\frac{1}{18}(18R - 11C - 7P_2) \geq 0$. We can write $18R - 11C - 7P_2 = bg_{105}(a/b)$, where

$$g_{105}(x) = \frac{(x^2 + 7x + 1)(x-1)^2}{(x+1)(x^2+1)}.$$

Since $g_{105}(x) \geq 0$, $\forall x > 0$, hence proving the required result.

106. For $\frac{11C+7P_2}{18} \leq P_5$: We have to show that $\frac{1}{18}(18P_5 - 11C - 7P_2) \geq 0$. We can write $18P_5 - 11C - 7P = bg_{106}(a/b)$, where

$$g_{106}(x) = \frac{u_{106}(x)}{(x+1)(x^2+1)(\sqrt{x}+1)^2}$$

with

$$u_{106}(x) = \begin{pmatrix} 7x^5 - 22x^{9/2} + 36x^4 - 14x^{7/2} + 29x^3 \\ +29x^2 - 72x^{5/2} + 36x - 14x^{3/2} - 22\sqrt{x} + 7 \end{pmatrix}.$$

After simplifications,

$$u_{106}(x) = (\sqrt{x}-1)^2 \begin{pmatrix} 3x^4 + 4(x^3+1)(\sqrt{x}-1)^2 + 9x^3 \\ +20x^{5/2} + 56x^2 + 20x^{3/2} + 9x + 3 \end{pmatrix}.$$

Since $u_{106}(x) \geq 0$, $\forall x > 0$, hence proving the required result.

107. For $\frac{H+20N_2}{21} \leq \frac{P_5+5N_1}{6}$: We have to show that $\frac{1}{42}(7P_5 + 35N_1 - 2H - 40N_2) \geq 0$. We can write $7P_5 + 35N_1 - 2H - 40N_2 = bg_{107}(a/b)$, where

$$g_{107}(x) = \frac{u_{107}(x)}{4(x+1)(\sqrt{x}+1)^2}$$

with

$$u_{107}(x) = \left(\begin{array}{l} 63x^3 + 140x^{5/2} + 313x^2 \\ +248x^{3/2} + 313x + 140\sqrt{x} + 63 \end{array} \right) \\ - 40\sqrt{2x+2}(x+1)(\sqrt{x}+1)^3.$$

Let's consider

$$v_{107}(x) = \left(\begin{array}{l} 63x^3 + 140x^{5/2} + 313x^2 \\ +248x^{3/2} + 313x + 140\sqrt{x} + 63 \end{array} \right)^2 \\ - \left[40\sqrt{2x+2}(x+1)(\sqrt{x}+1)^3 \right]^2.$$

After simplifications,

$$v_{107}(x) = (\sqrt{x}-1)^4 \left(\begin{array}{l} 769x^4 + 1516x^{7/2} + 2888x^3 \\ +2820x^{5/2} + 4494x^2 + 2820x^{3/2} \\ +2888x + 1516\sqrt{x} + 769 \end{array} \right).$$

In view of Note 1, $v_{107}(x) \geq 0$ implies $u_{107}(x) \geq 0$, $\forall x > 0$, hence proving the required result.

108. For $\frac{P_5 + 5P_1}{6} \leq \frac{7A + 2P_4}{9}$: We have to show that $\frac{1}{54}(42A + 12P_4 - 9P_5 - 45N_1) \geq 0$. We can write $42A + 12P_4 - 9P_5 - 45N_1 = bg_{108}(a/b)$, where $g_{108}(x) = \frac{3}{4}u_{108}(x)/(\sqrt{x}+1)^2$ with

$$u_{108}(x) = x^2 - 4x^{3/2} + 6x - 4\sqrt{x} + 1 = (\sqrt{x}-1)^4.$$

Since $u_{108}(x) \geq 0$, $\forall x > 0$, hence proving the required result.

109. For $\frac{7A + 2P_4}{9} \leq \frac{11C + 7P_2}{18}$: We have to show that $\frac{1}{18}(11C + 7P_2 - 14A - 4P_4) \geq 0$. We can write $11C + 7P_2 - 14A - 4P_4 = bg_{109}(a/b)$, where

$$g_{109}(x) = \frac{u_{109}(x)}{(x+1)(x^2+1)(\sqrt{x}+1)^2}$$

with

$$u_{109}(x) = \left(\begin{array}{l} 4x^5 - 19x^4 + 8x^{9/2} - x^3 - 14x^{7/2} - x^2 \\ +44x^{5/2} - 19x - 14x^{3/2} + 8\sqrt{x} + 4 \end{array} \right).$$

After simplifications,

$$u_{109}(x) = (\sqrt{x}-1)^4 \left(\begin{array}{l} 4x^3 + 24x^{5/2} + 53x^2 \\ +70x^{3/2} + 53x + 24\sqrt{x} + 4 \end{array} \right).$$

Since $u_{109}(x) \geq 0$, $\forall x > 0$, hence proving the required result.

110. For $\frac{P_2+44N_2}{45} \leq \frac{22P_5+5P_2}{27}$: We have to show that $\frac{22}{135}(5P_5 + P_2 -$
$6N_2) \geq 0$. We can write $5P_5 + P_2 - 6N_2 = bg_{110}(a/b)$, where

$$g_{110}(x) = \frac{u_{110}(x)}{2(\sqrt{x}+1)^2(x^2+1)}$$

with

$$u_{110}(x) = 10x^4 + 22x^3 + 4x^{5/2} + 24x^2 + 4x^{3/2} + 22x + 10$$
$$- 3\sqrt{2x+2}(x^2+1)(\sqrt{x}+1)^3.$$

Let's consider

$$v_{110}(x) = \left(10x^4 + 22x^3 + 4x^{5/2} + 24x^2 + 4x^{3/2} + 22x + 10\right)^2$$
$$- \left[3\sqrt{2x+2}(x^2+1)(\sqrt{x}+1)^3\right]^2.$$

After simplifications,

$$v_{110}(x) = 2(x+1)(\sqrt{x}-1)^4 \begin{pmatrix} 41x^5 + 110x^{9/2} + 229x^4 \\ +280x^{7/2} + 304x^3 + 268x^{5/2} \\ +304x^2 + 280x^{3/2} + 229x \\ +110\sqrt{x} + 41 \end{pmatrix}.$$

In view of Note 1, $v_{110}(x) \geq 0$ implies $u_{110}(x) \geq 0$, $\forall x > 0$, hence proving
the required result.

111. For $\frac{22P_5+5P_2}{27} \leq \textbf{P}_5$: We have to show that $\frac{5}{27}(P_5 - P_2) \geq 0$. It is
true by definition. See inequalities (30).

112. For $\frac{7A+2P_4}{9} \leq \textbf{A}$: We have to show that $\frac{2}{9}(A - P_4) \geq 0$. It is true
by definition. See inequalities (30).

Combining Results 99 to 112 we get the proof of Proposition 7.   □

Proposition 8. *The following inequalities hold:*

$$\textbf{N}_1 \leq \begin{cases} \frac{C+3G}{4} \leq \begin{cases} \frac{S+5N_1}{6} \\ \frac{9N_1+C}{10} \end{cases} \leq \frac{3R+11N_1}{14} \\ \textbf{N} \leq \frac{H+20N_2}{21} \\ \frac{H+S}{2} \leq \frac{P_4+S}{2} \end{cases} \leq \textbf{N}_2.$$

Proof. The proof is based on the following results.

113. For $N_1 \leq \frac{C+3G}{4}$: We have to show that $\frac{1}{4}(C+3G-4N_1) \geq 0$. We can write $C+3G-4N_1 = bg_{113}(a/b)$, where $g_{113}(x) = u_{113}(x)/(x+1)$ with

$$u_{113}(x) = x^{3/2} - 2x + \sqrt{x} = \sqrt{x}(\sqrt{x} - 1)^2.$$

Since $u_{113}(x) \geq 0$, $\forall x > 0$, hence proving the required result.

114. For $\frac{C+3G}{4} \leq \frac{S+5N_1}{6}$: We have to show that $\frac{1}{12}(2S + 10N_1 - 3C - 9G) \geq 0$. We can write $2S+10N_1 - 3C - 9G = bg_{114}(a/b)$, where $g_{114}(x) = \frac{1}{2}u_{114}(x)/(x+1)$ with

$$u_{114}(x) = 2\sqrt{2x^2 + 2}(x+1) - (x^2 + 4x^{3/2} - 2x + 4\sqrt{x} + 1).$$

After simplifications,

$$u_{114}(x) = 2\sqrt{2x^2 + 2}(x+1) \\ - \left[x^2 + 3x^{3/2} + \sqrt{x}(\sqrt{x} - 1)^2 + 3\sqrt{x} + 1\right].$$

Let's consider

$$v_{114}(x) = \left[2\sqrt{2x^2 + 2}(x+1)\right]^2 \\ - \left[x^2 + 3x^{3/2} + \sqrt{x}(\sqrt{x} - 1)^2 + 3\sqrt{x} + 1\right]^2.$$

After simplifications,

$$v_{114}(x) = (x-1)^2\left[3(x^2 + 1) + 4(x+1)(\sqrt{x} - 1)^2 + 10x\right].$$

In view of Note 1, $v_{114}(x) \geq 0$ implies $u_{114}(x) \geq 0$, $\forall x > 0$, hence proving the required result.

115. For $\frac{C+3G}{4} \leq \frac{9N_1+C}{10}$: We have to show that $\frac{1}{20}(18N_1 - 3C - 15G) \geq 0$. We can write $18N_1 - 3C - 15G = bg_{115}(a/b)$, where $g_{115}(x) = \frac{3}{2}u_{115}(x)/(x+1)$ with

$$u_{115}(x) = x^2 - 4x^{3/2} + 6x - 4\sqrt{x} + 1 = (\sqrt{x} - 1)^4.$$

Since $u_{115}(x) \geq 0$, $\forall x > 0$, hence proving the required result.

116. For $\frac{S+5N_1}{6} \leq \frac{3R+11N_1}{14}$: We have to show that $\frac{1}{42}(9R - 2N_1 - 7S) \geq 0$. We can write $9R - 2N_1 - 7S = bg_{116}(a/b)$, where $g_{116}(x) = \frac{1}{2}u_{116}(x)/(x+1)$ with

$$u_{116}(x) = 11x^2 - 2x^{3/2} + 10x - 2\sqrt{x} + 11 - 7\sqrt{2x^2 + 2}(x+1) \\ = 10(x^2 + 1) + (x+1)(\sqrt{x} - 1)^2 + 8x - 7\sqrt{2x^2 + 2}(x+1).$$

Let's consider

$$v_{116}(x) = \left[10(x^2 + 1) + (x + 1)(\sqrt{x} - 1)^2 + 8x\right]^2$$
$$- \left[7\sqrt{2x^2 + 2}(x + 1)\right]^2.$$

After simplifications,

$$v_{116}(x) = (\sqrt{x} - 1)^4(23x^2 + 48x^{3/2} + 82x + 48\sqrt{x} + 23).$$

In view of Note 1, $v_{116}(x) \geq 0$ implies $u_{116}(x) \geq 0$, $\forall x > 0$, hence proving the required result.

117. For $\frac{9N_1 + C}{10} \leq \frac{3R + 11N_1}{14}$: We have to show that $\frac{1}{70}(15R - 8N_1 - 7C) \geq 0$. We can write $15R - 8N_1 - 7C = bg_{117}(a/b)$, where $g_{117}(x) = u_{117}(x)/(x + 1)$ with

$$u_{117}(x) = x^2 - 4x^{3/2} + 6x - 4\sqrt{x} + 1 = (\sqrt{x} - 1)^4.$$

Since $u_{117}(x) \geq 0$, $\forall x > 0$, hence proving the required result.

118. For $\frac{H + 20N_2}{21} \leq \frac{3R + 11N_1}{14}$: We have to show that $\frac{1}{42}(9R + 33N_1 - 2H - 40N_2) \geq 0$. We can write $9R + 33N_1 - 2H - 40N_2 = bg_{118}(a/b)$, where $g_{118}(x) = \frac{1}{4}u_{118}(x)/(x + 1)$ with

$$u_{118}(x) = 57x^2 + 66x^{3/2} + 74x + 66\sqrt{x} + 57$$
$$- 40\sqrt{2x + 2}(\sqrt{x} + 1)(x + 1).$$

Let's consider

$$v_{118}(x) = \left(57x^2 + 66x^{3/2} + 74x + 66\sqrt{x} + 57\right)^2$$
$$- \left[40\sqrt{2x + 2}(\sqrt{x} + 1)(x + 1)\right]^2.$$

After simplifications,

$$v_{118}(x) = (\sqrt{x} - 1)^2 \left(\begin{array}{c} 49x^3 + 1222x^{5/2} + 2387x^2 \\ +1644x^{3/2} + 2387x + 1222\sqrt{x} + 49 \end{array}\right).$$

In view of Note 1, $v_{118}(x) \geq 0$ implies $u_{118}(x) \geq 0$, $\forall x > 0$, hence proving the required result.

119. For $\frac{3R + 11N_1}{14} \leq N_2$: We have to show that $\frac{1}{14}(14N_2 - 3R - 11N_1) \geq 0$. We can write $14N_2 - 3R - 11N_1 = bg_{119}(a/b)$, where $g_{119}(x) = \frac{1}{4}u_{119}(x)/(x + 1)$ with

$$u_{119}(x) = 14\sqrt{2x+2}(\sqrt{x}+1)(x+1)$$
$$- 19x^2 + 22x^{3/2} + 30x + 22\sqrt{x} + 19.$$

Let's consider

$$v_{119}(x) = \left[14\sqrt{2x+2}(\sqrt{x}+1)(x+1)\right]^2$$
$$- \left(19x^2 + 22x^{3/2} + 30x + 22\sqrt{x} + 19\right)^2.$$

After simplifications,

$$v_{119}(x) = (\sqrt{x}-1)^4\left(31x^2 + 72x^{3/2} + 46x + 72\sqrt{x} + 31\right).$$

In view of Note 1, $v_{119}(x) \geq 0$ implies $u_{119}(x) \geq 0$, $\forall x > 0$, hence proving the required result.

Remark 3. The results $N_1 \leq \frac{1}{2}(H + S)$ and $\frac{1}{2}(P_4 + S) \leq N_2$ are not shown here as they already appeared before in Results 58 and 96 respectively. Also $\frac{1}{2}(H+S) \leq \frac{1}{2}(P_4+S)$ is not written above as it is obvious in view of $P_4 \geq H$.

Combining Results 113 to 119 we get the proof of Proposition 8. □

Proposition 9. *The following inequalities hold:*

$$\mathbf{N_2} \leq \frac{11N + R}{12} \leq \frac{A + 3N}{4} \leq \frac{9N + P_5}{10}$$

$$\leq \left\{ \begin{array}{l} \frac{7A+H}{8} \leq \left\{ \begin{array}{l} \frac{11A+P_3}{12} \leq \left\{ \begin{array}{l} \frac{15A+P_2}{16} \leq \left\{ \begin{array}{l} \frac{23A+P_1}{24} \\ \frac{2S+H}{3} \end{array} \right. \\ \frac{9R+4P_4}{13} \end{array} \right. \\ \frac{2N+R}{3} \end{array} \right. \\ \left. \begin{array}{l} \frac{S+4N_2}{5} \\ \frac{S+3N}{4} \end{array} \right\} \leq \frac{9R+4P_4}{13} \end{array} \right\} \leq \mathbf{A}.$$

Proof. The proof is based on the following parts.

120. For $\mathbf{N_2} \leq \frac{\mathbf{11N+R}}{\mathbf{12}}$: We have to show that $\frac{1}{12}(11N + R - 12N_2) \geq 0$. We can write $11N + R - 12N_2 = bg_{120}(a/b)$, where $g_{120}(x) = \frac{1}{3}u_{120}(x)/(x+1)$ with

$$u_{120}(x) = 13x^2 + 11x^{3/2} + 24x + 11\sqrt{x} + 13$$
$$- 9\sqrt{2x+2}(\sqrt{x}+1)(x+1).$$

Let's consider

$$v_{120}(x) = \left(13x^2 + 11x^{3/2} + 24x + 11\sqrt{x} + 13\right)^2$$
$$- \left[9\sqrt{2x+2}(\sqrt{x}+1)(x+1)\right]^2.$$

After simplifications,

$$v_{120}(x) = (\sqrt{x}-1)^4\left[2(x^2+1) + 5(x+1)(\sqrt{x}-1)^2 + 5x\right].$$

In view of Note 1, $v_{120}(x) \geq 0$ implies $u_{120}(x) \geq 0$, $\forall x > 0$, hence proving the required result.

121. For $\frac{11N+R}{12} \leq \frac{A+3N}{4}$**:** We have to show that $\frac{1}{3}(3A - 2N - R) \geq 0$. We can write $3A - 2N - R = bg_{121}(a/b)$, where $g_{121}(x) = \frac{1}{6}u_{121}(x)/(x+1)$ with

$$u_{121}(x) = x^2 - 4x^{3/2} + 6x - 4\sqrt{x} + 1 = (\sqrt{x}-1)^4.$$

Since $u_{121}(x) \geq 0$, $\forall x > 0$, hence proving the required result.

122. For $\frac{A+3N}{4} \leq \frac{9N+P_5}{10}$**:** We have to show that $\frac{1}{20}(2P_5 + 3N - 5A) \geq 0$. We can write $2P_5 + 3N - 5A = bg_{122}(a/b)$, where $g_{122}(x) = \frac{1}{2}u_{122}(x)/(\sqrt{x}+1)^2$ with

$$u_{122}(x) = x^2 - 4x^{3/2} + 6x - 4\sqrt{x} + 1 = (\sqrt{x}-1)^4.$$

Since $u_{122}(x) \geq 0$, $\forall x > 0$, hence proving the required result.

123. For $\frac{9N+P_5}{10} \leq \frac{7A+H}{8}$**:** We have to show that $\frac{1}{40}(35A + 5H - 4P_5 - 36N) \geq 0$. We can write $35A + 5H - 4P_5 - 36N = bg_{123}(a/b)$, where

$$g_{123}(x) = \frac{u_{123}(x)}{2(x+1)(\sqrt{x}+1)^2}$$

with

$$u_{123}(x) = 3x^3 - 2x^{5/2} - 19x^2 + 36x^{3/2} - 19x - 2\sqrt{x} + 3$$
$$= (\sqrt{x}+3)(3\sqrt{x}+1)(\sqrt{x}-1)^4.$$

Since $u_{123}(x) \geq 0$, $\forall x > 0$, hence proving the required result.

124. For $\frac{7A+H}{8} \leq \frac{11A+P_3}{12}$**:** We have to show that $\frac{1}{24}(2P_3 + A - 3H) \geq 0$. We can write $2P_3 + A - 3H = bg_{124}(a/b)$, where

$$g_{124}(x) = \frac{u_{124}(x)}{2(x+1)(x-\sqrt{x}+1)}$$

with

$$u_{124}(x) = x^3 - x^{5/2} - 5x^2 + 10x^{3/2} - \sqrt{x} - 5x + 1$$
$$= (x + 3\sqrt{x} + 1)(\sqrt{x} - 1)^4.$$

Since $u_{124}(x) \geq 0$, $\forall x > 0$, hence proving the required result.

125. For $\frac{7A+H}{8} \leq \frac{2N+R}{3}$: We have to show that $\frac{1}{24}(16N + 8R - 21A - 3H) \geq 0$. We can write $16N + 8R - 21A - 3H = bg_{125}(a/b)$, where $g_{125}(x) = \frac{1}{6}u_{125}(x)/(x+1)$ with

$$u_{125}(x) = 32x^{3/2} + x^2 + 32\sqrt{x} - 66x + 1$$
$$= (x + 34\sqrt{x} + 1)(\sqrt{x} - 1)^2.$$

Since $u_{125}(x) \geq 0$, $\forall x > 0$, hence proving the required result.

126. For $\frac{11A+P_3}{12} \leq \frac{15A+P_2}{16}$: We have to show that $\frac{1}{48}(3P_2 + A - 4P_3) \geq 0$. We can write $3P_2 + A - 4P_3 = bg_{126}(a/b)$, where

$$g_{126}(x) = \frac{u_{126}(x)}{2(x^2 + 1)(x - \sqrt{x} + 1)}$$

with

$$u_{126}(x) = x^4 - x^{(7/2)} - 7x^{5/2} + 14x^2 - 7x^{3/2} - \sqrt{x} + 1$$
$$= (x^2 + 3x^{3/2} + 6x + 3\sqrt{x} + 1)(\sqrt{x} - 1)^4.$$

Since $u_{126}(x) \geq 0$, $\forall x > 0$, hence proving the required result.

127. For $\frac{11A+P_3}{12} \leq \frac{9R+4P_4}{13}$: We have to show that $\frac{1}{156}(108R + 48P_4 - 13P_3 - 143A) \geq 0$. We can write $108R + 48P_4 - 13P_3 - 143A = bg_{127}(a/b)$, where

$$g_{127}(x) = \frac{u_{127}(x)}{2(x+1)(\sqrt{x}+1)^2(x - \sqrt{x} + 1)}$$

with

$$u_{127}(x) = \begin{pmatrix} x^4 + 216x^3 + x^{7/2} + 718x^2 \\ -577x^{5/2} - 577x^{3/2} + \sqrt{x} + 216x + 1 \end{pmatrix}.$$

After simplifications,

$$u_{127}(x) = (\sqrt{x} - 1)^2 \begin{pmatrix} x^3 + 3x^{5/2} + 152x^2 + 152x \\ +69x(\sqrt{x} - 1)^2 + 3\sqrt{x} + 1 \end{pmatrix}.$$

Since $u_{127}(x) \geq 0$, $\forall x > 0$, hence proving the required result.

128. For $\frac{15A+P_2}{16} \le \frac{23A+P_1}{24}$**:** We have to show that $\frac{1}{48}(2P_1+A-3P_2) \ge 0$. We can write $2P_1 + A - 3P_2 = bg_{128}(a/b)$, where

$$g_{128}(x) = \frac{(x^2 + 3x + 1)(x - 1)^4}{2(x^3 + 1)(x^2 + 1)}.$$

Since $g_{128}(x) \ge 0$, $\forall x > 0$, hence proving the required result.

129. For $\frac{15A+P_2}{16} \le \frac{2S+H}{3}$**:** We have to show that $\frac{1}{48}(32S + 16H - 3P_2 - 45A) \ge 0$. We can write $32S + 16H - 3P_2 - 45A = bg_{129}(a/b)$, where

$$g_{129}(x) = \frac{u_{129}(x)}{2(x + 1)(x^2 + 1)},$$

with

$$u_{129}(x) = 32\sqrt{2x^2 + 2}(x + 1)(x^2 + 1) - (45x^4 + 32x^3 + 102x^2 + 32x + 45).$$

Let's consider

$$v_{129}(x) = \left[32\sqrt{2x^2 + 2}(x + 1)(x^2 + 1)\right]^2 - (45x^4 + 32x^3 + 102x^2 + 32x + 45)^2.$$

After simplifications,

$$v_{129}(x) = (x - 1)^2 \left(\begin{array}{c} 23x^6 + 1262x^5 + 489x^4 \\ +2596x^3 + 489x^2 + 1262x + 23 \end{array} \right).$$

In view of Note 1, $v_{129}(x) \ge 0$ implies $u_{129}(x) \ge 0$, $\forall x > 0$, hence proving the required result.

130. For $\frac{23A+P_1}{24} \le A$**:** We have to show that $\frac{1}{24}(A - P_1) \ge 0$. It is true by definition. See inequalities (30).

131. For $\frac{2S+H}{3} \le A$**:** We have to show that $\frac{1}{3}(3A - 2S - H) \ge 0$. It is true in view of Result 59 done before.

132. For $\frac{2N+R}{3} \le A$**:** We have to show that $\frac{1}{3}(3A - 2N - R) \ge 0$. It follows in view of Result 121 done above.

133. For $\frac{9N+P_5}{10} \le \frac{S+4N_2}{5}$**:** We have to show that $\frac{1}{10}(2S + 8N_2 - P_5 - 9N) \ge 0$. We can write $2S + 8N_2 - P_5 - 9N = bg_{133}(a/b)$, where $g_{133}(x) = u_{133}(x)/(\sqrt{x} + 1)^2$ with

$$u_{133}(x) = (\sqrt{x}+1)^2[2(\sqrt{x}+1)\sqrt{2x+2}+\sqrt{2x^2+2}]$$
$$-(4x^2+9x^{3/2}+14x+9\sqrt{x}+4).$$

Let's consider

$$v_{133}(x) = \{(\sqrt{x}+1)^2[2(\sqrt{x}+1)\sqrt{2x+2}+\sqrt{2x^2+2}]\}^2$$
$$-(4x^2+9x^{3/2}+14x+9\sqrt{x}+4)^2.$$

After simplifications,

$$v_{133}(x) = 4\sqrt{2x^2+2}\sqrt{2x+2}(\sqrt{x}+1)^5$$
$$-\begin{pmatrix} 6x^4+16x^{7/2}+53x^3+108x^{5/2} \\ +146x^2+108x^{3/2}+53x+16\sqrt{x}+6 \end{pmatrix}.$$

Still, we are unable to decide the nonnegativity of $v_{133}(x)$. Let's apply again Note 1. Let's consider

$$w_{133}(x) = \left[4\sqrt{2x^2+2}\sqrt{2x+2}(\sqrt{x}+1)^5\right]^2$$
$$-\begin{pmatrix} 6x^4+16x^{7/2}+53x^3+108x^{5/2} \\ +146x^2+108x^{3/2}+53x+16\sqrt{x}+6 \end{pmatrix}^2.$$

After simplifications,

$$w_{133}(x) = (\sqrt{x}-1)^2 \begin{pmatrix} 28x^7+504x^{13/2}+3032x^6 \\ +10888x^{11/2}+27111x^5+50366x^{9/2} \\ +72213x^4+81316x^{7/2}+72213x^3 \\ +50366x^{5/2}+27111x^2+10888x^{3/2} \\ +3032x+504\sqrt{x}+28 \end{pmatrix}.$$

In view of Note 1, the nonnegativity of $w_{133}(x)$ proves the nonnegativity of $v_{133}(x)$, giving $u_{133}(x) \geq 0$, $\forall x > 0$, hence proving the required result.

134. For $\frac{9N+P_5}{10} \leq \frac{S+3N}{4}$**:** We have to show that $\frac{1}{20}(5S-3N-2P_5) \geq 0$. We can write $5S - 3N - 2P_5 = bg_{134}(a/b)$, where $g_{134}(x) = \frac{1}{2}u_{134}(x)/(\sqrt{x}+1)^2$ with

$$u_{134}(x) = 5(\sqrt{x}+1)^2\sqrt{2x^2+2}-(6x^2+6x^{3/2}+16x+6\sqrt{x}+6).$$

Let's consider

$$v_{134}(x) = \left[5(\sqrt{x}+1)^2\sqrt{2x^2+2}\right]^2$$
$$-(6x^2+6x^{3/2}+16x+6\sqrt{x}+6)^2.$$

After simplifications,

$$v_{134}(x) = (\sqrt{x}-1)^2 \begin{pmatrix} 14x^3 + 156x^{5/2} + 370x^2 \\ +520x^{3/2} + 370x + 156\sqrt{x} + 14 \end{pmatrix}.$$

In view of Note 1, $v_{134}(x) \geq 0$ implies $u_{134}(x) \geq 0$, $\forall x > 0$, hence proving the required result.

135. For $\frac{S+4N_2}{5} \leq \frac{9R+4P_4}{13}$: We have to show that $\frac{1}{65}(45R+20P_4-13S-52N_2) \geq 0$. We can write $45R + 20P_4 - 13S - 52N_2 = bg_{135}(a/b)$, where

$$g_{135}(x) = \frac{u_{135}(x)}{2(x+1)(\sqrt{x}+1)^2}$$

with

$$u_{135}(x) = 20\big(3x^3 + 6x^{5/2} + 14x^2 + 6x^{3/2} + 14x + 6\sqrt{x} + 3\big)$$
$$- 13(x+1)(\sqrt{x}+1)^2\big[2(\sqrt{x}+1)\sqrt{2x+2} + \sqrt{2x^2+2}\big].$$

Let's consider

$$v_{135}(x) = \big[20(3x^3 + 6x^{5/2} + 14x^2 + 6x^{3/2} + 14x + 6\sqrt{x} + 3)\big]^2$$
$$- \big\{13(x+1)(\sqrt{x}+1)^2\big[2(\sqrt{x}+1)\sqrt{2x+2} + \sqrt{2x^2+2}\big]\big\}^2.$$

After simplifications,

$$v_{135}(x) = \begin{pmatrix} 1910x^6 + 4936x^{11/2} + 20960x^5 + 26168x^{9/2} \\ +50554x^4 + 29824x^{7/2} + 77408x^3 + 29824x^{5/2} \\ +50554x^2 + 26168x^{3/2} + 20960x + 4936\sqrt{x} + 1910 \end{pmatrix}$$
$$- 676\sqrt{2x^2+2}\sqrt{2x+2}(x+1)^2(\sqrt{x}+1)^5.$$

Still, we are unable to decide the nonnegativity of $v_{135}(x)$? Let's apply again Note 1 over the expression $v_{135}(x)$, and consider

$$w_{135}(x) = \begin{pmatrix} 1910x^6 + 4936x^{11/2} + 20960x^5 + 26168x^{9/2} \\ +50554x^4 + 29824x^{7/2} + 77408x^3 + 29824x^{5/2} \\ +50554x^2 + 26168x^{3/2} + 20960x + 4936\sqrt{x} + 1910 \end{pmatrix}^2$$
$$- \big[676\sqrt{2x^2+2}\sqrt{2x+2}(x+1)^2(\sqrt{x}+1)^5\big]^2.$$

After simplifications,

$$w_{135}(x) = 4(\sqrt{x}-1)^4 \times p_{135}(x),$$

where

$$p_{135}(x) = \begin{pmatrix} 455049x^{10} + 1964316x^{19/2} + 8385994x^9 \\ +22612076x^{17/2} + 66415353x^8 + 149447536x^{15/2} \\ +321040064x^7 + 521671696x^{13/2} + 782768958x^6 \\ +928143096x^{11/2} + 1053386604x^5 \\ +928143096x^{9/2} + 782768958x^4 + 521671696x^{7/2} \\ +321040064x^3 + 149447536x^{5/2} + 66415353x^2 \\ +22612076x^{3/2} + 8385994x + 1964316\sqrt{x} + 455049 \end{pmatrix}.$$

In view of Note 1, the nonnegativity of $w_{135}(x)$ proves the nonnegativity of $v_{135}(x)$, giving $u_{135}(x) \geq 0$, $\forall x > 0$, hence proving the required result.

136. For $\frac{S+3N}{4} \leq \frac{9R+4P_4}{13}$: We have to show that $\frac{1}{52}(36R + 16P_4 - 13S - 39N) \geq 0$. We can write $36R + 16P_4 - 13S - 39N = bg_{136}(a/b)$, where

$$g_{136}(x) = \frac{u_{136}(x)}{2(x+1)(\sqrt{x}+1)^2}$$

with

$$u_{136}(x) = 22x^3 + 18x^{5/2} + 94x^2 - 60x^{3/2} + 94x + 18\sqrt{x} + 22 \\ - 13\sqrt{2x^2 + 2}(x+1)(\sqrt{x}+1)^2.$$

After simplifications,

$$u_{136}(x) = \begin{pmatrix} 22x^3 + 18x^{5/2} + 64x^2 + 64x \\ +30x(\sqrt{x}-1)^2 + 18\sqrt{x} + 22 \end{pmatrix} \\ - 13\sqrt{2x^2 + 2}(x+1)(\sqrt{x}+1)^2.$$

Let's consider

$$v_{136}(x) = \begin{pmatrix} 22x^3 + 18x^{5/2} + 64x^2 + 64x \\ +30x(\sqrt{x}-1)^2 + 18\sqrt{x} + 22 \end{pmatrix}^2 \\ - \left[13\sqrt{2x^2 + 2}(x+1)(\sqrt{x}+1)^2\right]^2.$$

This gives

$$v_{136}(x) = 2(\sqrt{x}-1)^4 \begin{pmatrix} 73x^4 + 12x^{7/2} + 488x^3 + 516x^{5/2} \\ +1982x^2 + 516x^{3/2} + 488x + 12\sqrt{x} + 73 \end{pmatrix}.$$

In view of Note 1, $v_{136}(x) \geq 0$ implies $u_{136}(x) \geq 0$, $\forall x > 0$, hence proving the required result.

137. For $\frac{9R+4P_4}{13} \leq$ **A:** We have to show that $\frac{1}{13}(13A - 9R - 4P_4) \geq 0$. It follows in view of Result 70 done before.

Combining Results 120 to 137 we get the proof of Proposition 9. $\quad\square$

Proposition 10. *The following inequalities hold:*

$$A \leq \begin{cases} \left.\begin{array}{c} \frac{P_5+2N_2}{3} \\ \frac{3S+P_3}{4} \end{array}\right\} \leq \frac{P_4+3P_5}{4} \leq \frac{P_1+12P_5}{13} \leq P_5 \\[2em] \begin{array}{c} \frac{P_5+2N_2}{3} \leq \frac{6R+P_2}{7} \\[1em] \frac{3S+P_3}{4} \leq \frac{9R+2P_3}{11} \leq \left.\begin{array}{c} \frac{6R+P_2}{7} \\[1em] \frac{4S+P_2}{5} \end{array}\right\} \leq \begin{cases} \frac{P_1+12P_5}{13} \leq P_5 \\[1em] \frac{P_1+9R}{10} \leq (a) \end{cases} \end{array} \end{cases}$$

$$(a) \leq \begin{cases} \left.\begin{array}{c} \frac{4N+3C}{7} \\[1em] \frac{16N_2+11C}{27} \end{array}\right\} \leq R \\[2em] \frac{3H+5C}{8} \leq \frac{21R+P_3}{22} \leq \begin{cases} P_5 \\[1em] \frac{A+2S}{3} \leq \frac{11S+P_3}{12} \leq R. \end{cases} \end{cases}$$

Proof. The proof is based on the following results.

138. For $A \leq \frac{P_5+2N_2}{3}$**:** We have to show that $\frac{1}{3}(P_5 + 2N_2 - 3A) \geq 0$. We can write $P_5 + 2N_2 - 3A = bg_{138}(a/b)$, where $g_{138}(x) = \frac{1}{2}u_{138}(x)/(\sqrt{x}+1)^2$ with

$$u_{138}(x) = \sqrt{2x+2}(\sqrt{x}+1)^3 - (x+1)(x+6\sqrt{x}+1).$$

Let's consider

$$v_{138}(x) = \left[\sqrt{2x+2}(\sqrt{x}+1)^3\right]^2 - \left[(x+1)(x+6\sqrt{x}+1)\right]^2.$$

After simplifications,

$$v_{138}(x) = (x+1)(x+4\sqrt{x}+1)(\sqrt{x}-1)^4.$$

In view of Note 1, $v_{138}(x) \geq 0$ implies $u_{138}(x) \geq 0$, $\forall x > 0$, hence proving the required result.

139. For $A \leq \frac{3S+P_3}{4}$**:** We have to show that $\frac{1}{4}(3S + P_3 - 4A) \geq 0$. We can write $3S + P_3 - 4A = bg_{139}(a/b)$, where $g_{139}(x) = \frac{1}{2}u_{139}(x)/(x-\sqrt{x}+1)$

with

$$u_{139}(x) = 3\sqrt{2x^2 + 2}(x - \sqrt{x} + 1)$$
$$- 2(2x^2 - 2x^{3/2} + 3x - 2\sqrt{x} + 2).$$

After simplifications,

$$u_{139}(x) = 3\sqrt{2x^2 + 2}(x - \sqrt{x} + 1)$$
$$- 2[x^2 + (x+1)(\sqrt{x} - 1)^2 + x + 1].$$

Let's consider

$$v_{139}(x) = [3\sqrt{2x^2 + 2}(x - \sqrt{x} + 1)]^2$$
$$- 4[x^2 + (x+1)(\sqrt{x} - 1)^2 + x + 1]^2.$$

After simplifications,

$$v_{139}(x) = 2(\sqrt{x} - 1)^4[x^2 + 2\sqrt{x}(\sqrt{x} - 1)^2 + x + 1].$$

In view of Note 1, $v_{139}(x) \geq 0$ implies $u_{139}(x) \geq 0$, $\forall x > 0$, hence proving the required result.

140. **For** $\frac{P_5 + 2N_2}{3} \leq \frac{P_4 + 3P_5}{4}$: We have to show that $\frac{1}{12}(3P_4 + 5P_5 - 8N_2) \geq 0$. We can write $3P_4 + 5P_5 - 8N_2 = bg_{140}(a/b)$, where $g_{140}(x) = u_{140}(x)/(\sqrt{x} + 1)^2$ with

$$u_{140}(x) = 5x^2 + 22x + 5 - 2\sqrt{2x + 2}(\sqrt{x} + 1)^3.$$

Let's consider

$$v_{140}(x) = (5x^2 + 22x + 5)^2 - [2\sqrt{2x + 2}(\sqrt{x} + 1)^3]^2.$$

After simplifications,

$$v_{140}(x) = (\sqrt{x} - 1)^4(17x^2 + 20x^{3/2} + 70x + 20\sqrt{x} + 17).$$

In view of Note 1, $v_{140}(x) \geq 0$ implies $u_{140}(x) \geq 0$, $\forall x > 0$, hence proving the required result.

141. **For** $\frac{3S + P_3}{4} \leq \frac{P_4 + 3P_5}{4}$: We have to show that $\frac{1}{4}(P_4 + 3P_5 - 3S - P_3) \geq 0$. We can write $P_4 + 3P_5 - 3S - P_3 = bg_{141}(a/b)$, where

$$g_{141}(x) = \frac{3u_{141}(x)}{2(\sqrt{x} + 1)^2(x - \sqrt{x} + 1)}$$

with

$$u_{141}(x) = 2(x^3 - x^{5/2} + 4x^2 - 4x^{3/2} + 4x - \sqrt{x} + 1)$$
$$- 2\sqrt{2x^2 + 2}(x - \sqrt{x} + 1)(\sqrt{x} + 1)^2.$$

After simplifications,

$$u_{141}(x) = 2(x^2 + x^{3/2} + x + \sqrt{x} + 1)(\sqrt{x} - 1)^2 + 8x(x - \sqrt{x} + 1)$$
$$- 2\sqrt{2x^2 + 2}(x - \sqrt{x} + 1)(\sqrt{x} + 1)^2.$$

Let's consider

$$v_{141}(x) = \left[2(x^2 + x^{3/2} + x + \sqrt{x} + 1)(\sqrt{x} - 1)^2 + 8x(x - \sqrt{x} + 1)\right]^2$$
$$- \left[2\sqrt{2x^2 + 2}(x - \sqrt{x} + 1)(\sqrt{x} + 1)^2\right]^2.$$

This gives

$$v_{141}(x) = 2(\sqrt{x} - 1)^6\left[x^3 + x^2 + x(\sqrt{x} - 1)^2 + x + 1\right].$$

In view of Note 1, $v_{141}(x) \geq 0$ implies $u_{141}(x) \geq 0$, $\forall x > 0$, hence proving the required result.

142. For $\frac{3S+P_3}{4} \leq \frac{9R+2P_3}{11}$: We have to show that $\frac{1}{44}(36R - 3P_3 - 33S) \geq 0$. We can write $36R - 3P_3 - 33S = bg_{142}(a/b)$, where

$$g_{142}(x) = \frac{3u_{142}(x)}{2(x+1)(x - \sqrt{x} + 1)}$$

with

$$u_{142}(x) = 2(8x^3 - 8x^{5/2} + 15x^2 - 8x^{3/2} + 15x - 8\sqrt{x} + 8)$$
$$- 11\sqrt{2x^2 + 2}(x + 1)(x - \sqrt{x} + 1).$$

After simplifications,

$$u_{142}(x) = 2\left[4(\sqrt{x} - 1)^2(2x^2 + 2x^{3/2} + 3x + 2\sqrt{x} + 2) + 11x(x + 1)\right]$$
$$- 11\sqrt{2x^2 + 2}(x + 1)(x - \sqrt{x} + 1).$$

Let's consider

$$v_{142}(x) = 4\left(4(\sqrt{x} - 1)^2\begin{pmatrix}2x^2 + 2x^{3/2} \\ +3x + 2\sqrt{x} + 2\end{pmatrix} + 11x(x + 1)\right)^2$$
$$- \left[11\sqrt{2x^2 + 2}(x + 1)(x - \sqrt{x} + 1)\right]^2.$$

This gives

$$v_{142}(x) = 2(\sqrt{x} - 1)^4 \left(\begin{array}{l} 7x^4 + 14x^{7/2} + 17x^3 + 2x^{5/2} \\ +52x^2 + 2x^{3/2} + 17x + 14\sqrt{x} + 7 \end{array} \right).$$

In view of Note 1, $v_{142}(x) \geq 0$ implies $u_{142}(x) \geq 0$, $\forall x > 0$, hence proving the required result.

143. For $\frac{9R+2P_3}{11} \leq \frac{4S+P_2}{5}$: We have to show that $\frac{1}{55}(44S + 11P_2 - 45R - 10P_3) \geq 0$. We can write $44S + 11P_2 - 45R - 10P_3 = bg_{143}(a/b)$, where

$$g_{143}(x) = \frac{u_{143}(x)}{(x+1)(x^2+1)(x-\sqrt{x}+1)}$$

with

$$u_{143}(x) = 22\sqrt{2x^2 + 2}(x+1)(x-\sqrt{x}+1)(x^2+1)$$
$$- \left(\begin{array}{l} 30x^5 - 30x^{9/2} + 59x^4 - 19x^{7/2} + 67x^3 \\ -38x^{5/2} + 67x^2 - 19x^{3/2} + 59x - 30\sqrt{x} + 30 \end{array} \right).$$

After simplifications,

$$u_{143}(x) = 22\sqrt{2x^2 + 2}(x+1)(x-\sqrt{x}+1)(x^2+1)$$
$$- \left(\begin{array}{l} 10(x-\sqrt{x}+1)(3x^4 + 2x^3 + 4x^2 + 2x + 3) \\ +9x^4 + x^{7/2} + 7x^3 + 2x^{5/2} + 7x^2 + x^{3/2} + 9x \end{array} \right).$$

Let's consider

$$v_{143}(x) = \left[22\sqrt{2x^2 + 2}(x+1)(x-\sqrt{x}+1)(x^2+1) \right]^2$$
$$- \left(\begin{array}{l} 10(x-\sqrt{x}+1)(3x^4 + 2x^3 + 4x^2 + 2x + 3) \\ +9x^4 + x^{7/2} + 7x^3 + 2x^{5/2} + 7x^2 + x^{3/2} + 9x \end{array} \right)^2.$$

This gives

$$v_{143}(x) = (\sqrt{x} - 1)^4 \left(\begin{array}{l} 68x^8 + 136x^{15/2} + 536x^7 + 472x^{13/2} \\ +1155x^6 + 722x^{11/2} + 2103x^5 \\ +1058x^{9/2} + 1932x^4 + 1058x^{7/2} \\ +2103x^3 + 722x^{5/2} + 1155x^2 \\ +472x^{3/2} + 536x + 136\sqrt{x} + 68 \end{array} \right).$$

In view of Note 1, $v_{143}(x) \geq 0$ implies $u_{143}(x) \geq 0$, $\forall x > 0$, hence proving the required result.

144. For $\frac{P_5+2N_2}{3} \leq \frac{6R+P_2}{7}$**:** We have to show that $\frac{1}{21}(18R + 3P_2 - 7P_5 - 14N_2) \geq 0$. We can write $18R + 3P_2 - 7P_5 - 14N_2 = bg_{144}(a/b)$, where

$$g_{144}(x) = \frac{u_{144}(x)}{2(x+1)(x^2+1)(\sqrt{x}+1)^2}$$

with

$$u_{144}(x) = \begin{pmatrix} 10x^5 + 48x^{9/2} + 12x^4 + 60x^{7/2} + 34x^3 \\ +120x^{5/2} + 34x^2 + 60x^{3/2} + 12x + 48\sqrt{x} + 10 \end{pmatrix} \\ - 7\sqrt{2x+2}(x+1)(x^2+1)(\sqrt{x}+1)^3.$$

Let's consider

$$v_{144}(x) = \begin{pmatrix} 10x^5 + 48x^{9/2} + 12x^4 + 60x^{7/2} + 34x^3 \\ +120x^{5/2} + 34x^2 + 60x^{3/2} + 12x + 48\sqrt{x} + 10 \end{pmatrix}^2 \\ - \left[7\sqrt{2x+2}(x+1)(x^2+1)(\sqrt{x}+1)^3\right]^2.$$

After simplifications,

$$v_{144}(x) = 2(\sqrt{x}-1)^4 \begin{pmatrix} x^8 + 190x^{15/2} + 1144x^7 + 2754x^{13/2} \\ +5018x^6 + 6782x^{11/2} + 8928x^5 \\ +10210x^{9/2} + 11258x^4 + 10210x^{7/2} \\ +8928x^3 + 6782x^{5/2} + 5018x^2 \\ +2754x^{3/2} + 1144x + 190\sqrt{x} + 1 \end{pmatrix}.$$

In view of Note 1, $v_{144}(x) \geq 0$ implies $u_{144}(x) \geq 0$, $\forall x > 0$, hence proving the required result.

145. For $\frac{9R+2P_3}{11} \leq \frac{6R+P_2}{7}$**:** We have to show that $\frac{1}{77}(3R + 11P_2 - 14P_3) \geq 0$. We can write $3R + 11P_2 - 14P_3 = bg_{145}(a/b)$, where

$$g_{145}(x) = \frac{u_{145}(x)}{(x+1)(x^2+1)(x-\sqrt{x}+1)}$$

with

$$u_{145}(x) = \begin{pmatrix} 2x^5 - 2x^{9/2} + x^4 - 13x^{7/2} + 25x^3 - 26x^{5/2} \\ +25x^2 - 13x^{3/2} + x - 2\sqrt{x} + 2 \end{pmatrix}.$$

After simplifications,

$$u_{145}(x) = (\sqrt{x}-1)^4 \left(2x^3 + 6x^{5/2} + 13x^2 + 11x^{3/2} + 13x + 6\sqrt{x} + 2\right).$$

Since, $u_{145}(x) \geq 0$, $\forall x > 0$, hence proving the required result.

146. For $\frac{6R+P_2}{7} \leq \frac{P_1+12P_5}{13}$: We have to show that $\frac{1}{91}(7P_1 + 84P_5 - 78R - 13P_2) \geq 0$. We can write $7P_1 + 84P_5 - 78R - 13P_2 = bg_{146}(a/b)$, where

$$g_{146}(x) = \frac{u_{146}(x)}{(x^3+1)(x^2+1)(\sqrt{x}+1)^2}$$

with

$$u_{146}(x) = \begin{pmatrix} 32x^7 + 110x^6 - 104x^{13/2} + 45x^5 - 12x^{11/2} \\ +149x^4 - 234x^{(9/2)} + 149x^3 + 28x^{(7/2)} \\ +45x^2 - 234x^{5/2} + 110x - 12x^{3/2} - 104\sqrt{x} + 32 \end{pmatrix}.$$

After simplifications,

$$u_{146}(x) = (\sqrt{x}-1)^4 \begin{pmatrix} 32x^5 + 24x^{9/2} + 14x^4 + 28x^{7/2} \\ +137x^3 + 178x^{5/2} + 137x^2 \\ +28x^{3/2} + 14x + 24\sqrt{x} + 32 \end{pmatrix}.$$

Since, $u_{146}(x) \geq 0$, $\forall x > 0$, hence proving the required result.

147. For $\frac{4S+P_2}{5} \leq \frac{P_1+12P_5}{13}$: We have to show that $\frac{1}{65}(5P_1 + 60P_5 - 52S - 13P_2) \geq 0$. We can write $5P_1 + 60P_5 - 52S - 13P_2 = bg_{147}(a/b)$, where

$$g_{147}(x) = \frac{u_{147}(x)}{(x^3+1)(x^2+1)(\sqrt{x}+1)^2}$$

with

$$u_{147}(x) = \begin{pmatrix} 60x^7 + 112x^6 - 16x^{11/2} + 99x^5 - 26x^{9/2} \\ +177x^4 + 20x^{7/2} + 177x^3 - 26x^{5/2} \\ +99x^2 - 16x^{3/2} + 112x + 60 \end{pmatrix}$$
$$- 26\sqrt{2x^2 + 2}\left(x^3+1\right)\left(x^2+1\right)\left(\sqrt{x}+1\right)^2.$$

After simplifications,

$$u_{147}(x) = \begin{pmatrix} 60x^7 + 104x^6 + 78x^5 + 164x^4 \\ +[8x(x^4+1) + 13x^2(x^2+1)](\sqrt{x}-1)^2 \\ +20x^{7/2} + 164x^3 + 78x^2 + 104x + 60 \end{pmatrix}$$
$$- 26\sqrt{2x^2 + 2}\left(x^3+1\right)\left(x^2+1\right)\left(\sqrt{x}+1\right)^2.$$

Let's consider

$$v_{147}(x) = \left(\begin{array}{l} 60x^7 + 104x^6 + 78x^5 + 164x^4 \\ +[8x(x^4+1)+13x^2(x^2+1)](\sqrt{x}-1)^2 \\ +20x^{7/2}+164x^3+78x^2+104x+60 \end{array} \right)^2 \\ - \left[26\sqrt{2x^2+2}(x^3+1)(x^2+1)(\sqrt{x}+1)^2 \right]^2.$$

After simplifications,

$$v_{147}(x) = (\sqrt{x}-1)^4 \times w_{147}(x)$$

where

$$w_{147}(x) = \left(\begin{array}{l} 2248x^{12} + 3584x^{23/2} + 6176x^{11} + 4864x^{21/2} \\ +13504x^{10} + 23024x^{19/2} + 40984x^9 \\ +41312x^{17/2} + 45121x^8 + 37032x^{15/2} \\ +53124x^7 + 62152x^{13/2} + 79222x^6 \\ +62152x^{11/2} + 53124x^5 + 37032x^{9/2} \\ +45121x^4 + 41312x^{7/2} + 40984x^3 \\ +23024x^{5/2} + 13504x^2 + 4864x^{3/2} \\ +6176x + 3584\sqrt{x} + 2248 \end{array} \right).$$

In view of Note 1, $w_{147}(x) \geq 0$ implies $u_{147}(x) \geq 0$, $\forall x > 0$, hence proving the required result.

148. For $\frac{P_4 + 3P_5}{4} \leq \frac{P_1 + 12P_5}{13}$: We have to show that $\frac{1}{52}(4P_1 + 9P_5 - 13P_4) \geq 0$. We can write $4P_1 + 9P_5 - 13P_4 = bg_{148}(a/b)$, where

$$g_{148}(x) = \frac{u_{148}(x)}{(x^3+1)(\sqrt{x}+1)^2}$$

with

$$u_{148}(x) = 9x^5 - 30x^4 + 8x^{7/2} + 13x^3 + 13x^2 + 8x^{3/2} - 30x + 9.$$

After simplifications,

$$u_{148}(x) = (\sqrt{x}-1)^4 \left(\begin{array}{l} 9x^3 + 36x^{5/2} + 60x^2 \\ +68x^{3/2} + 60x + 36\sqrt{x} + 9 \end{array} \right).$$

Since $u_{148}(x) \geq 0$, $\forall x > 0$, hence proving the required result.

149. For $\frac{P_1+12P_5}{13} \leq P_5$: We have to show that $\frac{1}{13}(P_5 - P_1) \geq 0$. It is true by definition.

150. For $\frac{4S+P_2}{5} \leq \frac{P_1+9R}{10}$: We have to show that $\frac{1}{10}(P_1 + 9R - 8S - 2P_2) \geq 0$. We can write $P_1 + 9R - 8S - 2P = bg_{150}(a/b)$, where

$$g_{150}(x) = \frac{u_{150}(x)}{(x^2 + 1)(x^3 + 1)}$$

with

$$u_{150}(x) = 6x^6 - x^5 + 10x^4 + 2x^3 + 10x^2 - x + 6 \\ - 4\sqrt{2x^2 + 2}(x + 1)(x^3 + 1).$$

After simplifications,

$$u_{150}(x) = 5x^6 + (x^4 + 1)(x - 1)^2 + x^5 + 9x^4 + 2x^3 + 9x^2 + x + 5 \\ - 4\sqrt{2x^2 + 2}(x + 1)(x^3 + 1).$$

Let's consider

$$v_{150}(x) = \left[5x^6 + (x^4 + 1)(x - 1)^2 + x^5 + 9x^4 + 2x^3 + 9x^2 + x + 5\right]^2 \\ - \left[4\sqrt{2x^2 + 2}(x + 1)(x^3 + 1)\right]^2.$$

This gives

$$v_{150}(x) = (x - 1)^2\left(4x^8 + 4x^7 + 17x^6 + 30x^4 + 17x^2 + 4x + 4\right).$$

In view of Note 1, $v_{150}(x) \geq 0$ implies $u_{150}(x) \geq 0$, $\forall x > 0$, hence proving the required result.

151. For $\frac{P_2+6R}{7} \leq \frac{P_1+9R}{10}$: We have to show that $\frac{1}{70}(7P_1 + 3R - 10P_2) \geq 0$. We can write $7P_1 + 3R - 10P_2 = bg_{151}(a/b)$, where

$$g_{151}(x) = \frac{(x + 2)(2x + 1)(x - 1)^4}{(x^2 + 1)(x^3 + 1)}.$$

Since $g_{151}(x) \geq 0$, $\forall x > 0$, hence proving the required result.

152. For $\frac{P_1+9R}{10} \leq \frac{16N_2+11C}{27}$: We have to show that

$$\frac{1}{270}(160N_2 + 110C - 27P_1 - 243R) \geq 0.$$

We can write $160N_2 + 110C - 27P_1 - 243R = bg_{152}(a/b)$, where $g_{152}(x) = u_{152}(x)/(x^3 + 1)$ with

$$u_{152}(x) = 40\sqrt{2x+2}(\sqrt{x}+1)(x^3+1)$$
$$- (52x^4 + 137x^3 - 58x^2 + 137x + 52).$$

After simplifications,

$$u_{152}(x) = 40\sqrt{2x+2}(\sqrt{x}+1)(x^3+1)$$
$$- [52x^4 + 108x^3 + 29x(x-1)^2 + 108x + 52].$$

Let's consider

$$v_{152}(x) = \left[40\sqrt{2x+2}(\sqrt{x}+1)(x^3+1)\right]^2$$
$$- \left[52x^4 + 108x^3 + 29x(x-1)^2 + 108x + 52\right]^2.$$

This gives

$$v_{152}(x) = (\sqrt{x}-1)^2 \begin{pmatrix} 496x^7 + 7392x^{13/2} + 6440x^6 + 11888x^{11/2} \\ +7799x^5 + 3710x^{9/2} + 7665x^4 + 24420x^{7/2} \\ +7665x^3 + 3710x^{5/2} + 7799x^2 + 11888x^{3/2} \\ +6440x + 7392\sqrt{x} + 496 \end{pmatrix}.$$

In view of Note 1, $v_{152}(x) \geq 0$ implies $u_{152}(x) \geq 0$, $\forall x > 0$, hence proving the required result.

153. For $\frac{P_1+9R}{10} \leq \frac{4N+3C}{7}$: We have to show that $\frac{1}{70}(40N + 30C - 7P_1 - 63R) \geq 0$. We can write $40N + 30C - 7P_1 - 63R = bg_{153}(a/b)$, where $g_{153}(x) = \frac{1}{3}u_{153}(x)/(x^3+1)$ with

$$u_{153}(x) = 4x^4 + 40x^{7/2} - 71x^3 + 54x^2 - 71x + 40\sqrt{x} + 4.$$

After simplifications,

$$u_{153}(x) = (\sqrt{x}-1)^4 \begin{pmatrix} 4x^3 + 48x^{5/2} + 18x^2 \\ +3x(\sqrt{x}-1)^2 + 18x + 48\sqrt{x} + 4 \end{pmatrix}.$$

Since $u_{153}(x) \geq 0$, $\forall x > 0$, hence proving the required result.

154. For $\frac{P_1+9R}{10} \leq \frac{3H+5C}{8}$: We have to show that $\frac{1}{40}(15H + 25C - 4P_1 - 36R) \geq 0$. We can write $15H + 25C - 4P_1 - 36R = bg_{154}(a/b)$, where

$$g_{154}(x) = \frac{(x^2 + 3x + 1)(x-1)^2}{x^3+1}.$$

Since $g_{154}(x) \geq 0$, $\forall x > 0$, hence proving the required result.

155. For $\frac{3H+5C}{8} \leq \frac{21R+P_3}{22}$: We have to show that $\frac{1}{88}(84R+4P_3-33H-55C) \geq 0$. We can write $84R+4P_3-33H-55C = bg_{155}(a/b)$, where

$$g_{155}(x) = \frac{u_{155}(x)}{(x+1)(x-\sqrt{x}+1)}$$

with

$$u_{155}(x) = x^3 - x^{5/2} - 5x^2 + 10x^{3/2} - 5x - \sqrt{x} + 1$$
$$= (x+3\sqrt{x}+1)(\sqrt{x}-1)^2.$$

Since $u_{155}(x) \geq 0$, $\forall x > 0$, hence proving the required result.

156. For $\frac{21R+P_3}{22} \leq \frac{A+2S}{3}$: We have to show that $\frac{1}{66}(22A+44S-63R-3P_3) \geq 0$. We can write $22A+44S-63R-3P_3 = bg_{156}(a/b)$, where

$$g_{156}(x) = \frac{u_{156}(x)}{(x+1)(x-\sqrt{x}+1)}$$

with

$$u_{156}(x) = 22\sqrt{2x^2+2}(x+1)(x-\sqrt{x}+1)$$
$$- \left(31x^3 - 31x^{5/2} + 54x^2 - 20x^{3/2} + 54x - 31\sqrt{x} + 31\right).$$

After simplifications,

$$u_{156}(x) = 22\sqrt{2x^2+2}(x+1)(x-\sqrt{x}+1)$$
$$- \left[(31x^2 + 20x + 31)(x-\sqrt{x}+1) + 3x(x+1)\right].$$

Let's consider

$$v_{156}(x) = \left[22\sqrt{2x^2+2}(x+1)(x-\sqrt{x}+1)\right]^2$$
$$- \left[(31x^2 + 20x + 31)(x-\sqrt{x}+1) + 3x(x+1)\right]^2.$$

This gives

$$v_{156}(x) = (\sqrt{x}-1)^2 \left(\begin{array}{l} 7x^5 + 438x^4 + 86x(x^2+1)(\sqrt{x}-1)^2 \\ +254x^3 + 538x^{5/2} + 254x^2 + 438x + 7 \end{array}\right).$$

In view of Note 1, $v_{156}(x) \geq 0$ implies $u_{156}(x) \geq 0$, $\forall x > 0$, hence proving the required result.

157. For $\frac{A+2S}{3} \leq \frac{11S+P_3}{12}$**:** We have to show that $\frac{1}{12}(3S + P_3 - 4A) \geq 0$. We can write $3S + P_3 - 4A = bg_{157}(a/b)$, where $g_{157}(x) = \frac{1}{2}u_{157}(x)/(x - \sqrt{x} + 1)$ with

$$u_{157}(x) = 3\sqrt{2x^2 + 2}(x - \sqrt{x} + 1)$$
$$- 2(2x^2 - 2x^{3/2} + 3x - 2\sqrt{x} + 2).$$

After simplifications,

$$u_{157}(x) = 3\sqrt{2x^2 + 2}(x - \sqrt{x} + 1) - 2[(x - \sqrt{x} + 1)^2 + x^2 + 1].$$

Let's consider

$$v_{157}(x) = [3\sqrt{2x^2 + 2}(x - \sqrt{x} + 1)]^2$$
$$- 4[(x - \sqrt{x} + 1)^2 + x^2 + 1]^2.$$

This gives

$$v_{157}(x) = 2(\sqrt{x} - 1)^4[x^2 + 2\sqrt{x}(\sqrt{x} - 1)^2 + x + 1].$$

In view of Note 1, $v_{157}(x) \geq 0$ implies $u_{157}(x) \geq 0$, $\forall x > 0$, hence proving the required result.

158. For $\frac{11S+P_3}{12} \leq \mathbf{R}$**:** We have to show that $\frac{1}{12}(12R - 11S - P_3) \geq 0$. We can write $12R - 11S - P_3 = bg_{158}(a/b)$, where

$$g_{158}(x) = \frac{u_{158}(x)}{2(x + 1)(x - \sqrt{x} + 1)}$$

with

$$u_{158}(x) = 2(8x^3 - 8x^{5/2} + 15x^2 - 8x^{3/2} + 15x - 8\sqrt{x} + 8)$$
$$- 11\sqrt{2x^2 + 2}(x + 1)(x - \sqrt{x} + 1).$$

After simplifications,

$$u_{158}(x) = 2[4(x^3 + 1) + 4(x^2 + x + 1)(\sqrt{x} - 1)^2 + 7x(x + 1)]$$
$$- 11\sqrt{2x^2 + 2}(x + 1)(x - \sqrt{x} + 1).$$

Let's consider

$$v_{158}(x) = 4[4(x^3 + 1) + 4(x^2 + x + 1)(\sqrt{x} - 1)^2 + 7x(x + 1)]^2$$
$$- [11\sqrt{2x^2 + 2}(x + 1)(x - \sqrt{x} + 1)]^2.$$

This gives

$$v_{158}(x) = 2(\sqrt{x} - 1)^4 \left(\begin{array}{l} 7x^4 + 14x^{7/2} + 17x^3 + 2x^{5/2} \\ +52x^2 + 2x^{3/2} + 17x + 14\sqrt{x} + 7 \end{array} \right).$$

In view of Note 1, $v_{158}(x) \geq 0$ implies $u_{158}(x) \geq 0$, $\forall x > 0$, hence proving the required result.

159. For $\frac{4N+3C}{7} \leq R$: We have to show that $\frac{1}{7}(7R - 4N - 3C) \geq 0$. We can write $7R - 4N - 3C = bg_{159}(a/b)$, where $g_{159}(x) = \frac{1}{3}u_{159}(x)/(x+1)$ with

$$u_{159}(x) = x^2 - 4x^{3/2} + 6x - 4\sqrt{x} + 1 = (\sqrt{x} - 1)^4.$$

Since $u_{159}(x) \geq 0$, $\forall x > 0$, hence proving the required result.

160. For $\frac{16N_2+11C}{27} \leq R$: We have to show that $\frac{1}{27}(27R - 16N_2 - 11C) \geq 0$. We can write $27R - 16N_2 - 11C = bg_{160}(a/b)$, where $g_{160}(x) = u_{160}(x)/(x+1)$ with

$$u_{160}(x) = 7x^2 + 18x + 7 - 4\sqrt{2x + 2}(x + 1)(\sqrt{x} + 1).$$

Let's consider

$$v_{160}(x) = \left(7x^2 + 18x + 7\right)^2 - \left[4\sqrt{2x + 2}(x + 1)(\sqrt{x} + 1)\right]^2.$$

After simplifications,

$$v_{160}(x) = (\sqrt{x} - 1)^4 \left(17x^2 + 4x^{3/2} + 38x + 4\sqrt{x} + 17\right).$$

In view of Note 1, $v_{160}(x) \geq 0$ implies $u_{160}(x) \geq 0$, $\forall x > 0$, hence proving the required result.

161. For $\frac{21R+P_3}{22} \leq P_5$: We have to show that $\frac{1}{22}(22P_5 - 21R - P_3) \geq 0$. We can write $22P_5 - 21R - P_3 = bg_{161}(a/b)$, where

$$g_{161}(x) = \frac{u_{161}(x)}{(x+1)(x - \sqrt{x} + 1)(\sqrt{x} + 1)^2}$$

with

$$u_{161}(x) = \left(\begin{array}{l} 8x^4 + 73x^3 - 36x^{7/2} + 102x^2 \\ -96x^{5/2} - 96x^{3/2} + 73x - 36\sqrt{x} + 8 \end{array} \right).$$

After simplifications,

$$u_{161}(x) = (\sqrt{x} - 1)^4 \left[2(x+1)(\sqrt{x} - 1)^2 + 6x^2 + 5x + 6 \right].$$

Since $u_{161}(x) \geq 0$, $\forall x > 0$, hence proving the required result.

Combining Results 138 to 161 we get the proof of Proposition 10. $\square$

Proposition 11. *The following inequalities hold:*

$$\mathbf{R} \leq \frac{P_1 + 20S}{21} \leq \mathbf{S} \leq \left\{ \begin{matrix} \frac{5C+2P_4}{7} \\ \frac{5C+4N_2}{9} \end{matrix} \right\} \leq \frac{7C + P_1}{8} \leq \mathbf{C}.$$

Proof. The proof is based on the following results.

162. For $\mathbf{R} \leq \frac{P_1 + 20S}{21}$: We have to show that $\frac{1}{21}(P_1 + 20S - 21R) \geq 0$.
We can write $P_1 + 20S - 21R = b g_{162}(a/b)$, where $g_{162}(x) = u_{162}(x)/(x^3 + 1)$
with

$$u_{162}(x) = 10\sqrt{2x^2 + 2}(x^3 + 1) - (14x^4 - x^3 + 14x^2 - x + 14).$$

After simplifications,

$$\begin{aligned} u_{162}(x) = {} & 10\sqrt{2x^2 + 2}(x^3 + 1) \\ & - \left[(x^2 + 1)(x - 1)^2 + (13x^2 + 14x + 13)(x^2 - x + 1) \right]. \end{aligned}$$

Let's consider

$$\begin{aligned} v_{162}(x) = {} & \left[10\sqrt{2x^2 + 2}(x^3 + 1) \right]^2 \\ & - \left[(x^2 + 1)(x - 1)^2 + (13x^2 + 14x + 13)(x^2 - x + 1) \right]^2. \end{aligned}$$

This gives

$$v_{162}(x) = (x - 1)^4 \left[4x^4 + 23x^3 + 21x(x - 1)^2 + x^2 + 23x + 4 \right].$$

In view of Note 1, $v_{162}(x) \geq 0$ implies $u_{162}(x) \geq 0$, $\forall x > 0$, hence proving
the required result.

163. For $\frac{P_1 + 20S}{21} \leq \mathbf{S}$: We have to show that $\frac{1}{21}(S - P_1) \geq 0$. It is true by
definition. See inequalities (30).

164. For $\mathbf{S} \leq \frac{5C+2P_4}{7}$: We have to show that $\frac{1}{7}(5C + 2P_4 - 7S) \geq 0$. We
can write $5C + 2P_4 - 7S = b g_{164}(a/b)$, where

$$g_{164}(x) = \frac{u_{164}(x)}{2(x+1)(\sqrt{x} + 1)^2}$$

with

$$u_{164}(x) = 2\left(5x^3 + 10x^{5/2} + 13x^2 + 13x + 10\sqrt{x} + 5\right)$$
$$- 7\sqrt{2x^2 + 2}(x+1)(\sqrt{x}+1)^2.$$

Let's consider

$$v_{164}(x) = 4\left(5x^3 + 10x^{5/2} + 13x^2 + 13x + 10\sqrt{x} + 5\right)^2$$
$$- \left[7\sqrt{2x^2 + 2}(x+1)(\sqrt{x}+1)^2\right]^2.$$

After simplifications,

$$v_{164}(x) = 2(\sqrt{x} - 1)^4 \begin{pmatrix} x^4 + 8x^{7/2} + 94x^3 + 264x^{5/2} \\ +386x^2 + 264x^{3/2} + 94x + 8\sqrt{x} + 1 \end{pmatrix}.$$

In view of Note 1, $v_{164}(x) \geq 0$ implies $u_{164}(x) \geq 0$, $\forall x > 0$, hence proving the required result.

165. For S $\leq \frac{5C+4N_2}{9}$: We have to show that $\frac{1}{9}(5C + 4N_2 - 9S) \geq 0$. We can write $5C + 4N_2 - 9S = bg_{165}(a/b)$, where $g_{165}(x) = \frac{1}{2}u_{165}(x)/(x+1)$ with

$$u_{165}(x) = 10\left(x^2 + 1\right) - (x+1)\left[9\sqrt{2x^2 + 2} - 2(\sqrt{x}+1)\sqrt{2x+2}\right].$$

Before proceeding further, we shall show that the second expression in the first line of $u_{165}(x)$ is nonnegative. For simplicity, let's write

$$p_{165}(x) = 9\sqrt{2x^2 + 2} - 2(\sqrt{x}+1)\sqrt{2x+2}.$$

Let's consider

$$h_{165}(x) = \left[9\sqrt{2x^2 + 2}\right]^2 - \left[2(\sqrt{x}+1)\sqrt{2x+2}\right]^2.$$

After simplifications,

$$h_{165}(x) = 16(2x + 3\sqrt{x} + 2)(\sqrt{x} - 1)^2 + 2\left(61x^2 + 8x + 61\right).$$

In view of Note 1, $h_{165}(x) > 0$ implies $p_{165}(x) > 0$, $\forall x > 0$. Now we shall show that $u_{165}(x) \geq 0$, $\forall x > 0$. Let's consider

$$v_{165}(x) = \left[10\left(x^2 + 1\right)\right]^2$$
$$- \left\{(x+1)\left[9\sqrt{2x^2 + 2} - 2(\sqrt{x}+1)\sqrt{2x+2}\right]\right\}^2.$$

After simplifications,

$$v_{165}(x) = 36\sqrt{2x^2 + 2}\sqrt{2x + 2}(\sqrt{x} + 1)(x + 1)^2$$
$$- 2(7x + 10\sqrt{x} + 7)\begin{pmatrix}2(x^3 + 1) + 10x(x + 1) \\ +(3x^2 + 16x + 3)(\sqrt{x} - 1)^2\end{pmatrix}.$$

Still, we are unable to decide the nonnegativity of $v_{165}(x)$. Let's apply again Note 1 over the expression $v_{165}(x)$, and consider

$$w_{165}(x) = \left[36\sqrt{2x^2 + 2}\sqrt{2x + 2}(\sqrt{x} + 1)(x + 1)^2\right]^2$$
$$- \left\{2(7x + 14\sqrt{x} + 7)\begin{pmatrix}2(x^3 + 1) + 10x(x + 1) \\ +(3x^2 + 16x + 3)(\sqrt{x} - 1)^2\end{pmatrix}\right\}^2.$$

This gives

$$w_{165}(x) = 4(\sqrt{x} - 1)^4 \begin{pmatrix}71x^6 + 2316x^{11/2} + 4090x^5 \\ +11180x^{9/2} + 12021x^4 + 11960x^{7/2} \\ +6004x^3 + 11960x^{5/2} + 12021x^2 \\ +11180x^{3/2} + 4090x + 2316\sqrt{x} + 71\end{pmatrix}.$$

In view of Note 1, the nonnegativity of $w_{165}(x)$ proves the nonnegativity of $v_{165}(x)$, giving $u_{165}(x) \geq 0$, $\forall x > 0$, hence proving the required result.

166. For $\frac{5C + 2P_4}{7} \leq \frac{7C + P_1}{8}$: We have to show that $\frac{1}{56}(9C + 7P_1 - 16P_4) \geq 0$. We can write $9C + 7P_1 - 16P_4 = bg_{166}(a/b)$, where

$$g_{166}(x) = \frac{u_{166}(x)}{(x^3 + 1)(\sqrt{x} + 1)^2}$$

with

$$u_{166}(x) = \begin{pmatrix}9x^5 - 57x^4 + 18x^{9/2} + 16x^3 - 4x^{7/2} \\ +16x^2 + 36x^{5/2} - 57x - 4x^{3/2} + 18\sqrt{x} + 9\end{pmatrix}.$$

After simplifications,

$$u_{166}(x) = (\sqrt{x} - 1)^4 \begin{pmatrix}9x^3 + 54x^{5/2} + 105x^2 \\ +128x^{3/2} + 105x + 54\sqrt{x} + 9\end{pmatrix}.$$

Since $u_{166}(x) \geq 0$, $\forall x > 0$, hence proving the required result.

167. For $\frac{5C + 4N_2}{9} \leq \frac{7C + P_1}{8}$: We have to show that $\frac{1}{72}(23C + 9P_1 - 32N_2) \geq 0$. We can write $23C + 9P_1 - 32N_2 = bg_{167}(a/b)$, where $g_{167}(x) =$

$u_{167}(x)/(x+1)$ with

$$u_{167}(x) = 23x^4 - 14x^3 + 46x^2 - 14x + 23$$
$$- 8\sqrt{2x+2}(x^3+1)(\sqrt{x}+1)^2.$$

After simplifications,

$$u_{167}(x) = (x^2+1)\big[16(x^2+1)+7(x-1)^2\big]$$
$$- 8\sqrt{2x+2}(x^3+1)(\sqrt{x}+1)^2.$$

Let's consider

$$v_{167}(x) = \big\{(x^2+1)\big[16(x^2+1)+7(x-1)^2\big]\big\}^2$$
$$- \big[8\sqrt{2x+2}(x^3+1)(\sqrt{x}+1)^2\big]^2.$$

This gives

$$v_{167}(x) = (\sqrt{x}-1)^4 \begin{pmatrix} 101x^6 + 1348x^{11/2} + 2086x^5 \\ +1604x^{9/2} + 1075x^4 + 1672x^{7/2} \\ +2380x^3 + 1672x^{5/2} + 1075x^2 \\ +1604x^{3/2} + 2086x + 1348\sqrt{x} + 401 \end{pmatrix}.$$

In view of Note 1, $v_{167}(x) \geq 0$ implies $u_{167}(x) \geq 0$, $\forall x > 0$, hence proving the required result.

168. For $\frac{7C+P_1}{8} \leq C$: We have to show that $\frac{1}{8}(C - P_1) \geq 0$. It is true by definition. See inequalities (30).

Combining Results 162 to 168 we get the proof of Proposition 11. $\square$

3.1 Equality Relations

Based on Results 1–168 given in Section 3, we have some equality relations. These are given as follows:

(i) In view of Results 34, 49, 124, and 155, the following equalities hold:

$$4P_3 + C - 5H$$
$$= 2(2P_3 + 9R - 11A) = 2(2P_3 + A - 3H)$$
$$= 84R + 4P_3 - 33H - 55C.$$

It leads us to the following equalities among the means given in (30):

(a) $C + 22A = 5H + 18R$;

(d) $A = \frac{3R+H}{4}$;

(b) $A = \frac{C+H}{2}$;

(e) $3H + 5C = 6R + 2A$;

(c) $R = \frac{2C+H}{3}$;

(f) $R = \frac{2A+27H+55C}{84}$.

(ii) In view of Results 24 and 26, the following equalities hold:

$$2(P_2 + 6R - 7A) = 2P_2 + 7C - 9R.$$

It leads us to the following equalities among the means given in (30):

$$R = \frac{2A + C}{3}.$$

(iii) In view of Results 71, 108, and 122, the following equalities hold:

$$P_4 + 3P_5 - 4A$$
$$= \frac{4}{3}(42A + 12P_4 - 9P_5 - 45N_1)$$
$$= 2(2P_5 + 3N - 5A).$$

It leads us to the following equalities among the means given in (30):

(a) $P_4 + 6A = P_5 + 6N$;

(b) $P_4 + 4A = P_5 + 4N_1$;

(c) $8P_4 + 33A = 8P_5 + 3N + 30N_1$.

(iv) In view of Results 57, 66, 76, 115, 117, 121, and 159, the following equalities hold:

$$2(H + 2N_1 - 3G)$$
$$= 2(10N_1 - 3R - 7G) = 9R - 4G - 5C$$
$$= \frac{2}{3}(18N_1 - 3C - 15G) = 15R - 8N_1 - 7C$$
$$= 6(3A - 2N - R)(121) = 3(7R - 4N - 3C).$$

It leads us to the following equalities among the means given in (30):

(a) $N_1 = \frac{H+4G+3R}{8}$;

(f) $7G + 9A = 10N_1 + 6N$;

(b) $N_1 = \frac{H+2G+C}{4}$;

(g) $6N + C = 4N_1 + 3R$;

(c) $R = \frac{2A+C}{3}$;

(h) $3N + C = G + 3R$;

(d) $2G + 9R = 2H + 4N_1 + 5C$;

(i) $21R + 10G = 12N + 7C + 12N_1$;

(e) $2G + 3R = 4N_1 + C$;

(j) $27R + 14G = 20N_1 + 12N + 9C$;

(k) $12N + 21R = 18A + 8N_1 + 7C$;

(l) $2N_1 + 2N + R = \frac{5G+9A+C}{3}$;

(m) $4N + 7R = \frac{8N_1+18A+7C}{3}$;

(n) $4N + 5R = \frac{4G+18A+5C}{3}$;

(o) $2G + 7R = \frac{2H+4N_1+12N+9C}{3}$;

(p) $G + 3A = \frac{H+2N_1+6N+3R}{3}$;

(q) $2G + 5R = \frac{2H+12N_1+7C}{3}$.

(v) In view of Results 69 and 70, the following equalities hold:

$$21R - 8P_4 - 13C = 2(13A - 4P_4 - 9R).$$

It leads us to the following equalities among the means given in (30):

$$R = \frac{2A + C}{3}.$$

(vi) In view of Results 86 and 113, the following equalities hold:

$$\tfrac{1}{5}(2C + 28N_1 - 27N - 3H) = (C + 3G - 4N_1).$$

It leads us to the following equalities among the means given in (30):

$$N_1 = \frac{H + 5G + 9N + C}{16}.$$

(vii) In view of Results 20 and 43, the following equalities hold:

$$2(P_2 + 6N_1 - 7G) = 2P_2 + 9N - 11G.$$

It leads us to the following equalities among the means given in (30):

$$N_1 = \frac{G + 3N}{4}.$$

(viii) In view of Results 56 and 98, the following equalities hold:

$$G + H - 2P_4 = 3N - C - 2P_4.$$

It leads us to the following equalities among the means given in (30):

$$N = \frac{G + H + C}{3}.$$

(ix) In view of Remark 1, the following equalities among the means given in (30) hold:

(a) $A = \frac{H+3R}{4} = \frac{C+H}{2};$ **(c)** $N = \frac{2A+G}{3} = \frac{G+H+C}{3};$

(b) $R = \frac{2A+C}{3} = \frac{2C+H}{3};$ **(d)** $2N_1 + H + C = 3A + G;$

                                                  (e) $2A + P_4 = 2G + P_5.$

Remark 4. There are same results in many of the above cases.

3.2 Unification Results

Above we have proved 11 propositions with two means. Let's write in a simplified way the measures given in Group I (Section 2.5):

$$n_1 := \frac{7P_2 + 3R}{10} \qquad n_{15} := \frac{C + 5P_2}{6} \qquad n_{29} := \frac{A + 2P_3}{3}$$

$$n_2 := \frac{P_1 + 2P_3}{3} \qquad n_{16} := \frac{2P_4 + 3P_2}{5} \qquad n_{30} := \frac{4P_3 + C}{5}$$

$$n_3 := \frac{H + P_1}{2} \qquad n_{17} := \frac{A + 3P_2}{4} \qquad n_{31} := \frac{3P_3 + 2N_1}{5}$$

$$n_4 := \frac{3P_1 + 2G}{5} \qquad n_{18} := \frac{4P_2 + S}{5} \qquad n_{32} := \frac{7P_3 + 4N_2}{11}$$

$$n_5 := \frac{3P_1 + C}{4} \qquad n_{19} := \frac{11P_2 + 3R}{14} \qquad n_{33} := \frac{5P_3 + 2P_5}{7}$$

$$n_6 := \frac{5P_1 + 4P_4}{9} \qquad n_{20} := \frac{5P_2 + 2N_1}{7} \qquad n_{34} := \frac{8P_3 + 3R}{11}$$

$$n_7 := \frac{15P_1 + 8N_2}{23} \qquad n_{21} := \frac{2P_2 + G}{3} \qquad n_{35} := \frac{P_1 + 9G}{10}$$

$$n_8 := \frac{7P_1 + 4N_1}{11} \qquad n_{22} := \frac{11P_2 + 4N_2}{15} \qquad n_{36} := \frac{G + H}{2}$$

$$n_9 := \frac{9P_1 + 4P_5}{13} \qquad n_{23} := \frac{4P_4 + P_2}{5} \qquad n_{37} := \frac{3G + P_3}{4}$$

$$n_{10} := \frac{7P_3 + 2P_5}{9} \qquad n_{24} := \frac{2G + P_2}{3} \qquad n_{38} := \frac{5H + S}{6}$$

$$n_{11} := \frac{5P_1 + 2S}{7} \qquad n_{25} := \frac{2P_4 + P_3}{3} \qquad n_{39} := \frac{4H + P_5}{5}$$

$$n_{12} := \frac{A + 2P_1}{3} \qquad n_{26} := \frac{8P_4 + P_1}{9} \qquad n_{40} := \frac{2H + N_1}{3}$$

$$n_{13} := \frac{P_1 + 3H}{4} \qquad n_{27} := \frac{5P_2 + 6N}{11} \qquad n_{41} := \frac{A + 3H}{4}$$

$$n_{14} := \frac{H + P_2}{2} \qquad n_{28} := \frac{P_3 + G}{2} \qquad n_{42} := \frac{6N_1 + P_2}{7}$$

$$n_{43} := \frac{C + 6P_4}{7}$$

$$n_{44} := \frac{P_1 + 10N_1}{11}$$

$$n_{45} := \frac{A + 2P_4}{3}$$

$$n_{46} := \frac{4P_4 + S}{5}$$

$$n_{47} := \frac{3P_4 + P_5}{4}$$

$$n_{48} := \frac{10P_4 + 3R}{13}$$

$$n_{49} := \frac{S + 3G}{4}$$

$$n_{50} := \frac{C + 5G}{6}$$

$$n_{51} := \frac{G + 2N_2}{3}$$

$$n_{52} := \frac{P_2 + 21N}{22}$$

$$n_{53} := \frac{P_1 + 33N}{34}$$

$$n_{54} := \frac{4N_2 + P_4}{5}$$

$$n_{55} := \frac{9N + H}{10}$$

$$n_{56} := \frac{P_5 + 2G}{3}$$

$$n_{57} := \frac{P_3 + 15N}{16}$$

$$n_{58} := \frac{7G + 3R}{10}$$

$$n_{59} := \frac{H + S}{2}$$

$$n_{60} := \frac{C + 14N_1}{15}$$

$$n_{61} := \frac{C + 2P_4}{3}$$

$$n_{62} := \frac{P_4 + 14N_2}{15}$$

$$n_{63} := \frac{S + 8N_1}{9}$$

$$n_{64} := \frac{8P_4 + 7S}{15}$$

$$n_{65} := \frac{11C + 7P_2}{18}$$

$$n_{66} := \frac{22P_5 + 5P_2}{27}$$

$$n_{67} := \frac{P_1 + 68N_2}{69}$$

$$n_{68} := \frac{P_2 + 44N_2}{45}$$

$$n_{69} := \frac{P_5 + 5N_1}{6}$$

$$n_{70} := \frac{H + 20N_2}{21}$$

$$n_{71} := \frac{P_3 + 32N_2}{33}$$

$$n_{72} := \frac{7A + 2P_4}{9}$$

$$n_{73} := \frac{3R + 11N_1}{14}$$

$$n_{74} := \frac{C + 3G}{4}$$

$$n_{75} := \frac{S + 5N_1}{6}$$

$$n_{76} := \frac{7A + H}{8}$$

$$n_{77} := \frac{P_1 + 23A}{24}$$

$$n_{78} := \frac{P_2 + 15A}{16}$$

$$n_{79} := \frac{P_5 + 9N}{10}$$

$$n_{80} := \frac{P_3 + 11A}{12}$$

$$n_{81} := \frac{A + 3N}{4}$$

$$n_{82} := \frac{11N + R}{12}$$

$$n_{83} := \frac{9N_1 + C}{10}$$

$$n_{84} := \frac{P_4 + S}{2}$$

$$n_{85} := \frac{9R + 4P_4}{13}$$

$$n_{86} := \frac{2S + H}{3}$$

$$n_{87} := \frac{S + 4N_2}{5}$$

$$n_{88} := \frac{S + 3N}{4}$$

$$n_{89} := \frac{3S + P_3}{4}$$

$$n_{90} := \frac{P_5 + 2N_2}{3}$$

$$n_{91} := \frac{4S + P_2}{5}$$

$$n_{92} := \frac{6R + P_2}{7}$$

$$n_{93} := \frac{P_1 + 12P_5}{13}$$

$$n_{94} := \frac{P_1 + 9R}{10}$$

$$n_{95} := \frac{P_4 + 3P_5}{4}$$

$$n_{96} := \frac{9R + 2P_3}{11}$$

$$n_{97} := \frac{2N+R}{3} \qquad n_{101} := \frac{21R+P_3}{22} \qquad n_{105} := \frac{A+2S}{3}$$

$$n_{98} := \frac{5C+2P_4}{7} \qquad n_{102} := \frac{3H+5C}{8} \qquad n_{106} := \frac{4N+3C}{7}$$

$$n_{99} := \frac{5C+4N_2}{9} \qquad n_{103} := \frac{11S+P_3}{12} \qquad n_{107} := \frac{16N_2+11C}{27}$$

$$n_{100} := \frac{7C+P_1}{8} \qquad n_{104} := \frac{P_1+20S}{21}$$

Based on the above notations, the results appearing in Propositions 1–11 are summarized in a theorem below.

Theorem 4. *The following inequalities hold:*

(i) $\quad \mathbf{P_1} \le \mathbf{P_2} \le n_2 \le \begin{cases} \mathbf{P_3} \\ n_3 \le \begin{cases} \mathbf{H} \\ n_6 \le \begin{cases} \mathbf{P_4} \\ n_4 \le \begin{cases} \mathbf{G} \\ n_8 \le n_7 \le n_{12} \le \begin{cases} n_{11} \le n_5 \le \mathbf{N_1} \\ n_9 \le \mathbf{N}. \end{cases} \end{cases} \end{cases} \end{cases} \end{cases}$

(ii) $\quad \mathbf{P_3} \le n_{14}$

$\le \begin{cases} n_{13} \le \mathbf{H} \le n_{30} \le \mathbf{N_1} \\ n_{16} \le \begin{cases} \mathbf{P_4} \\ n_{21} \le \begin{cases} \mathbf{G} \\ n_{20} \le n_{22} \le n_{17} \le \begin{cases} n_{18} \\ n_{19} \end{cases} \le n_{15} \le \begin{cases} n_{30} \\ n_1 \\ n_{10} \end{cases} \le \mathbf{N_1}. \end{cases} \end{cases} \end{cases}$

(iii) $\quad \mathbf{H} \le n_{25} \le \begin{cases} n_{23} \le \begin{cases} n_{24} \\ n_{26} \end{cases} \le \begin{cases} \mathbf{G} \\ n_{27} \le \mathbf{N_1} \\ n_{26} \le \mathbf{P_4} \end{cases} \\ \begin{cases} n_{23} \\ n_{28} \end{cases} \le \begin{cases} \mathbf{G} \\ n_{31} \le n_{32} \le n_{29} \le n_{34} \le n_{33} \le \mathbf{N}. \end{cases} \end{cases}$

(iv) $\quad \mathbf{P_4} \le n_{36} \le \begin{cases} n_{37} \le n_{35} \le \mathbf{G} \\ n_{40} \le n_{38} \le n_{41} \le n_{39} \end{cases} \le n_{58} \le \mathbf{N_1}.$

(v) $\quad \mathbf{G} \le n_{46} \le n_{43} \le n_{48} \le n_{45} \le \begin{cases} n_{47} \\ n_{42} \le n_{44} \\ \begin{cases} n_{49} \\ n_{50} \end{cases} \le n_{58} \end{cases} n_{51} \le \mathbf{N_1}.$

$$(\text{vi}) \quad \mathbf{N_1} \le n_{54} \le \left\{ \begin{array}{l} n_{55} \le \left\{ \begin{array}{l} \left. \begin{array}{l} n_{57} \le n_{52} \le n_{53} \\ n_{60} \\ n_{63} \\ n_{64} \end{array} \right\} \le n_{62} \end{array} \right. \\ \left. \begin{array}{l} n_{56} \\ n_{64} \end{array} \right\} \le n_{61} \end{array} \right\} \le \mathbf{N_1}.$$

$$(\text{vii}) \quad \mathbf{N} \le n_{70} \le \left\{ \begin{array}{l} n_{71} \le n_{68} \le n_{67} \le \left\{ \begin{array}{l} \mathbf{N_2} \\ n_{65} \le \left\{ \begin{array}{l} \mathbf{P_5} \\ \mathbf{R} \end{array} \right. \end{array} \right. \\ \left. \begin{array}{l} n_{71} \le n_{68} \\ n_{69} \le n_{72} \end{array} \right\} \le \left\{ \begin{array}{l} \mathbf{A} \\ n_{66} \le \mathbf{P_5}. \end{array} \right. \end{array} \right.$$

$$(\text{viii}) \quad \mathbf{N_1} \le \left\{ \begin{array}{l} n_{74} \le \left\{ \begin{array}{l} n_{75} \\ n_{83} \end{array} \right\} \le n_{73} \\ \mathbf{N} \le n_{70} \\ n_{59} \le n_{84} \end{array} \right\} \le \mathbf{N_2}.$$

$$(\text{ix}) \quad \mathbf{N_2} \le n_{82} \le n_{81} \le n_{79} \le \left\{ \begin{array}{l} n_{76} \le \left\{ \begin{array}{l} n_{80} \le \left\{ \begin{array}{l} n_{78} \le \left\{ \begin{array}{l} n_{77} \\ n_{86} \end{array} \right. \\ n_{85} \end{array} \right. \\ n_{97} \end{array} \right. \\ \left. \begin{array}{l} n_{87} \\ n_{88} \end{array} \right\} \le n_{85} \end{array} \right\} \le \mathbf{A}.$$

$$(\text{x}) \quad \mathbf{A} \le \left\{ \begin{array}{l} \left. \begin{array}{l} n_{90} \\ n_{89} \end{array} \right\} \le n_{95} \le n_{93} \le \mathbf{P_5} \\ n_{90} \le n_{92} \\ n_{89} \le n_{96} \le \left\{ \begin{array}{l} n_{92} \\ n_{91} \end{array} \right\} \end{array} \right\} \le \left\{ \begin{array}{l} n_{93} \le \mathbf{P_5} \\ n_{94} \le \left\{ \begin{array}{l} \left. \begin{array}{l} n_{106} \\ n_{107} \end{array} \right\} \le \mathbf{R} \\ n_{102} \le n_{101} \le \left\{ \begin{array}{l} \mathbf{P_5} \\ n_{105} \le n_{103} \le \mathbf{R}. \end{array} \right. \end{array} \right. \end{array} \right.$$

$$(\text{xi}) \quad \mathbf{R} \le n_{104} \le \mathbf{S} \le \left\{ \begin{array}{l} n_{98} \\ n_{99} \end{array} \right\} \le n_{100} \le \mathbf{C}.$$

According to 11 parts given above, we shall write the inequalities in a theorem below in a continued form giving refinement over the inequalities (30).

Theorem 5. *The following inequalities give refinement over the inequalities (30) in terms of two means:*

$$\mathbf{P_1} \leq \mathbf{P_2} \leq n_2 \leq \mathbf{P_3} \leq n_{14} \leq n_{13} \leq \mathbf{H} \leq n_{25} \leq n_{23} \leq n_{26} \leq \mathbf{P_4}, \tag{37}$$

$$\mathbf{P_4} \leq n_{36} \leq n_{37} \leq n_{35} \leq \mathbf{G} \leq n_{46} \leq n_{43} \leq n_{48} \leq n_{45} \leq \left\{ \begin{array}{c} n_{47} \\ \left. \begin{array}{c} n_{42} \leq n_{44} \end{array} \right\} \\ \left. \begin{array}{c} n_{49} \\ n_{50} \end{array} \right\} \leq n_{58} \end{array} \right\} \leq n_{51} \right\}$$

$$\leq \mathbf{N_1}, \tag{38}$$

$$\mathbf{G} \leq n_{46} \leq n_{43} \leq n_{48} \leq n_{45} \leq \left\{ \begin{array}{c} n_{47} \\ \left. \begin{array}{c} n_{42} \leq n_{44} \end{array} \right\} \\ \left. \begin{array}{c} n_{49} \\ n_{50} \end{array} \right\} \leq n_{58} \end{array} \right\} \leq n_{51} \right\} \leq \mathbf{N_1}, \tag{39}$$

$$\mathbf{N_1} \leq n_{54} \leq \left\{ \begin{array}{c} n_{55} \leq \left\{ \begin{array}{c} n_{57} \leq n_{52} \leq n_{53} \\ \left. \begin{array}{c} n_{60} \\ n_{63} \\ n_{64} \end{array} \right\} \leq n_{62} \end{array} \right\} \\ \left. \begin{array}{c} n_{56} \\ n_{64} \end{array} \right\} \leq n_{61} \end{array} \right\} \leq \mathbf{N} \leq n_{70} \leq \left\{ \begin{array}{c} n_{71} \leq n_{68} \leq n_{67} \\ n_{73} \end{array} \right\}$$

$$\leq \mathbf{N_2}, \tag{40}$$

$$\mathbf{N_2} \leq n_{82} \leq n_{81} \leq n_{79} \leq \left\{ \begin{array}{c} n_{76} \leq \left\{ \begin{array}{c} n_{80} \leq \left\{ \begin{array}{c} n_{78} \leq \left\{ \begin{array}{c} n_{77} \\ n_{86} \end{array} \right. \\ n_{85} \end{array} \right. \\ n_{97} \end{array} \right. \\ \left. \begin{array}{c} n_{87} \\ n_{88} \end{array} \right\} \leq n_{85} \end{array} \right\} \leq \mathbf{A}, \tag{41}$$

$$\mathbf{A} \leq \left\{ \begin{array}{c} \left. \begin{array}{c} n_{90} \\ n_{89} \end{array} \right\} \leq n_{95} \leq n_{93} \leq \mathbf{P_5} \\ n_{90} \leq n_{92} \\ n_{89} \leq n_{96} \leq \left\{ \begin{array}{c} n_{92} \\ n_{91} \end{array} \right\} \end{array} \right\} \leq \left\{ \begin{array}{c} n_{93} \leq \mathbf{P_5} \\ n_{102} \leq n_{101} \leq \left\{ \begin{array}{c} \mathbf{P_5} \\ n_{105} \leq n_{103} \leq \mathbf{R} \end{array} \right. \\ \left. \begin{array}{c} n_{106} \\ n_{107} \end{array} \right\} \leq \mathbf{R}, \end{array} \right. \tag{42}$$

and

$$\mathbf{R} \le n_{104} \le \mathbf{S} \le \begin{Bmatrix} n_{98} \\ n_{99} \end{Bmatrix} \le n_{100} \le \mathbf{C}. \tag{43}$$

Remark 5. Theorem 5 improves Theorem 1, but with more means. The difference is that Theorem 1 is a mixture of two and three means at the same time, while Theorem 5 is only with two means inequalities. Results for the three means are given in the following section.

4. THREE MEANS INEQUALITIES

This section deals with the improvement of inequalities given in (30). This is based on the three means in each case as given in Group 2, Section 2.6. The results are divided in 9 propositions. The argument given in Note 1 shall be used frequently.

Proposition 12. *The following inequalities hold:*

$$\mathbf{P}_2 \le \begin{cases} \dfrac{23N_1 + 14P_1 - 14N_2}{23} \le \dfrac{18N + 11P_1 - 11A}{18} \le \mathbf{N_2} \\ C + 3P_3 - 3G \le P_5 + 3P_3 - 3P_4 \le \mathbf{P_5} \\ \dfrac{16P_5 + 9P_1 - 9C}{16} \le \mathbf{A}. \end{cases}$$

Proof. The proof is based on the following results.

169. For $\mathbf{P}_2 \le \frac{23\mathbf{N_1} + 14\mathbf{P_1} - 14\mathbf{N_2}}{23}$: We have to show that

$$\frac{1}{23}(23N_1 + 14P_1 - 23P_2 - 14N_2) \ge 0.$$

We can write $23N_1 + 14P_1 - 23P_2 - 14N_2 = bg_{169}(a/b)$, where

$$g_{169}(x) = \frac{u_{169}(x)}{4(x^2 + 1)(x^3 + 1)}$$

with

$$u_{169}(x) = \begin{pmatrix} 23x^6 + 46x^{11/2} + 13x^5 + 69x^4 + 46x^{7/2} \\ +158x^3 + 46x^{5/2} - 69x^2 - 13x + 46\sqrt{x} + 23 \end{pmatrix}$$
$$- 14\sqrt{2x + 2}(\sqrt{x} + 1)(x^2 + 1)(x^3 + 1).$$

After simplifications,

$$u_{169}(x) = (\sqrt{x} - 1)^2 \begin{pmatrix} 7x^5 + 52x^{9/2} + 83x^4 + 114x^{7/2} \\ +76x^3 + 76x^{5/2} + 76x^2 \\ 114x^{3/2} + 83x + 52\sqrt{x} + 7 \end{pmatrix}$$
$$+ \begin{pmatrix} 16x^6 + 8x^{11/2} + x^5 + 8x^{7/2} \\ +158x^3 + 8x^{5/2} + x + 8\sqrt{x} + 16 \end{pmatrix}$$
$$- 14\sqrt{2x + 2}(\sqrt{x} + 1)(x^2 + 1)(x^3 + 1).$$

Let's consider

$$v_{169}(x) = \left\{ (\sqrt{x} - 1)^2 \begin{pmatrix} 7x^5 + 52x^{9/2} + 83x^4 + 114x^{7/2} \\ +76x^3 + 76x^{5/2} + 76x^2 \\ 114x^{3/2} + 83x + 52\sqrt{x} + 7 \end{pmatrix} \right.$$
$$\left. + \begin{pmatrix} 16x^6 + 8x^{11/2} + x^5 + 8x^{7/2} \\ +158x^3 + 8x^{5/2} + x + 8\sqrt{x} + 16 \end{pmatrix} \right\}^2$$
$$- \left[14\sqrt{2x + 2}\left(\sqrt{x} + 1\right)(x^2 + 1)(x^3 + 1) \right]^2.$$

After simplifications,

$$v_{169}(x) = 2(\sqrt{x} - 1)^4 \times w_{169}(x),$$

where

$$w_{169}(x) = \begin{pmatrix} 137x^{10} + 1880x^{19/2} + 7432x^9 + 17016x^{17/2} \\ +26674x^8 + 26648x^{15/2} + 18122x^7 + 14600x^{13/2} \\ +28553x^6 + 56208x^{11/2} + 72164x^5 + 56208x^{9/2} \\ +28553x^4 + 14600x^{7/2} + 18122x^3 + 26648x^{5/2} \\ +26674x^2 + 17016x^{3/2} + 7432x + 1880\sqrt{x} + 137 \end{pmatrix}.$$

In view of Note 1, $v_{169}(x) \geq 0$ implies $u_{169}(x) \geq 0$, $\forall x > 0$, hence proving the required result.

170. For $P_2 \leq C + 3P_3 - 3G$: We have to show that $C + 3P_3 - 3G - P_2 \geq 0$. We can write $C + 3P_3 - 3G - P_2 = bg_{170}(a/b)$, where

$$g_{170}(x) = \frac{u_{170}(x)}{(x^2 + 1)(x + 1)(x - \sqrt{x} + 1)}$$

with

$$u_{170}(x) = \begin{pmatrix} x^5 - 4x^{9/2} + 6x^4 - 5x^{7/2} + 5x^3 - 6x^{5/2} \\ +5x^2 - 5x^{3/2} + 6x - 4\sqrt{x} + 1 \end{pmatrix}.$$

After simplifications,

$$u_{170}(x) = (\sqrt{x} - 1)^4 \big[(\sqrt{x} - 1)^2 (x - \sqrt{x} + 1)^2 + x^{3/2} \big].$$

Since, $u_{170}(x) \geq 0$, $\forall x > 0$, hence proving the required result.

171. For $P_2 \leq \frac{16P_5 + 9P_1 - 9C}{16}$: We have to show that $\frac{1}{16}(16P_5 + 9P_1 - 16P_2 - 9C) \geq 0$. We can write $16P_5 + 9P_1 - 16P_2 - 9C = bg_{171}(a/b)$, where

$$g_{171}(x) = \frac{u_{171}(x)}{(x^2 + 1)(x^3 + 1)(\sqrt{x} + 1)^2}$$

with

$$u_{171}(x) = \begin{pmatrix} 7x^7 + 25x^6 - 18x^{13/2} - 9x^5 + 4x^{11/2} \\ +41x^4 - 86x^{9/2} + 11x^3 + 72x^{7/2} \\ -9x^2 - 86x^{5/2} + 25x + 4x^{3/2} - 18\sqrt{x} + 7 \end{pmatrix}.$$

After simplifications,

$$u_{171}(x) = (\sqrt{x} - 1)^4 \begin{pmatrix} 7x^5 + 10x^{9/2} + 23x^4 + 64x^{7/2} \\ +142x^3 + 180x^{5/2} + 142x^2 \\ +64x^{3/2} + 23x + 10\sqrt{x} + 7 \end{pmatrix}.$$

Since, $u_{171}(x) \geq 0$, $\forall x > 0$, hence proving the required result.

172. For $C + 3P_3 - 3G \leq P_5 + 3P_3 - 3P_4$: We have to show that $P_5 - 3P_4 - C + 3G \geq 0$. We can write $P_5 - 3P_4 - C + 3G = bg_{172}(a/b)$, where

$$g_{172}(x) = \frac{x^{5/2} + 6x^{3/2} - 4x^2 - 4x + \sqrt{x}}{(x+1)(\sqrt{x} + 1)^2} = \frac{\sqrt{x}(\sqrt{x} - 1)^4}{(x+1)(\sqrt{x} + 1)^2}.$$

Since, $g_{172}(x) \geq 0$, $\forall x > 0$, hence proving the required result.

173. For $P_5 + 3P_3 - 3P_4 \leq P_5$: We have to show that $3(P_4 - P_3) \geq 0$. It is true by definition. See inequalities (30).

174. For $\frac{23N_1 + 14P_1 - 14N_2}{23} \leq \frac{18N + 11P_1 - 11N_2}{18}$: We have to show that

$$\frac{1}{414}(414N + P_1 - 414N_1 - N_2) \geq 0.$$

We can write $414N + P_1 - 414N_1 - N_2 = bg_{174}(a/b)$, where $g_{174}(x) = \frac{1}{4}u_{174}(x)/(x^3 + 1)$ with

$$u_{174}(x) = 138x^4 - 276x^{7/2} + 142x^3 + 142x - 276\sqrt{x} + 138$$
$$- \sqrt{2x+2}(\sqrt{x}+1)(x^3+1).$$

After simplifications,

$$u_{174}(x) = 138(x^3 + 1)(\sqrt{x} - 1)^2 + 4x(x^2 + 1)$$
$$- \sqrt{2x+2}(\sqrt{x}+1)(x^3+1).$$

Let's consider

$$v_{174}(x) = \left[138(x^3 + 1)(\sqrt{x} - 1)^2 + 4x(x^2 + 1)\right]^2$$
$$- \left[\sqrt{2x+2}(\sqrt{x}+1)(x^3+1)\right]^2.$$

After simplifications,

$$v_{174}(x) = 2(\sqrt{x} - 1)^2 \times w_{174}(x),$$

where

$$w_{174}(x) = \begin{pmatrix} 9521x^7 - 19048x^{13/2} + 10065x^6 - 16x^{11/2} \\ -16x^5 - 16x^{9/2} + 19578x^4 - 38112x^{7/2} \\ +19578x^3 - 16x^{5/2} - 16x^2 - 16x^{3/2} \\ +10065x - 19048\sqrt{x} + 9521 \end{pmatrix}.$$

Still, we need to show that $w_{174}(x) \geq 0$, $\forall x > 0$. Let's write

$$k_{174}(t) := w_{174}(t^2) = \begin{pmatrix} 9521t^{14} - 19048t^{13} + 10065t^{12} - 16t^{11} \\ -16t^{10} - 16t^9 + 19578t^8 - 38112t^7 \\ +19578t^6 - 16t^5 - 16t^4 - 16t^3 \\ +10065t^2 - 19048t + 9521 \end{pmatrix}.$$

The polynomial equation $k_{174}(t) = 0$ is of 14th degree. It admits 14 solutions. All these 14 solutions are complex (not written here), i.e., it doesn't have any real solution. Since we are working with $t > 0$, then there are two possibilities, one whole the graph of $k_{174}(t)$ is either in first quadrant or in 4th quadrant. Since $k_{174}(1) = 2024$, this means that $k_{174}(t) > 0$, for all $t > 0$. Thus we have $w_{174}(x) > 0$, for all $x > 0$ giving $v_{174}(x) \geq 0$, for all $x > 0$. Finally, proving the required result.

175. For $\frac{18N+11P_1-11N_2}{18} \leq \mathbf{N_2}$**:** We have to show that $\frac{1}{18}(29N_2 - 18N - 11P_1 t) \geq 0$. We can write $29N_2 - 18N - 11P_1 = bg_{175}(a/b)$, where $g_{175}(x) = \frac{1}{4}u_{175}(x)/(x^3 + 1)$ with

$$u_{175}(x) = 29\sqrt{2x+2}(\sqrt{x}+1)(x^3+1)$$
$$- 4(6x^4 + 6x^{7/2} + 17x^3 + 17x + 6\sqrt{x} + 6).$$

Let us consider

$$v_{175}(x) = \left[29\sqrt{2x+2}(\sqrt{x}+1)(x^3+1)\right]^2$$
$$- \left[4(6x^4 + 6x^{7/2} + 17x^3 + 17x + 6\sqrt{x} + 6)\right]^2.$$

After simplifications,

$$v_{175}(x) = (\sqrt{x}-1)^2 \begin{pmatrix} 1106x^7 + 4424x^{13/2} + 7266x^6 \\ +10208x^{11/2} + 10208x^5 + 10208x^{9/2} \\ +10308x^4 + 12720x^{7/2} + 10308x^3 \\ +10208x^{5/2} + 10208x^2 + 10208x^{3/2} \\ +7266x + 4424\sqrt{x} + 1106 \end{pmatrix}.$$

In view of Note 1, $v_{175}(x) \geq 0$ implies $u_{175}(x) \geq 0$, $\forall x > 0$, hence proving the required result.

176. For $\frac{16P_5+9P_1-9C}{16} \leq \mathbf{A}$**:** We have to show that $\frac{1}{16}(16A - 16P_5 - 9P_1 + 9C) \geq 0$. We can write $16A - 16P_5 - 9P_1 + 9C = bg_{176}(a/b)$, where

$$g_{176}(x) = \frac{u_{176}(x)}{(\sqrt{x}+1)^2(x^3+1)}$$

with

$$u_{176}(x) = \begin{pmatrix} x^5 - 25x^4 + 34x^{9/2} - 8x^3 - 20x^{7/2} - 8x^2 \\ +36x^{5/2} - 25x - 20x^{3/2} + 34\sqrt{x} + 1 \end{pmatrix}.$$

After simplifications,

$$u_{176}(x) = (\sqrt{x}-1)^2 \begin{pmatrix} x^4 + 36x^{7/2} + 46x^3 + 36x^{5/2} \\ +18x^2 + 36x^{3/2} + 46x + 36\sqrt{x} + 1 \end{pmatrix}.$$

Since, $u_{176}(x) \geq 0$, $\forall x > 0$, hence proving the required result.

Combining Results 169 to 176 we get the proof of Proposition 12.   □

Proposition 13. *The following inequalities hold:*

$$P_3 \leq N + 8R - 8S \leq \frac{3N_1 + 2P_2 - 2N_2}{3} \leq \frac{13A + 6P_1 - 6P_5}{13} \leq N_1.$$

Proof. Proof is based on the following results.

177. For $P_3 \leq N + 8R - 8S$: We have to show that $N + 8R - 8S - P_3 \geq 0$. We can write $N + 8R - 8S - P_3 = bg_{177}(a/b)$, where

$$g_{177}(x) = \frac{u_{177}(x)}{(x+1)\left(x - \sqrt{x} + 1\right)}$$

with

$$u_{177}(x) = 17x^3 - 16x^{5/2} + 31x^2 - 16x^{3/2} + 31x - 16\sqrt{x} + 17$$
$$- 12\sqrt{2x^2 + 2}(x+1)(x - \sqrt{x} + 1).$$

After simplifications,

$$u_{177}(x) = 8\left(x^2 + x + 1\right)(\sqrt{x} - 1)^2 + 9x^3 + 15x^2 + 15x + 9$$
$$- 12\sqrt{2x^2 + 2}(x+1)(x - \sqrt{x} + 1).$$

Let us consider

$$v_{177}(x) = \left[8\left(x^2 + x + 1\right)(\sqrt{x} - 1)^2 + 9x^3 + 15x^2 + 15x + 9\right]^2$$
$$- \left[12\sqrt{2x^2 + 2}(x+1)(x - \sqrt{x} + 1)\right]^2.$$

This gives

$$v_{177}(x) = (\sqrt{x} - 1)^4 \left(\begin{array}{c} x^3 + 3x^{5/2} + 31\sqrt{x}(x+1)(\sqrt{x} - 1)^2 \\ +38x^{3/2} + x^2 + x + 3\sqrt{x} + 1 \end{array} \right).$$

In view of Note 1, $v_{177}(x) \geq 0$ implies $u_{177}(x) \geq 0$, $\forall x > 0$, hence proving the required result.

178. For $N + 8R - 8S \leq \frac{3N_1 + 2P_2 - 2N_2}{3}$: We have to show that

$$\frac{1}{3}(3N_1 + 2P_2 - 2N_2 - 3N - 24R + 24S) \geq 0.$$

We can write $3N_1 + 2P_2 - 2N_2 - 3N - 24R + 24S = bg_{178}(a/b)$, where

$$g_{178}(x) = \frac{u_{178}(x)}{4\left(x^2 + 1\right)(x+1)}$$

with

$$u_{178}(x) = 2(x+1)(x^2+1)\left[24\sqrt{2x^2+2} - (\sqrt{x}+1)\sqrt{2x^2+2}\right]$$
$$- \left(\begin{array}{l}65x^4 - 2x^{7/2} + 58x^3 - 2x^{5/2}\\ +114x^2 - 2x^{3/2} + 58x - 2\sqrt{x} + 65\end{array}\right).$$

After simplifications,

$$u_{178}(x) = 2(x+1)(x^2+1)[24\sqrt{2x^2+2} - (\sqrt{x}+1)\sqrt{2x+2}]$$
$$- \left(\begin{array}{l}(x^3+x^2+x+1)(\sqrt{x}-1)^2\\ +64x^4 + 56x^3 + 112x^2 + 56x + 64\end{array}\right).$$

Before proceeding further we need to show that the expression $24\sqrt{2x^2+2} - (\sqrt{x}+1)\sqrt{2x+2}$ is nonnegative for all $x > 0$. It is true in view of Result 165 given in Group 1. Now we shall show that $u_{178}(x) \geq 0$, $\forall x > 0$. Let's consider

$$v_{178}(x) = \{2(x+1)(x^2+1)[24\sqrt{2x^2+2} - (\sqrt{x}+1)\sqrt{2x+2}]\}^2$$
$$- \left(\begin{array}{l}(x^3+x^2+x+1)(\sqrt{x}-1)^2\\ +64x^4 + 56x^3 + 112x^2 + 56x + 64\end{array}\right)^2.$$

After simplifications,

$$h_{178}(x) = \left(\begin{array}{l}391x^8 + 276x^{15/2} + 1704x^7 + 540x^{13/2} + 304x^6\\ 1028x^{11/2} + 6968x^5 + 1077x^{9/2} + 1077x^{7/2}\\ +215x^{7/2}(\sqrt{x}-1)^2 + 6968x^3 + 1028x^{5/2}\\ +304x^2 + 540x^{3/2} + 1704x + 276\sqrt{x} + 391\end{array}\right)$$
$$- 192\sqrt{2x+2}\sqrt{2x^2+2}(\sqrt{x}+1)(x+1)^2(x^2+1)^2.$$

Still, we are unable to decide the nonnegativity of $u_{178}(x)$. Let's consider

$$v_{178}(x) = \left(\begin{array}{l}391x^8 + 276x^{15/2} + 1704x^7 + 540x^{13/2} + 304x^6\\ 1028x^{11/2} + 6968x^5 + 1077x^{9/2} + 1077x^{7/2}\\ +215x^{7/2}(\sqrt{x}-1)^2 + 6968x^3 + 1028x^{5/2}\\ +304x^2 + 540x^{3/2} + 1704x + 276\sqrt{x} + 391\end{array}\right)^2$$
$$- \left[192\sqrt{2x+2}\sqrt{2x^2+2}(\sqrt{x}+1)(x+1)^2(x^2+1)^2\right]^2.$$

After simplifications,

$$h_{178}(x) = (\sqrt{x}-1)^4 \times w_{178}(x),$$

where

$$w_{178}(x) = \begin{pmatrix} 5425 + 261898x + 17852305x^4 + 11383048x^{7/2} \\ +10560340x^3 + 7248200x^{5/2} + 3891571x^2 \\ -57380\sqrt{x} + 1301900x^{3/2} + 7248200x^{23/2} \\ +10560340x^{11} + 11383048x^{21/2} + 17852305x^{10} \\ +32261396x^{19/2} + 43811318x^9 + 35617668x^{17/2} \\ +32891851x^8 + 50557872x^{(15/2)} + 67124056x^7 \\ +50557872x^{13/2} + 32891851x^6 + 35617668x^{11/2} \\ +43811318x^5 + 32261396x^{9/2} + 5425x^{14} - 57380x^{27/2} \\ +261898x^{13} + 1301900x^{25/2} + 3891571x^{12} \end{pmatrix}.$$

For simplicity, let's write

$$k_{178}(t) := w_{178}(t^2) = \begin{pmatrix} 5425t^{28} - 57380t^{27} + 261898t^{26} + 1301900t^{25} \\ +3891571t^{24} + 7248200t^{23} + 10560340t^{22} \\ +11383048t^{21} + 17852305t^{20} + 32261396t^{19} \\ +43811318t^{18} + 35617668t^{17} + 32891851t^{16} \\ +50557872t^{15} + 67124056t^{14} + 50557872t^{13} \\ +32891851t^{12} + 35617668t^{11} + 43811318t^{10} \\ +32261396t^9 + 17852305t^8 + 11383048t^7 \\ +10560340t^6 + 7248200t^5 + 3891571t^4 \\ +1301900t^3 + 261898t^2 - 57380t + 5425 \end{pmatrix}.$$

The polynomial equation $k_{178}(t) = 0$ is of 28th degree. It admits 28 solutions. All these 28 solutions are complex (not written here), i.e., it doesn't have any real solution. Since $t > 0$, then there are two possibilities, one when the graph of $k_{178}(t)$ is either in first quadrant or in 4th quadrant. Since $k_{178}(1) = 562298880$, this means that $k_{178}(t) > 0$, for all $t > 0$. Thus we have $w_{178}(x) > 0$, for all $x > 0$. Consequently proving that $h_{178}(x) \geq 0$, for all $x > 0$. In view of Note 1, $w_{178}(x) \geq 0$ implies $v_{178}(x) \geq 0$ implies $v_{178}(x) \geq 0$, $\forall x > 0$, hence proving the required result.

179. For $\frac{3N_1 + 2P_2 - 2N_2}{3} \leq \frac{13A + 6P_1 - 6P_5}{13}$: We have to show that

$$\frac{1}{39}(39A + 18P_1 - 18P_5 - 39N_1 - 26P_2 + 26N_2) \geq 0.$$

We can write $39A + 18P_1 - 18P_5 - 39N_1 - 26P_2 + 26N_2 = bg_{179}(x)(a/b)$, where

$$g_{179}(x) = \frac{u_{179}(x)}{4(x^3 + 1)(x^2 + 1)(\sqrt{x} + 1)^2}$$

with

$$u_{179}(x) = 26\sqrt{2x+2}\left(x^3+1\right)\left(x^2+1\right)\left(\sqrt{x}+1\right)^3$$
$$- \begin{pmatrix} 33x^7 + 254x^6 + 64x^{11/2} + 202x^5 + 208x^{9/2} \\ +71x^4 + 144x^3(\sqrt{x}-1)^2 + 71x^3 + 208x^{5/2} \\ +202x^2 + 64x^{3/2} + 254x + 33 \end{pmatrix}.$$

Let us consider

$$v_{179}(x) = \left[26\sqrt{2x+2}\left(x^3+1\right)\left(x^2+1\right)\left(\sqrt{x}+1\right)^3\right]^2$$
$$- \begin{pmatrix} 33x^7 + 254x^6 + 64x^{11/2} + 202x^5 + 208x^{9/2} \\ +71x^4 + 144x^3(\sqrt{x}-1)^2 + 71x^3 + 208x^{5/2} \\ +202x^2 + 64x^{3/2} + 254x + 33 \end{pmatrix}^2.$$

After simplifications,

$$v_{179}(x) = (\sqrt{x}-1)^4 \times w_{179}(x),$$

where

$$w_{179}(x) = \begin{pmatrix} 263x^{12} + 9164x^{23/2} + 39946x^{11} + 136780x^{21/2} \\ +309253x^{10} + 572088x^{19/2} + 886706x^9 \\ +1196656x^{17/2} + 1381737x^8 + 1491492x^{15/2} \\ +1511076x^7 + 1469276x^{13/2} + 1410006x^6 \\ +1469276x^{11/2} + 1511076x^5 + 1491492x^{9/2} \\ +1381737x^4 + 1196656x^{7/2} + 886706x^3 \\ +572088x^{5/2} + 309253x^2 + 136780x^{3/2} \\ +39946x + 9164\sqrt{x} + 263 \end{pmatrix}.$$

In view of Note 1, $v_{179}(x) \geq 0$ implies $u_{179}(x) \geq 0$, $\forall x > 0$, hence proving the required result.

180. For $\frac{13A+6P_1-6P_5}{13} \leq N_1$: We have to show that $\frac{1}{13}(13N_1 - 13A - 6P_1 + 6P_5) \geq 0$. We can write $13N_1 - 13A - 6P_1 + 6P_5 = bg_{180}(a/b)$, where

$$g_{180}(x) = \frac{u_{180}(x)}{4(\sqrt{x}+1)^2(x^3+1)}$$

with

$$u_{180}(x) = \begin{pmatrix} 11x^5 + 50x^4 - 48x^{7/2} - 13x^3 \\ -13x^2 - 48x^{3/2} + 50x + 11 \end{pmatrix}.$$

After simplifications,

$$u_{180}(x) = (\sqrt{x}-1)^2 \left(\begin{matrix} 11x^4 + 22x^{7/2} + 83x^3 + 96x^{5/2} \\ +96x^2 + 96x^{3/2} + 83x + 22\sqrt{x} + 11 \end{matrix} \right).$$

Since, $u_{180}(x) \geq 0$, $\forall x > 0$, hence proving the required result.

Combining Results 177–180, we get the proof of Proposition 13. $\square$

Proposition 14. *The following inequalities hold:*

$$H \leq \begin{cases} \frac{11N+5P_3-5R}{11} \leq \begin{cases} P_3 + 3S - 3R \leq N \\ \frac{3P_5+2H-2S}{3} \leq \begin{cases} \mathbf{P_5} \\ \mathbf{R} \end{cases} \end{cases} \\ \frac{2N_1+P_3-A}{2} \leq \begin{cases} \mathbf{G} \\ P_3 + 3S - 3R \leq N \end{cases} \\ \left. \begin{matrix} \frac{4P_5+5G-5S}{4} \\ \frac{11P_5+30N_2-30R}{11} \end{matrix} \right\} \leq \begin{cases} \mathbf{N_1} \\ S + 2P_5 - 2C \leq \mathbf{P_5} \end{cases} \\ \frac{4P_2+C-P_1}{4} \leq \mathbf{N_1} \\ A + 4P_4 - 4G \leq \begin{cases} \mathbf{A} \\ S + 2P_5 - 2C \leq \mathbf{P_5}. \end{cases} \end{cases}$$

Proof. The proof is based on the following results.

181. For $H \leq \frac{11N+5P_3-5R}{11}$: We have to show that $\frac{1}{11}(11N + 5P_3 - 5R - 11H) \geq 0$. We can write $11N + 5P_3 - 5R - 11H = bg_{181}(a/b)$, where

$$g_{181}(x) = \frac{u_{181}(x)}{3(x+1)\left(x-\sqrt{x}+1\right)}$$

with

$$u_{181}(x) = 10x^{5/2} + x^3 + 76x^{3/2} - 49x^2 - 49x + 10\sqrt{x} + 1$$
$$= (x + 14\sqrt{x} + 1)(\sqrt{x} - 1)^4.$$

Since, $u_{181}(x) \geq 0$, $\forall x > 0$, hence proving the required result.

182. For $H \leq \frac{2N_1+P_3-A}{2}$: We have to show that $\frac{1}{2}(2N_1 + P_3 - A - 2H) \geq 0$. We can write $2N_1 + P_3 - A - 2H = bg_{182}(a/b)$, where

$$g_{182}(x) = \frac{u_{182}(x)}{(x+1)(x-\sqrt{x}+1)}$$

with

$$u_{182}(x) = x^{5/2} + 6x^{3/2} - 4x^2 + \sqrt{x} - 4x = \sqrt{x}(\sqrt{x} - 1)^4.$$

Since, $u_{182}(x) \geq 0$, $\forall x > 0$, hence proving the required result.

183. For H $\leq \frac{4P_5 + 5G - 5S}{4}$: We have to show that $\frac{1}{4}(4P_5 + 5G - 5S - 4H) \geq 0$. We can write $4P_5 + 5G - 5S - 4H = bg_{183}(a/b)$, where

$$g_{183}(x) = \frac{u_{183}(x)}{2(x+1)(\sqrt{x}+1)^2}$$

with

$$u_{183}(x) = 8x^3 + 10x^{5/2} + 22x^2 + 6x(\sqrt{x} - 1)^2 + 22x + 10\sqrt{x} + 8$$
$$- 5\sqrt{2x^2 + 2}(\sqrt{x} + 1)^2(x + 1).$$

Let us consider

$$v_{183}(x) = \left[8x^3 + 10x^{5/2} + 22x^2 + 6x(\sqrt{x} - 1)^2 + 22x + 10\sqrt{x} + 8\right]^2$$
$$- \left[5\sqrt{2x^2 + 2}(\sqrt{x} + 1)^2(x + 1)\right]^2.$$

After simplifications,

$$v_{183}(x) = (\sqrt{x} - 1)^4 \begin{pmatrix} 14x^4 + 16x^{7/2} + 128x^3 + 240x^{5/2} \\ +484x^2 + 240x^{3/2} + 128x + 16\sqrt{x} + 14 \end{pmatrix}.$$

In view of Note 1, $v_{183}(x) \geq 0$ implies $u_{183}(x) \geq 0$, $\forall x > 0$, hence proving the required result.

184. For H $\leq \frac{11P_5 + 30N_2 - 30R}{11}$: We have to show that $\frac{1}{11}(11P_5 + 30N_2 - 30R - 11H) \geq 0$. We can write $11P_5 + 30N_2 - 30R - 11H = bg_{184}(a/b)$, where

$$g_{184}(x) = \frac{u_{184}(x)}{2(x+1)(\sqrt{x}+1)^2}$$

with

$$u_{184}(x) = 15\sqrt{2x + 2}(x + 1)(\sqrt{x} + 1)^3$$
$$- 2(9x^3 + 40x^{5/2} + 29x^2 + 84x^{3/2} + 29x + 40\sqrt{x} + 9).$$

Let us consider

$$v_{184}(x) = \left[15\sqrt{2x + 2}(x + 1)(\sqrt{x} + 1)^3\right]^2$$

$$- 4\left(9x^3 + 40x^{5/2} + 29x^2 + 84x^{3/2} + 29x + 40\sqrt{x} + 9\right)^2.$$

After simplifications,

$$v_{184}(x) = 2(\sqrt{x} - 1)^4 \left(\begin{array}{l} 63x^4 + 162x^{7/2} + 76x^3 + 470x^{5/2} \\ +18x^2 + 470x^{3/2} + 76x + 1624\sqrt{x} + 63 \end{array} \right).$$

In view of Note 1, $v_{184}(x) \geq 0$ implies $u_{184}(x) \geq 0$, $\forall x > 0$, hence proving the required result.

185. For $H \leq A + 4P_4 - 4G$: We have to show that $A + 4P_4 - 4G - H \geq 0$. We can write $A + 4P_4 - 4G - H = bg_{185}(a/b)$, where

$$g_{185}(x) = \frac{u_{185}(x)}{2(x+1)(\sqrt{x}+1)^2}$$

with

$$u_{185}(x) = x^3 - 6x^{5/2} + 15x^2 - 20x^{3/2} + 15x - 6\sqrt{x} + 1 = (\sqrt{x} - 1)^6.$$

Since $u_{185}(x) \geq 0$, $\forall x > 0$, hence proving the required result.

186. For $H \leq \frac{4P_2 + C - P_1}{4}$: We have to show that $\frac{1}{4}(4P_2 + C - P_1 - 4H) \geq 0$. We can write $4P_2 + C - P_1 - 4H = bg_{186}(a/b)$, where

$$g_{186}(x) = \frac{(x-1)^6}{(x^3+1)(x^2+1)}.$$

Since, $g_{186}(x) \geq 0$, $\forall x > 0$, hence proving the required result.

187. For $\frac{11N + 5P_3 - 5R}{11} \leq P_3 + 3S - 3R$: We have to show that

$$\frac{1}{11}(6P_3 + 33S - 28R - 11N) \geq 0.$$

We can write $6P_3 + 33S - 28R - 11N = bg_{187}(a/b)$, where

$$g_{187}(x) = \frac{u_{187}(x)}{6(x+1)(x - \sqrt{x} + 1)}$$

with

$$u_{187}(x) = 99\sqrt{2x^2 + 2}(x^3 + 1)$$
$$- 2(67x^3 - 56x^{5/2} + 116x^2 - 56x^{3/2} + 116x - 56\sqrt{x} + 67).$$

After simplifications,

$$u_{187}(x) = 99\sqrt{2x^2 + 2}(x^3 + 1) - 2\left(\frac{28\left(x^2 + x + 1\right)\left(\sqrt{x} - 1\right)^2}{+39x^3 + 60x^2 + 60x + 39}\right).$$

Let us consider

$$u_{187}(x) = \left[99\sqrt{2x^2 + 2}(x^3 + 1)\right]^2$$
$$- 4\left(\frac{28(x^2 + x + 1)(\sqrt{x} - 1)^2}{+39x^3 + 60x^2 + 60x + 39}\right)^2.$$

This gives

$$v_{187}(x) = 2(\sqrt{x} - 1)^4 \times w_{187}(x),$$

where

$$w_{187}(x) = \left(\frac{823x^4 - 1302x^{7/2} + 1499x^3 - 714x^{5/2}}{-216x^2 - 714x^{3/2} + 1499x - 1302\sqrt{x} + 823}\right).$$

Still, we need to show that $w_{187}(x)$ is nonnegative. For simplicity, let's write

$$k_{187}(t) := w_{187}\left(t^2\right) = \left(\frac{823t^8 - 1302t^7 + 1499t^6 - 714t^5}{-216t^4 - 714t^3 + 1499t^2 - 1302t + 823}\right).$$

The polynomial equation $k_{187}(t) = 0$ is of 8th degree. It admits 8 solutions. All these 8 solutions are complex (not written here), i.e., it doesn't have any real solution. Since, $t > 0$, then there are two possibilities, one when the graph of $k_{187}(t)$ is either in first quadrant or in 4th quadrant. Since $k_{187}(1) = 396$, this means that $k_{187}(t) > 0$, for all $t > 0$. Thus, we have $w_{187}(x) > 0$, for all $x > 0$. Consequently proving that $v_{187}(x) \geq 0$, for all $x > 0$. In view of Note 1, $v_{187}(x) \geq 0$ implies $u_{187}(x) \geq 0$, $\forall x > 0$, hence proving the required result.

188. For $P_3 + 3S - 3R \leq N$: We have to show that $N - P_3 - 3S + 3R \geq 0$. We can write $N - P_3 - 3S + 3R = bg_{188}(a/b)$, where

$$g_{188}(x) = \frac{u_{188}(x)}{6(x + 1)(x - \sqrt{x} + 1)}$$

with

$$u_{188}(x) = 14x^3 - 12x^{5/2} + 22x^2 - 12x^{3/2} + 22x - 12\sqrt{x} + 14$$

$$- 9\sqrt{2x^2 + 2}(x+1)(x - \sqrt{x} + 1).$$

After simplifications,

$$u_{188}(x) = 8x^3 + 6(x^2 + x + 1)(\sqrt{x} - 1)^2 + 10x^2 + 10x + 8$$
$$- 9\sqrt{2x^2 + 2}(x+1)(x - \sqrt{x} + 1).$$

Let us consider

$$v_{188}(x) = \left[8x^3 + 6(x^2 + x + 1)(\sqrt{x} - 1)^2 + 10x^2 + 10x + 8\right]^2$$
$$- \left[9\sqrt{2x^2 + 2}(x+1)\left(x - \sqrt{x} + 1\right)\right]^2.$$

This gives

$$v_{188}(x) = 2(x-1)^2 \begin{pmatrix} 14x^4 + \left(3x^3 + 17x^{3/2} + 3\right)\left(\sqrt{x} - 1\right)^2 \\ +6x^3 + 25x^{5/2} + 25x^{3/2} + 6x + 14 \end{pmatrix}.$$

In view of Note 1, $v_{188}(x) \geq 0$ implies $u_{188}(x) \geq 0$, $\forall x > 0$, hence proving the required result.

189. For $P_3 + 3S - 3R \leq \frac{3P_5 + 2H - 2S}{3}$: We have to show that

$$\frac{1}{3}(3P_5 + 2H - 11S - 3P_3 + 9R) \geq 0.$$

We can write $3P_5 + 2H - 11S - 3P_3 + 9R = bg_{189}(a/b)$, where

$$g_{189}(x) = \frac{u_{189}(x)}{2(x+1)(\sqrt{x} + 1)^2(x - \sqrt{x} + 1)}$$

with

$$u_{189}(x) = \begin{pmatrix} 18x^4 + 6x^{7/2} + 38x^3 + 2x^{5/2} \\ +48x^2 + 2x^{3/2} + 38x + 6\sqrt{x} + 18 \end{pmatrix}$$
$$- 11\sqrt{2x^2 + 2}(x+1)(x - \sqrt{x} + 1)(\sqrt{x} + 1)^2.$$

Let us consider

$$v_{189}(x) = \begin{pmatrix} 18x^4 + 6x^{7/2} + 38x^3 + 2x^{5/2} \\ +48x^2 + 2x^{3/2} + 38x + 6\sqrt{x} + 18 \end{pmatrix}^2$$
$$- \left[11\sqrt{2x^2 + 2}(x+1)(x - \sqrt{x} + 1)(\sqrt{x} + 1)^2\right]^2.$$

After simplifications,

$$v_{189}(x) = 2(\sqrt{x} - 1)^2 \times w_{189}(x),$$

where

$$w_{189}(x) = (\sqrt{x} - 1)^2 \begin{pmatrix} 26x^6 + 11x^5 + 2x^4 \\ +42x^3 + 2x^2 + 11x + 26 \end{pmatrix}$$
$$+ \begin{pmatrix} 15x^7 + 157x^6 + 379x^5 + 505x^4 \\ +505x^3 + 379x^2 + 157x + 15 \end{pmatrix}.$$

In view of Note 1, $v_{189}(x) \geq 0$ implies $u_{189}(x) \geq 0$, $\forall x > 0$, hence proving the required result.

190. For $\frac{3P_5 + 2H - 2S}{3} \leq R$: We have to show that $\frac{1}{3}(3R - 3P_5 - 2H + 2S)$. We can write $3R - 3P_5 - 2H + 2S = bg_{190}(a/b)$, where

$$g_{190}(x) = \frac{u_{190}(x)}{(x + 1)(\sqrt{x} + 1)^2}$$

with

$$u_{190}(x) = \sqrt{2x^2 + 2}(x + 1)(\sqrt{x} + 1)^2$$
$$- (x^3 - 4x^{5/2} + 9x^2 + 4x^{3/2} + 9x - 4\sqrt{x} + 1).$$

After simplifications,

$$u_{190}(x) = \sqrt{2x^2 + 2}(x + 1)(\sqrt{x} + 1)^2$$
$$- [x^2(\sqrt{x} - 2)^2 + 5x^2 + 4x^{3/2} + 5x + (2\sqrt{x} - 1)^2].$$

Let us consider

$$v_{190}(x) = [\sqrt{2x^2 + 2}(x + 1)(\sqrt{x} + 1)^2]^2$$
$$- [x^2(\sqrt{x} - 2)^2 + 5x^2 + 4x^{3/2} + 5x + (2\sqrt{x} - 1)^2]^2.$$

This gives

$$v_{190}(x) = (\sqrt{x} - 1)^2 \begin{pmatrix} x^5 + 18x^{9/2} + 17x^4 + 104x^{7/2} \\ +154x^3 + 244x^{5/2} + 154x^2 \\ +104x^{3/2} + 17x + 18\sqrt{x} + 1 \end{pmatrix}.$$

In view of Note 1, $v_{190}(x) \geq 0$ implies $u_{190}(x) \geq 0$, $\forall x > 0$, hence proving the required result.

191. For $\frac{3P_5 + 2H - 2S}{3} \leq P_5$**:** We have to show that $\frac{2}{3}(S - H) \geq 0$. It is true by definition. See inequalities (30).

192. For $\frac{2N_1 + P_3 - A}{2} \leq G$**:** We have to show that $\frac{1}{2}(2G - 2N_1 - P_3 + A) \geq 0$. We can write $2G - 2N_1 - P_3 + A = bg_{192}(a/b)$, where

$$g_{192}(x) = \frac{x^{3/2} + \sqrt{x} - 2x}{x - \sqrt{x} + 1} = \frac{\sqrt{x}(\sqrt{x} - 1)^2}{x - \sqrt{x} + 1}.$$

Since $g_{192}(x) \geq 0$, $\forall x > 0$, hence proving the required result.

193. For $\frac{2N_1 + P_3 - A}{2} \leq P_3 + 3S - 3R$ **:** We have to show that

$$\frac{1}{2}(P_3 + 6S - 6R - 2N_1 + A) \geq 0.$$

We can write $P_3 + 6S - 6R - 2N_1 + A = bg_{193}(a/b)$, where

$$g_{193}(x) = \frac{u_{193}(x)}{(x + 1)(x - \sqrt{x} + 1)}$$

with

$$u_{193}(x) = 3\sqrt{2x^2 + 2}(x - \sqrt{x} + 1)(x + 1) \\ - \left(4x^3 - 3x^{5/2} + 6x^2 - 2x^{3/2} + 6x - 3\sqrt{x} + 4\right).$$

After simplifications,

$$u_{193}(x) = 3\sqrt{2x^2 + 2}(x + 1)(x - \sqrt{x} + 1) \\ - \left(\begin{array}{c} 2x^3 + 2(x^2 + x + 1)(\sqrt{x} - 1)^2 \\ + x^{5/2} + 2x^2 + 2x^{3/2} + 2x + \sqrt{x}s + 2 \end{array} \right).$$

Let's consider

$$u_{193}(x) = \left[3\sqrt{2x^2 + 2}(x + 1)(x - \sqrt{x} + 1)\right]^2 \\ - \left(\begin{array}{c} 2x^3 + 2(x^2 + x + 1)(\sqrt{x} - 1)^2 \\ + x^{5/2} + 2x^2 + 2x^{3/2} + 2x + \sqrt{x}s + 2 \end{array} \right)^2.$$

This gives

$$v_{193}(x) = (\sqrt{x} - 1)^6 (2x^3 + 3x^2 + 2x^{3/2} + 3x + 2).$$

In view of Note 1, $v_{193}(x) \geq 0$ implies $u_{193}(x) \geq 0$, $\forall x > 0$, hence proving the required result.

194. For $\frac{4P_5+5G-5S}{4} \leq S+2P_5-2C$**:** We have to show that $\frac{1}{4}(9S+4P_5-8C-5G) \geq 0$. We can write $9S+4P_5-8C-5G = bg_{194}(a/b)$, where

$$g_{194}(x) = \frac{u_{194}(x)}{2(x+1)(\sqrt{x}+1)^2}$$

with

$$u_{194}(x) = 9\sqrt{2x^2+2}(x+1)(\sqrt{x}+1)^2$$
$$- \left(8x^3 + 42x^{5/2} + 12x^2 + 20x^{3/2} + 12x + 42\sqrt{x} + 8\right).$$

Let's consider

$$v_{194}(x) = \left[9\sqrt{2x^2+2}(x+1)(\sqrt{x}+1)^2\right]^2$$
$$- \left(8x^3 + 42x^{5/2} + 12x^2 + 20x^{3/2} + 12x + 42\sqrt{x} + 8\right)^2.$$

After simplifications,

$$v_{194}(x) = 2(\sqrt{x}-1)^6 \left(\begin{array}{c} 49x^3 + 282x^{5/2} + 627x^2 \\ +820x^{3/2} + 627x + 282\sqrt{x} + 49 \end{array} \right).$$

In view of Note 1, $v_{194}(x) \geq 0$ implies $u_{194}(x) \geq 0$, $\forall x > 0$, hence proving the required result.

195. For $\frac{11P_5+30N_2-30R}{11} \leq S + 2P_5 - 2C$**:** We have to show that

$$\frac{1}{11}(11S + 11P_5 - 22C - 30N_2 + 30R) \geq 0.$$

We can write $11S + 11P_5 - 22C - 30N_2 + 30R = bg_{195}(a/b)$, where

$$g_{195}(x) = \frac{u_{195}(x)}{2(x+1)(\sqrt{x}+1)^2}$$

with

$$u_{195}(x) = \left(18x^3 - 8x^{5/2} + 102x^2 + 80x^{3/2} + 102x - 8\sqrt{x} + 18\right)$$
$$- (x+1)(\sqrt{x}+1)^2\left[15(\sqrt{x}+1)\sqrt{2x+2} - 11\sqrt{2x^2+2}\right].$$

After simplifications,

$$u_{195}(x) = 14x^3 + 4(x^2 + 1)(\sqrt{x} - 1)^2 + 98x^2 + 80x^{3/2} + 98x + 14$$
$$- (x+1)(\sqrt{x}+1)^2\left[15(\sqrt{x}+1)\sqrt{2x+2} - 11\sqrt{2x^2+2}\right].$$

Before proceeding further we shall prove that the expression $15(\sqrt{x} + 1)$ $\sqrt{2x+2} - 11\sqrt{2x^2+2}$ is nonnegative for all $x > 0$. Let's consider

$$p_{195}(x) = \left[15(\sqrt{x}+1)\sqrt{2x+2}\right]^2 - \left[11\sqrt{2x^2+2}\right]^2.$$

After simplifications,

$$p_{195}(x) = 4\left(52x^2 + 225x^{3/2} + 225x + 225\sqrt{x} + 52\right).$$

In view of Note 1, it proves that the expression $15(\sqrt{x}+1)\sqrt{2x+2} - 11\sqrt{2x^2+2}$ is positive for all $x > 0$. Now we shall show that $u_{195}(x) \geq 0$, $\forall x > 0$. Let's consider

$$v_{195}(x) = \left[14x^3 + 4(x^2+1)(\sqrt{x}-1)^2 + 98x^2 + 80x^{3/2} + 98x + 14\right]^2$$
$$- \left\{(x+1)(\sqrt{x}+1)^2\left[15(\sqrt{x}+1)\sqrt{2x+2} - 11\sqrt{2x^2+2}\right]\right\}^2.$$

This gives

$$v_{195}(x) = 330\sqrt{2x+2}\sqrt{2x^2+2}(x+1)^2(\sqrt{x}+1)^5$$
$$- 2\begin{pmatrix}184x^6 + 1978x^{11/2} + 3150x^5 + 9378x^{9/2} \\ +9592x^4 + 13636x^{7/2} + 8644x^3 + 13636x^{5/2} \\ +9592x^2 + 9378x^{3/2} + 3150x + 1978\sqrt{x} + 184\end{pmatrix}.$$

Still, we are unable to decide the nonnegativity of $v_{195}(x)$. Let's apply again the argument given in Note 1. Let's consider

$$w_{195}(x) = \left[330\sqrt{2x+2}\sqrt{2x^2+2}(x+1)^2(\sqrt{x}+1)^5\right]^2$$
$$- 4\begin{pmatrix}184x^6 + 1978x^{11/2} + 3150x^5 + 9378x^{9/2} \\ +9592x^4 + 13636x^{7/2} + 8644x^3 + 13636x^{5/2} \\ +9592x^2 + 9378x^{3/2} + 3150x + 1978\sqrt{x} + 184\end{pmatrix}^2.$$

After simplifications,

$$w_{195}(x) = 16(\sqrt{x} - 1)^4 \times h_{195}(x),$$

where

$$h_{195}(x) = \begin{pmatrix} 18761x^{10} + 165318x^{19/2} + 642035x^9 \\ +2301400x^{17/2} + 5500320x^8 + 11274802x^{15/2} \\ +17473730x^7 + 23280222x^{13/2} + 24440183x^6 \\ +24107282x^{11/2} + 22359894x^5 + 24107282x^{9/2} \\ +24440183x^4 + 23280222x^{7/2} + 17473730x^3 \\ +11274802x^{5/2} + 5500320x^2 + 2301400x^{3/2} \\ +642035x + 165318\sqrt{x} + 18761 \end{pmatrix}.$$

In view of Note 1, $w_{195}(x) \geq 0$ implies $v_{195}(x) \geq 0$ implies $u_{195}(x) \geq 0$, $\forall x > 0$, hence proving the required result.

196. For $S + 2P_5 - 2C \leq P_5$: We have to show that $2C - P_5 - S \geq 0$. We can write $2C - P_5 - S = bg_{196}(a/b)$, where

$$g_{196}(x) = \frac{u_{196}(x)}{2(x+1)(\sqrt{x}+1)^2}$$

with

$$u_{196}(x) = 2\left(x^3 + 4x^{5/2} - x^2 - x + 4\sqrt{x} + 1\right) - \sqrt{2x^2 + 2}(x+1)(\sqrt{x}+1)^2.$$

After simplifications,

$$u_{196}(x) = \begin{pmatrix} \sqrt{x}(\sqrt{x}-1)^2[(\sqrt{x}+1)^3 + 2] \\ +x^3 + 7x^{5/2} + 5\sqrt{x} + x + 2 \end{pmatrix} - \sqrt{2x^2 + 2}(x+1)(\sqrt{x}+1)^2.$$

Let's consider

$$v_{196}(x) = \begin{pmatrix} \sqrt{x}(\sqrt{x}-1)^2[(\sqrt{x}+1)^3 + 2] \\ +x^3 + 7x^{5/2} + 5\sqrt{x} + x + 2 \end{pmatrix}^2 - \left[\sqrt{2x^2 + 2}(x+1)(\sqrt{x}+1)^2\right]^2.$$

This gives

$$v_{196}(x) = 2(\sqrt{x}-1)^2 \begin{pmatrix} x^5 + 14x^{9/2} + 47x^4 + 52x^{7/2} \\ +40x^3 + 12x^{5/2} + 40x^2 \\ +52x^{3/2} + 47x + 14\sqrt{x} + 1 \end{pmatrix}.$$

In view of Note 1, $v_{196}(x) \geq 0$ implies $u_{196}(x) \geq 0$, $\forall x > 0$, hence proving the required result.

197. For $\frac{4P_5+5G-5S}{4} \leq N_1$: We have to show that $\frac{1}{4}(5S + 4N_1 - 4P_5 - 5G) \geq 0$. We can write $5S + 4N_1 - 4P_5 - 5G = bg_{197}(a/b)$, where $g_{197}(x) = \frac{1}{2}u_{197}(x)/(\sqrt{x}+1)^2$ with

$$u_{197}(x) = 5\sqrt{2x^2 + 2}(\sqrt{x}+1)^2 - \left(6x^2 + 2x^{3/2} + 24x + 2\sqrt{x} + 6\right).$$

Let's consider

$$v_{197}(x) = \left[5\sqrt{2x^2 + 2}(\sqrt{x}+1)^2\right]^2 - \left(6x^2 + 2x^{3/2} + 24x + 2\sqrt{x} + 6\right)^2.$$

After simplifications,

$$v_{197}(x) = 2(\sqrt{x}-1)^2 \left(\begin{array}{c} 7x^3 + 102x^{5/2} + 201x^2 \\ +340x^{3/2} + 201x + 102\sqrt{x} + 7 \end{array}\right).$$

In view of Note 1, $v_{197}(x) \geq 0$ implies $u_{197}(x) \geq 0$, $\forall x > 0$, hence proving the required result.

198. For $\frac{11P_5+30N_2-30R}{11} \leq N_1$: We have to show that

$$\frac{1}{11}(11N_1 - 11P_5 - 30N_2 + 30R) \geq 0.$$

We can write $11N_1 - 11P_5 - 30N_2 + 30R = bg_{198}(a/b)$, where

$$g_{198}(x) = \frac{u_{198}(x)}{4(x+1)(\sqrt{x}+1)^2}$$

with

$$u_{198}(x) = \left(\begin{array}{c} 47x^3 + 204x^{5/2} + 105x^2 \\ +248x^{3/2} + 105x + 204\sqrt{x} + 47 \end{array}\right) - 30\sqrt{2x+2}(x+1)(\sqrt{x}+1)^3.$$

Let's consider

$$v_{198}(x) = \left(\begin{array}{c} 47x^3 + 204x^{5/2} + 105x^2 \\ +248x^{3/2} + 105x + 204\sqrt{x} + 47 \end{array}\right)^2 - \left[30\sqrt{2x+2}(x+1)(\sqrt{x}+1)^3\right]^2.$$

After simplifications,

$$v_{198}(x) = (\sqrt{x} - 1)^2 \begin{pmatrix} 409x^5 + 9194x^{9/2} + 37065x^4 + 62688x^{7/2} \\ +96990x^3 + 94188x^{5/2} + 96990x^2 \\ +62688x^{3/2} + 37065x + 9194\sqrt{x} + 409 \end{pmatrix}.$$

In view of Note 1, $v_{198}(x) \geq 0$ implies $u_{198}(x) \geq 0$, $\forall x > 0$, hence proving the required result.

199. For $A + 4P_4 - 4G \leq S + 2P_5 - 2C$: We have to show that

$$S + 2P_5 - 2C - A - 4P_4 + 4G \geq 0.$$

We can write $S + 2P_5 - 2C - A - 4P_4 + 4G = bg_{199}(a/b)$, where

$$g_{199}(x) = \frac{u_{199}(x)}{2(x+1)(\sqrt{x}+1)^2}$$

with

$$u_{199}(x) = \sqrt{2x+2}(x+1)(\sqrt{x}+1)^2$$
$$- \left(x^3 + 2x^{5/2} + 11x^2 - 12x^{3/2} + 11x + 2\sqrt{x} + 1\right).$$

After simplifications,

$$u_{199}(x) = \sqrt{2x+2}(x+1)(\sqrt{x}+1)^2$$
$$- \left(x^3 + 2x^{5/2} + 5x^2 + 6x(\sqrt{x}-1)^2 + 5x + 2\sqrt{x} + 1\right).$$

Let's consider

$$v_{199}(x) = \left[\sqrt{2x+2}(x+1)(\sqrt{x}+1)^2\right]^2$$
$$- \left(x^3 + 2x^{5/2} + 5x^2 + 6x(\sqrt{x}-1)^2 + 5x + 2\sqrt{x} + 1\right)^2.$$

This gives

$$v_{199}(x) = (\sqrt{x} - 1)^4 \begin{pmatrix} x^4 + 8x^{7/2} + 16x^3 + 7x^{5/2} + 7x^{3/2} \\ +17x^{3/2}(\sqrt{x}-1)^2 + 16x + 8\sqrt{x} + 1 \end{pmatrix}.$$

In view of Note 1, $v_{199}(x) \geq 0$ implies $u_{199}(x) \geq 0$, $\forall x > 0$, hence proving the required result.

200. For $A + 4P_4 - 4G \leq A$: We have to show that $4(G - P_4) \geq 0$. It is true by definition. See inequalities (30).

201. For $\frac{4P_2+C-P_1}{4} \leq \mathbf{N_1}$**:** We have to show that $\frac{1}{4}(4N_1 - 4P_2 - C + P_1) \geq 0$. We can write $4N_1 - 4P_2 - C + P_1 = bg_{201}(a/b)$, where

$$g_{201}(x) = \frac{u_{201}(x)}{(x^2+1)(x^3+1)}$$

with

$$u_{201}(x) = 2x^{11/2} - 6x^4 - x^5 + 6x^3 + 2x^{7/2} + 2x^{5/2} - x - 6x^2 + 2\sqrt{x}.$$

After simplifications,

$$u_{201}(x) = \sqrt{x}(\sqrt{x}-1)^2$$
$$\times \begin{pmatrix} 2\sqrt{x}(x+\sqrt{x}+1)(x^2+\sqrt{x}(\sqrt{x}-1)^2+1) \\ +(x-1)^2(x+1)(2x+\sqrt{x}+2) \end{pmatrix}.$$

Since $u_{201}(x) \geq 0$, $\forall x > 0$, hence proving the required result.
Combining Results 181 to 201 we get the proof of Proposition 14. $\quad\square$

Proposition 15. *The following inequalities hold:*

$$P_4 \leq \begin{cases} \frac{4S+5H-5A}{4} \leq \begin{cases} \mathbf{N_1} \\ P_5 + 2S - 2C \leq \mathbf{A} \end{cases} \\ \frac{3G+P_5-C}{3} \leq \frac{2N_1+H-A}{2} \leq \frac{3H+C-P_5}{3} \leq \text{(a)} \end{cases}$$

$$\text{(a)} \leq \begin{cases} \mathbf{G} \\ \frac{27R+26N_2-26C}{27} \leq \frac{4C+7H-7A}{4} \leq \mathbf{N_1}. \end{cases}$$

Proof. The proof is based on the following results.

202. For $P_4 \leq \frac{3G+P_5-C}{3}$**:** We have to show that $\frac{1}{3}(3G+P_5-C-3P_4) \geq 0$. It is the same as shown in Result 172.

203. For $P_4 \leq \frac{4S+5H-5A}{4}$**:** We have to show that $\frac{1}{4}(4S+5H-5A-4P_4) \geq 0$. We can write $4S+5H-5A-4P_4 = bg_{203}(a/b)$, where

$$g_{203}(x) = \frac{u_{203}(x)}{2(x+1)(\sqrt{x}+1)^2}$$

with

$$u_{203}(x) = 4\sqrt{2x^2+2}(\sqrt{x}+1)^2(x+1)$$
$$- (5x^3 + 10x^{5/2} + 27x^2 - 20x^{3/2} + 27x + 10\sqrt{x} + 5).$$

After simplifications,

$$u_{203}(x) = 4\sqrt{2x^2 + 2}(\sqrt{x} + 1)^2(x + 1)$$
$$- \left(5x^3 + 10x^{5/2} + 17x^2 + 10x(\sqrt{x} - 1)^2 + 17x + 10\sqrt{x} + 5\right).$$

Let's consider

$$v_{203}(x) = \left[4\sqrt{2x^2 + 2}(\sqrt{x} + 1)^2(x + 1)\right]^2$$
$$- \left(\begin{array}{c} 5x^3 + 10x^{5/2} + 17x^2 + 17x \\ +10x(\sqrt{x} - 1)^2 + 10\sqrt{x} + 5 \end{array}\right)^2.$$

This gives

$$v_{203}(x) = (\sqrt{x} - 1)^6 \left(\begin{array}{c} 7x^3 + 70x^{5/2} + 201x^2 \\ +340x^{3/2} + 201x + 70\sqrt{x} + 7 \end{array}\right).$$

In view of Note 1, $v_{203}(x) \geq 0$ implies $u_{203}(x) \geq 0$, $\forall x > 0$, hence proving the required result.

204. For $\frac{3G+P_5-C}{3} \leq \frac{2N_1+H-A}{2}$: We have to show that

$$\frac{1}{6}(6N_1 + 3H - 3A - 6G - 2P_5 + 2C) \geq 0.$$

We can write $6N_1 + 3H - 3A - 6G - 2P_5 + 2C = bg_{204}(a/b)$, where

$$g_{204}(x) = \frac{u_{204}(x)}{(x + 1)(\sqrt{x} + 1)^2}$$

with

$$u_{204}(x) = x^{5/2} + 6x^{3/2} - 4x^2 - 4x + \sqrt{x} = \sqrt{x}(\sqrt{x} - 1)^4.$$

Since $u_{204}(x) \geq 0$, $\forall x > 0$, hence proving the required result.

205. For $\frac{2N_1+H-A}{2} \leq \frac{3H+C-P_5}{3}$: We have to show that

$$\frac{1}{6}(3H + 2C - 2P_5 - 6N_1 + 3A) \geq 0.$$

We can write $3H + 2C - 2P_5 - 6N_1 + 3A = bg_{205}(a/b)$, where

$$g_{205}(x) = \frac{u_{205}(x)}{(x + 1)(\sqrt{x} + 1)^2}$$

with

$$u_{205}(x) = x^{5/2} + 6x^{3/2} - 4x^2 - 4x + \sqrt{x} = \sqrt{x}(\sqrt{x} - 1)^4.$$

Since $u_{205}(x) \geq 0$, $\forall x > 0$, hence proving the required result.

206. For $\frac{3H+C-P_5}{3} \leq \frac{27R+26N_2-26C}{27}$**: We have to show that**

$$\frac{1}{27}(27R + 26N_2 - 35C - 27H + 9P_5) \geq 0.$$

We can write $27R + 26N_2 - 35C - 27H + 9P_5 = bg_{206}(a/b)$, where

$$g_{206}(x) = \frac{u_{206}(x)}{2(x+1)(\sqrt{x}+1)^2}$$

with

$$u_{206}(x) = 13(x+1)\sqrt{2x+2}(\sqrt{x}+1)^3$$
$$- \left(16x^3 + 68x^{5/2} + 52x^2 + 144x^{3/2} + 52x + 68\sqrt{x} + 16\right).$$

Let's consider

$$v_{206}(x) = \left[13(x+1)\sqrt{2x+2}(\sqrt{x}+1)^3\right]^2$$
$$- \left(16x^3 + 68x^{5/2} + 52x^2 + 144x^{3/2} + 52x + 68\sqrt{x} + 16\right)^2.$$

After simplifications,

$$v_{206}(x) = 2(\sqrt{x}-1)^4 \begin{pmatrix} 41x^4 + 90x^{7/2} + 12x^3 \\ +221x^{5/2} + 33x^{3/2}(\sqrt{x}-1)^2 \\ +221x^{3/2} + 12x + 90\sqrt{x} + 41 \end{pmatrix}.$$

In view of Note 1, $v_{206}(x) \geq 0$ implies $u_{206}(x) \geq 0$, $\forall x > 0$, hence proving the required result.

207. For $\frac{27R+26N_2-26C}{27} \leq \frac{4C+7H-7A}{4}$**: We have to show that**

$$\frac{1}{108}(212C + 189H - 189A - 108R - 104N_2) \geq 0.$$

We can write $212C + 189H - 189A - 108R - 104N_2 = bg_{207}(a/b)$, where
$g_{207}(x) = \frac{13}{2}u_{207}(x)/(x+1)$ with

$$u_{207}(x) = 7x^2 + 18x + 7 - 4\sqrt{2x+2}(\sqrt{x}+1)(x+1).$$

Let's consider

$$v_{207}(x) = \left(7x^2 + 18x + 7\right)^2 - \left[4\sqrt{2x+2}(\sqrt{x}+1)(x+1)\right]^2.$$

After simplifications,

$$v_{207}(x) = (\sqrt{x}-1)^4\left(17x^2 + 4x^{3/2} + 38x + 4\sqrt{x} + 17\right).$$

In view of Note 1, $v_{207}(x) \geq 0$ implies $u_{207}(x) \geq 0$, $\forall x > 0$, hence proving the required result.

208. For $\frac{4C+7H-7A}{4} \leq N_1$: We have to show that $\frac{1}{4}(4N_1 - 4C - 7H + 7A) \geq 0$. We can write $4N_1 - 4C - 7H + 7A = bg_{208}(a/b)$, where $g_{208}(x) = \frac{1}{2}u_{208}(x)/(x+1)$ with

$$u_{208}(x) = x^2 + 4x^{3/2} - 10x + 4\sqrt{x} + 1 = (x + 6\sqrt{x} + 1)(\sqrt{x}-1)^2.$$

Since $u_{208}(x) \geq 0$, $\forall x > 0$, hence proving the required result.

209. For $\frac{3H+C-P_5}{3} \leq G$: We have to show that $\frac{1}{3}(3G - 3H - C + P_5) \geq 0$. We can write $3G - 3H - C + P_5 = bg_{209}(a/b)$, where

$$g_{209}(x) = \frac{u_{209}(x)}{(\sqrt{x}+1)^2(x+1)}$$

with

$$u_{209}(x) = x^{5/2} + 2x - 6x^{3/2} + 2x^2 + \sqrt{x}$$
$$= \sqrt{x}(x + 4\sqrt{x} + 1)(\sqrt{x}-1)^2.$$

Since $u_{209}(x) \geq 0$, $\forall x > 0$, hence proving the required result.

210. For $\frac{4S+5H-5A}{4} \leq P_5 + 2S - 2C$: We have to show that

$$\frac{1}{4}(4P_5 + 4S - 8C - 5H + 5A) \geq 0.$$

We can write $4P_5 + 4S - 8C - 5H + 5A = bg_{210}(a/b)$, where

$$g_{210}(x) = \frac{u_{210}(x)}{2(x+1)(\sqrt{x}+1)^2}$$

with

$$u_{210}(x) = 4\sqrt{2x^2 + 2}(x+1)(\sqrt{x}+1)^2$$
$$- \left(3x^3 + 22x^{5/2} - 3x^2 + 20x^{3/2} - 3x + 22\sqrt{x} + 3\right).$$

After simplifications,

$$u_{210}(x) = 4\sqrt{2x^2 + 2}(x+1)(\sqrt{x}+1)^2$$
$$- \left(\frac{3x^3 + 20x^{5/2} + 2\sqrt{x}(x+1)(\sqrt{x}-1)^2}{+x^2 + 16x^{3/2} + x + 20\sqrt{x} + 3} \right).$$

Let's consider

$$v_{210}(x) = \left[4\sqrt{2x^2 + 2}(x+1)(\sqrt{x}+1)^2 \right]^2$$
$$- \left(\frac{3x^3 + 20x^{5/2} + 2\sqrt{x}(x+1)(\sqrt{x}-1)^2}{+x^2 + 16x^{3/2} + x + 20\sqrt{x} + 3} \right)^2.$$

This gives

$$v_{210}(x) = (\sqrt{x}-1)^6 \left(\frac{23x^3 + 134x^{5/2} + 249x^2}{+340x^{3/2} + 249x + 134\sqrt{x} + 23} \right).$$

In view of Note 1, $v_{210}(x) \geq 0$ implies $u_{210}(x) \geq 0$, $\forall x > 0$, hence proving the required result.

211. For $P_5 + 2S - 2C \leq A$: We have to show that $A - P_5 - 2S + 2C \geq 0$. We can write $A - P_5 - 2S + 2C = bg_{211}(a/b)$, where

$$g_{211}(x) = \frac{u_{211}(x)}{2(x+1)(\sqrt{x}+1)^2}$$

with

$$u_{211}(x) = 3x^3 + 10x^{5/2} + x^2 + 4x^{3/2} + x + 10\sqrt{x} + 3$$
$$- 2\sqrt{2x^2 + 2}(x+1)(\sqrt{x}+1)^2.$$

Let's consider

$$v_{211}(x) = \left(3x^3 + 10x^{5/2} + x^2 + 4x^{3/2} + x + 10\sqrt{x} + 3 \right)^2$$
$$- \left[2\sqrt{2x^2 + 2}(x+1)(\sqrt{x}+1)^2 \right]^2.$$

After simplifications,

$$v_{211}(x) = (\sqrt{x}-1)^2 \left(\begin{array}{l} x^5 + 30x^{9/2} + 101x^4 + 120x^{7/2} \\ +106x^3 + 52x^{5/2} + 106x^2 \\ +120x^{3/2} + 101x + 30\sqrt{x} + 1 \end{array} \right).$$

In view of Note 1, $v_{211}(x) \geq 0$ implies $u_{211}(x) \geq 0$, $\forall x > 0$, hence proving the required result.

212. For $\frac{4S+5H-5A}{4} \leq N_1$**:** We have to show that $\frac{1}{4}(4N_1 - 4S - 5H + 5A) \geq 0$. We can write $4N_1 - 4S - 5H + 5A = bg_{212}(a/b)$, where $g_{212}(x) = \frac{1}{2}u_{212}(x)/(x+1)$ with

$$u_{212}(x) = \left(7x^2 + 4x^{3/2} - 6x + 4\sqrt{x} + 7\right) - 4\sqrt{2x^2 + 2}(x+1).$$

After simplifications,

$$u_{212}(x) = \left(3\sqrt{x}(\sqrt{x} - 1)^2 + 7x^2 + x^{3/2} + \sqrt{x} + 7\right) - 4\sqrt{2x^2 + 2}(x+1).$$

Let's consider

$$v_{212}(x) = \left[3\sqrt{x}(\sqrt{x} - 1)^2 + 7x^2 + x^{3/2} + \sqrt{x} + 7\right]^2$$
$$- \left[4\sqrt{2x^2 + 2}(x+1)\right]^2.$$

After simplifications,

$$v_{212}(x) = (\sqrt{x} - 1)^2 \left(\begin{array}{l} 17x^3 + 90x^{5/2} + 21x^2 + 21x \\ +10x(\sqrt{x} - 1)^2 + 90\sqrt{x} + 17 \end{array}\right).$$

In view of Note 1, $v_{212}(x) \geq 0$ implies $u_{212}(x) \geq 0$, $\forall x > 0$, hence proving the required result.

Combining Results 202 to 212 we get the proof of Proposition 15.   □

Proposition 16. *The following inequalities hold:*

$$G \leq \begin{cases} \frac{3H+2C-2P_5}{3} \leq P_2 + 2C - 2P_5 \leq \mathbf{A} \\ \frac{3H+2C-2P_5}{3} \leq C + 2N_1 - 2S \\ \frac{2N_1+S-C}{2} \leq \frac{5H+C-P_3}{5} \\ P_4 + 6N_2 - 6N \\ \frac{4P_4+3R-3A}{4} \end{cases} \leq \mathbf{N_1}.$$

Proof. The proof is based on the following results.

213. For $G \leq \frac{3H+2C-2P_5}{3}$**:** We have to show that $\frac{1}{3}(3H + 2C - 2P_5 - 3G) \geq 0$. We can write $3H + 2C - 2P_5 - 3G = bg_{213}(a/b)$, where

$$g_{213}(x) = \frac{u_{213}(x)}{(\sqrt{x} + 1)^2(x+1)}$$

with

$$u_{213}(x) = x^{5/2} + 6x^{3/2} - 4x^2 + \sqrt{x} - 4x = \sqrt{x}(\sqrt{x} - 1)^2.$$

Since $u_{213}(x) \geq 0$, $\forall x > 0$, hence proving the required result.

214. For $G \leq 2N_1 + S - C$: We have to show that $\frac{1}{2}(2N_1 + S - C - 2G) \geq 0$. We can write $2N_1 + S - C - 2G = bg_{214}(a/b)$, where $g_{214}(x) = \frac{1}{2}u_{214}(x)/(x+1)$ with

$$u_{214}(x) = \sqrt{2x^2 + 2}(x + 1) - (x^2 + 2x^{3/2} - 2x + 2\sqrt{x} + 1),$$

i.e.,

$$u_{214}(x) = \sqrt{2x^2 + 2}(x + 1) - \left[x^2 + x^{3/2} + \sqrt{x}(\sqrt{x} - 1)^2 + \sqrt{x} + 1\right].$$

Let's consider

$$u_{214}(x) = \left[\sqrt{2x^2 + 2}(x + 1)\right]^2 - \left[x^2 + x^{3/2} + \sqrt{x}(\sqrt{x} - 1)^2 + \sqrt{x} + 1\right]^2.$$

After simplifications,

$$v_{214}(x) = (\sqrt{x} - 1)^6(\sqrt{x} + 1)^2.$$

In view of Note 1, $v_{214}(x) \geq 0$ implies $u_{214}(x) \geq 0$, $\forall x > 0$, hence proving the required result.

215. For $G \leq P_4 + 6N_2 - 6N$: We have to show that $P_4 + 6N_2 - 6N - G \geq 0$. We can write $P_4 + 6N_2 - 6N - G = bg_{215}(a/b)$, where $g_{215}(x) = \frac{1}{2}u_{215}(x)/(\sqrt{x} + 1)^2$ with

$$u_{215}(x) = 3\sqrt{2x + 2}(\sqrt{x} + 1)^3 - (4x^2 + 14x^{3/2} + 12x + 14\sqrt{x} + 4).$$

Let's consider

$$v_{215}(x) = \left[3\sqrt{2x + 2}(\sqrt{x} + 1)^3\right]^2$$
$$- (4x^2 + 14x^{3/2} + 12x + 14\sqrt{x} + 4)^2.$$

After simplifications,

$$v_{215}(x) = 2(\sqrt{x} - 1)^4(x^2 + 2x^{3/2} + 2\sqrt{x} + 1).$$

In view of Note 1, $v_{215}(x) \geq 0$ implies $u_{215}(x) \geq 0$, $\forall x > 0$, hence proving the required result.

216. For $G \le \frac{4P_4 + 3R - 3A}{4}$: We have to show that $\frac{1}{4}(4P_4 + 3R - 3A - 4G) \ge 0$. We can write $4P_4 + 3R - 3A - 4G = bg_{216}(a/b)$, where

$$g_{216}(x) = \frac{u_{216}(x)}{2(x+1)(\sqrt{x}+1)^2}$$

with

$$u_{216}(x) = x^3 - 6x^{5/2} + 15x^2 - 20x^{3/2} + 15x - 6\sqrt{x} + 1 = (\sqrt{x} - 1)^6.$$

Since $u_{216}(x) \ge 0$, $\forall x > 0$, hence proving the required result.

217. For $\frac{3H + 2C - 2P_5}{3} \le P_2 + 2C - 2P_5$: We have to show that

$$\frac{1}{3}(3P_2 + 4C - 4P_5 - 3H) \ge 0.$$

We can write $3P_2 + 4C - 4P_5 - 3H = bg_{217}(a/b)$, where

$$g_{217}(x) = \frac{u_{217}(x)}{(x+1)(x^2+1)(\sqrt{x}+1)^2}$$

with

$$\begin{aligned}
u_{217}(x) &= \begin{pmatrix} 8x^{9/2} - 11x^4 - 6x^{7/2} - 5x^3 \\ +28x^{5/2} - 5x^2 - 11x - 6x^{3/2} + 8\sqrt{x} \end{pmatrix} \\
&= \sqrt{x}(\sqrt{x} - 1)^4 (8x^2 + 21x^{3/2} + 30x + 21\sqrt{x} + 8).
\end{aligned}$$

Since $u_{217}(x) \ge 0$, $\forall x > 0$, hence proving the required result.

218. For $P_2 + 2C - 2P_5 \le A$: We have to show that $A - P_2 - 2C + 2P_5 \ge 0$. We can write $A - P_2 - 2C + 2P_5 = bg_{218}(a/b)$, where

$$g_{218}(x) = \frac{u_{218}(x)}{2(x+1)(x^2+1)(\sqrt{x}+1)^2}$$

with

$$\begin{aligned}
u_{218}(x) &= x^5 - 6x^{9/2} + 9x^4 + 6x^3 - 20x^{5/2} + 6x^2 + 9x - 6\sqrt{x} + 1 \\
&= (\sqrt{x} - 1)^4 (x^2 - 2x^{3/2} - 2x - 2\sqrt{x} + 1)^2.
\end{aligned}$$

Since $u_{218}(x) \ge 0$, $\forall x > 0$, hence proving the required result.

219. For $\frac{3H+2C-2P_5}{3} \leq C + 2N_1 - 2S$: We have to show that

$$\frac{1}{3}(C + 6N_1 - 6S - 3H + 2P_5) \geq 0.$$

We can write $C + 6N_1 - 6S - 3H + 2P_5 = bg_{219}(a/b)$, where

$$g_{219}(x) = \frac{u_{219}(x)}{2(x+1)(\sqrt{x}+1)^2}$$

with

$$u_{219}(x) = 9x^3 + 16x^{5/2} + 23x^2 + 23x + 16\sqrt{x} + 9$$
$$- 6\sqrt{2x^2 + 2}(x+1)(\sqrt{x}+1)^2.$$

Let's consider

$$v_{219}(x) = \left(9x^3 + 16x^{5/2} + 23x^2 + 23x + 16\sqrt{x} + 9\right)^2$$
$$- \left[6\sqrt{2x^2 + 2}(x+1)(\sqrt{x}+1)^2\right]^2.$$

After simplifications,

$$v_{219}(x) = (\sqrt{x} - 1)^4 \left(\begin{array}{c} 9x^4 + 36x^{7/2} + 184x^3 + 428x^{5/2} \\ +606x^2 + 428x^{3/2} + 184x + 36\sqrt{x} + 9 \end{array}\right).$$

In view of Note 1, $v_{219}(x) \geq 0$ implies $u_{219}(x) \geq 0$, $\forall x > 0$, hence proving the required result.

220. For $\frac{2N_1+S-C}{2} \leq \frac{5H+C-P_3}{5}$: We have to show that

$$\frac{1}{10}(10H + 7C - 2P_3 - 10N_1 - 5S) \geq 0.$$

We can write $10H + 7C - 2P_3 - 10N_1 - 5S = bg_{220}(a/b)$, where

$$g_{220}(x) = \frac{u_{220}(x)}{2(x+1)(x - \sqrt{x} + 1)}$$

with

$$u_{220}(x) = 9x^3 - 19x^{5/2} + 45x^2 - 50x^{3/2} + 45x - 19\sqrt{x} + 9$$
$$- 5\sqrt{2x^2 + 2}(x+1)(x - \sqrt{x} + 1).$$

After simplifications,

$$u_{220}(x) = \begin{pmatrix} 5(\sqrt{x}-1)^2(x-\sqrt{x}+1)^2 \\ +4x^3+x^{5/2}+5x^2+5x+\sqrt{x}+4 \end{pmatrix}$$
$$- 5\sqrt{2x^2+2}(x+1)(x-\sqrt{x}+1).$$

Let's consider

$$u_{220}(x) = \begin{pmatrix} 5(\sqrt{x}-1)^2(x-\sqrt{x}+1)^2 \\ +4x^3+x^{5/2}+5x^2+5x+\sqrt{x}+4 \end{pmatrix}^2$$
$$- \left[5\sqrt{2x^2+2}(x+1)(x-\sqrt{x}+1)\right]^2.$$

This gives

$$v_{220}(x) = (\sqrt{x}-1)^6 \begin{pmatrix} (28x^2+65x+28)(\sqrt{x}-1)^2 \\ +3x^3+27x^2+27x+3 \end{pmatrix}.$$

In view of Note 1, $v_{220}(x) \geq 0$ implies $u_{220}(x) \geq 0$, $\forall x > 0$, hence proving the required result.

221. For $C + 2N_1 - 2S \leq N_1$: We have to show that $2S - N_1 - C \geq 0$. We can write $2S - N_1 - C = bg_{221}(a/b)$, where $g_{221}(x) = \frac{1}{4}u_{221}(x)/(x+1)$ with

$$u_{221}(x) = 4\sqrt{2x^2+2}(x+1) - \left(5x^2+2x^{3/2}+2x+2\sqrt{x}+5\right).$$

Let's consider

$$v_{221}(x) = \left[4\sqrt{2x^2+2}(x+1)\right]^2 - \left(5x^2+2x^{3/2}+2x+2\sqrt{x}+5\right)^2.$$

After simplifications,

$$v_{221}(x) = (\sqrt{x}-1)^2 \begin{pmatrix} 4x^3+3(x^2+1)(\sqrt{x}-1)^2 \\ +18x^2+20x^{3/2}+18x+4 \end{pmatrix}.$$

In view of Note 1, $v_{221}(x) \geq 0$ implies $u_{221}(x) \geq 0$, $\forall x > 0$, hence proving the required result.

222. For $\frac{5H+C-P_3}{5} \leq N_1$: We have to show that $\frac{1}{5}(5N_1 - 5H - C + P_3) \geq 0$. We can write $5N_1 - 5H - C + P_3 = bg_{222}(a/b)$, where

$$g_{222}(x) = \frac{u_{222}(x)}{4(x+1)(x-\sqrt{x}+1)}$$

with

$$u_{222}(x) = x^3 - 35x^2 + 9x^{5/2} + 50x^{3/2} - 35x + 9\sqrt{x} + 1$$
$$= (\sqrt{x} - 1)^2 [x^2 + 4x^{3/2} + 7\sqrt{x}(\sqrt{x} - 1)^2 + 4\sqrt{x} + 1].$$

Since $u_{222}(x) \geq 0$, $\forall x > 0$, hence proving the required result.

223. For $P_4 + 6N_2 - 6N \leq N_1$: We have to show that $N_1 - P_4 - 6N_2 + 6N \geq 0$. We can write $N_1 - P_4 - 6N_2 + 6N = bg_{223}(a/b)$, where $g_{223}(x) = \frac{1}{4} u_{223}(x)/(\sqrt{x} + 1)$ with

$$u_{223}(x) = 9x^2 + 28x^{3/2} + 22x + 28\sqrt{x} + 9 - 6\sqrt{2x + 2}(\sqrt{x} + 1)^3.$$

Let's consider

$$v_{223}(x) = \left(9x^2 + 28x^{3/2} + 22x + 28\sqrt{x} + 9\right)^2 - \left[6\sqrt{2x + 2}(\sqrt{x} + 1)^3\right]^2.$$

After simplifications,

$$v_{223}(x) = (\sqrt{x} - 1)^2 \left(\begin{array}{c} 9x^3 + 90x^{5/2} + 199x^2 \\ +172x^{3/2} + 199x + 90\sqrt{x} + 9 \end{array}\right).$$

In view of Note 1, $v_{223}(x) \geq 0$ implies $u_{223}(x) \geq 0$, $\forall x > 0$, hence proving the required result.

224. For $\frac{4P_4 + 3R - 3A}{4} \leq N_1$: We have to show that $\frac{1}{4}(4N_1 - 4P_4 - 3R + 3A) \geq 0$. We can write $4N_1 - 4P_4 - 3R + 3A = bg_{224}(a/b)$, where

$$g_{224}(x) = \frac{u_{224}(x)}{2(x + 1)(\sqrt{x} + 1)^2}$$

with

$$u_{224}(x) = x^3 - 17x^2 + 6x^{5/2} + 20x^{3/2} - 17x + 6\sqrt{x} + 1$$
$$= (\sqrt{x} - 1)^2 [x^2 + 7x^{3/2} + \sqrt{x}(\sqrt{x} - 1)^2 + 7\sqrt{x} + 1].$$

Since $u_{224}(x) \geq 0$, $\forall x > 0$, hence proving the required result.

Combining Results 213 to 224 we get the proof of Proposition 16. □

Proposition 17. *The following inequalities hold:*

$$N_1 \leq \frac{20N_2 + 3G - 3R}{20} \leq \left\{\begin{array}{c} C + P_4 - S \\ \frac{12R + 7G - 7S}{12} \end{array}\right\} \leq \frac{N_1 + 12N - N_2}{12} \leq N.$$

Proof. Proof is based on the following results.

225. For $N_1 \leq \frac{20N_2+3G-3R}{20}$: We have to show that $\frac{1}{20}(20N_2 + 3G - 3R - 20N_1) \geq 0$. We can write $20N_2 + 3G - 3R - 20N_1 = bg_{225}(a/b)$, where $g_{225}(x) = u_{225}(x)/(x+1)$ with

$$u_{225}(x) = 5\sqrt{2x+2}(\sqrt{x}+1)(x+1) \\ - \left(7x^2 + 7x^{3/2} + 12x + 7\sqrt{x} + 7\right).$$

Let's consider

$$v_{225}(x) = \left[5\sqrt{2x+2}(\sqrt{x}+1)(x+1)\right]^2 \\ - \left(7x^2 + 7x^{3/2} + 12x + 7\sqrt{x} + 7\right)^2.$$

After simplifications,

$$v_{225}(x) = (\sqrt{x}-1)^4\left(x^2 + 6x^{3/2} + x + 6\sqrt{x} + 1\right).$$

In view of Note 1, $v_{225}(x) \geq 0$ implies $u_{225}(x) \geq 0$, $\forall x > 0$, hence proving the required result.

226. For $\frac{20N_2+3G-3R}{20} \leq \frac{12R+7G-7S}{12}$: We have to show that

$$\frac{1}{60}(69R + 26G - 35S - 60N_2) \geq 0.$$

We can write $69R + 26G - 35S - 60N_2 = bg_{226}(a/b)$, where $g_{226}(x) = \frac{1}{2}u_{226}(x)/(x+1)$ with

$$u_{226}(x) = 92x^2 + 52x^{3/2} + 92x + 52\sqrt{x} + 92 \\ - 5(x+1)\left[6(\sqrt{x}+1)\sqrt{2x+2} + 7\sqrt{2x^2+2}\right].$$

Let's consider

$$v_{226}(x) = \left(92x^2 + 52x^{3/2} + 92x + 52\sqrt{x} + 92\right)^2 \\ - \left\{5(x+1)\left[6(\sqrt{x}+1)\sqrt{2x+2} + 7\sqrt{2x^2+2}\right]\right\}^2.$$

After simplifications,

$$v_{226}(x) = \begin{pmatrix} 4214x^4 + 5968x^{7/2} + 7532x^3 + 8336x^{5/2} \\ +15100x^2 + 8336x^{3/2} + 7532x + 5968\sqrt{x} + 4214 \end{pmatrix} \\ - 2100\sqrt{2x+2}\sqrt{2x^2+2}(\sqrt{x}+1)(x+1)^2.$$

Since we are unable to decide the nonnegativity of $v_{226}(x)$, we shall apply again the argument given in Note 1. Let's consider

$$w_{226}(x) = \left(\begin{array}{l} 4214x^4 + 5968x^{7/2} + 7532x^3 + 8336x^{5/2} \\ +15100x^2 + 8336x^{3/2} + 7532x + 5968\sqrt{x} + 4214 \end{array}\right)^2$$
$$- \left[2100\sqrt{2x+2}\sqrt{2x^2+2}(\sqrt{x}+1)(x+1)^2\right]^2.$$

After simplifications,

$$w_{226}(x) = 4(\sqrt{x} - 1)^4 \times h_{226}(x)$$

where

$$h_{226}(x) = \left(\begin{array}{l} 29449x^6 + 3872372x^{11/2} + 13626974x^5 \\ +27330900x^{9/2} + 43334875x^4 + 56985352x^{7/2} \\ +63952156x^3 + 56985352x^{5/2} + 43334875x^2 \\ +27330900x^{3/2} + 13626974x + 3872372\sqrt{x} + 29449 \end{array}\right).$$

In view of Note 1, $w_{226}(x) \geq 0$ implies $v_{226}(x) \geq 0$ implies $u_{226}(x) \geq 0$, $\forall x > 0$, hence proving the required result.

227. For $\frac{20N_2 + 3G - 3R}{20} \leq C + P_4 - S$: We have to show that

$$\frac{1}{20}(20C + 20P_4 - 20S - 20N_2 - 3G + 3R) \geq 0.$$

We can write $20C + 20P_4 - 20S - 20N_2 - 3G + 3R = bg_{227}(a/b)$, where

$$g_{227}(x) = \frac{u_{277}(x)}{(\sqrt{x}+1)^2(x+1)}$$

with

$$u_{227}(x) = 22x^3 + 41x^{5/2} + 98x^2 - 2x^{3/2} + 98x + 41\sqrt{x} + 22$$
$$- 5(x+1)(\sqrt{x}+1)^2[(\sqrt{x}+1)\sqrt{2x+2} + 2\sqrt{2x^2+2}].$$

After simplifications,

$$u_{227}(x) = 22x^3 + 41x^{5/2} + 97x^2 + x(\sqrt{x}-1)^2 + 97x + 41\sqrt{x} + 22$$
$$- 5(x+1)(\sqrt{x}+1)^2[(\sqrt{x}+1)\sqrt{2x+2} + 2\sqrt{2x^2+2}].$$

In this case also we shall apply twice the argument given in Note 1. Let's consider

$$v_{227}(x) = \left(\begin{array}{l} 22x^3 + 41x^{5/2} + 97x^2 \\ +x(\sqrt{x}-1)^2 + 97x + 41\sqrt{x} + 22 \end{array}\right)^2$$
$$- \{5(x+1)(\sqrt{x}+1)^2[(\sqrt{x}+1)\sqrt{2x+2} + 2\sqrt{2x^2+2}]\}^2.$$

After simplifications,

$$v_{227}(x) = \left(\begin{array}{l} 234x^6 + 704x^{11/2} + 3493x^5 + 3648x^{9/2} \\ +7602x^4 + 2048x^{7/2} + 15742x^3 + 2048x^{5/2} \\ +7602x^2 + 3648x^{3/2} + 3493x + 704\sqrt{x} + 234 \end{array}\right)$$
$$- 100\sqrt{2x^2+2}\sqrt{2x+2}(x+1)^2(\sqrt{x}+1)^5.$$

Still, we are unable to decide the nonnegativity of the expression $u_{227}(x)$. Let's consider

$$h_{227}(x) = \left(\begin{array}{l} 234x^6 + 704x^{11/2} + 3493x^5 + 3648x^{9/2} \\ +7602x^4 + 2048x^{7/2} + 15742x^3 + 2048x^{5/2} \\ +7602x^2 + 3648x^{3/2} + 3493x + 704\sqrt{x} + 234 \end{array}\right)^2$$
$$- \left[100\sqrt{2x^2+2}\sqrt{2x+2}(x+1)^2(\sqrt{x}+1)^5\right]^2.$$

After simplifications,

$$h_{227}(x) = (\sqrt{x}-1)^4 \times w_{227}(x),$$

where

$$w_{227}(x) = \left(\begin{array}{l} 14756x^{10} - 11504x^{19/2} - 4212x^9 - 63392x^{17/2} \\ +2766101x^8 + 10106420x^{15/2} + 29446034x^7 \\ +47168132x^{13/2} + 75003447x^6 \\ +84112344x^{11/2} + 109131748x^5 \\ +84112344x^{9/2} + 75003447x^4 + 47168132x^{7/2} \\ +29446034x^3 + 10106420x^{5/2} + 2766101x^2 \\ -63392x^{3/2} - 4212x - 11504\sqrt{x} + 14756 \end{array}\right).$$

Let's consider

$$k_{227}(t) := w_{227}(t^2) = \begin{pmatrix} 14756t^{20} - 11504t^{19} - 4212t^{18} - 63392t^{17} \\ +2766101t^{16} + 10106420t^{15} + 29446034t^{14} \\ +47168132t^{13} + 75003447t^{12} + 84112344t^{11} \\ +109131748t^{10} + 84112344t^{9} + 75003447t^{8} \\ +47168132t^{7} + 29446034t^{6} + 10106420t^{5} \\ +2766101t^{4} - 63392t^{3} - 4212t^{2} \\ -11504t + 14756 \end{pmatrix}.$$

The polynomial equation $k_{227}(t) = 0$ of 20th degree admits 20 solutions. All of them are convex (not written here). This means that there are no real positive solutions of the equation $k_{227}(t) = 0$. Thus we conclude that either $k_{227}(t) > 0$ or $k_{227}(t) < 0$, for all $t > 0$. Since $k_{227}(1) = 606208000$, this gives that $k_{227}(t) > 0$ for all $t > 0$, consequently, $w_{227}(x) \geq 0$, for all $x > 0$ giving $v_{227}(x) \geq 0$ and $u_{227}(x)$ nonnegative for all $x > 0$. Finally, we have the required result.

228. For $C + P_4 - S \leq \frac{N_1 + 12N - N_2}{12}$: We have to show that

$$\frac{1}{12}(N_1 + 12N - N_2 - 12C - 12P_4 + 12S) \geq 0.$$

We can write $N_1 + 12N - N_2 - 12C - 12P_4 + 12S = bg_{228}(a/b)$, where

$$g_{228}(x) = \frac{u_{228}(x)}{4(\sqrt{x} + 1)^2(x + 1)}$$

with

$$u_{228}(x) = (x + 1)(\sqrt{x} + 1)^2[24\sqrt{2x^2 + 2} - (\sqrt{x} + 1)\sqrt{2x + 2}]$$
$$- \begin{pmatrix} 31x^3 + 44x^{5/2} + 101x^2 + 101x \\ +52x(\sqrt{x} - 1)^2 + 44\sqrt{x} + 31 \end{pmatrix}.$$

Before proceeding further we shall prove that the expression $24\sqrt{2x^2 + 2} - (\sqrt{x} + 1)\sqrt{2x + 2}$ is nonnegative for all $x > 0$. It is true in view of Result 165 – Group 1. Now we shall show that $u_{228}(x) \geq 0$, $\forall x > 0$. Let's consider

$$v_{228}(x) = \{(x + 1)(\sqrt{x} + 1)^2[24\sqrt{2x^2 + 2} - (\sqrt{x} + 1)\sqrt{2x + 2}]\}^2$$
$$- \begin{pmatrix} 31x^3 + 44x^{5/2} + 101x^2 + 101x \\ +52x(\sqrt{x} - 1)^2 + 44\sqrt{x} + 31 \end{pmatrix}^2.$$

After simplifications,

$$v_{228}(x) = \begin{pmatrix} \sqrt{x}(\sqrt{x}-1)^2(1085x^4 + 3169x^3 \\ +22406x^2 + 3169x + 1085) + 193x^6 \\ +807x^{11/2} + 2630x^{9/2} + x^4 + 8657x^{7/2} \\ +8657x^{5/2} + x^2 + 2630x^{3/2} + 807\sqrt{x} + 193 \end{pmatrix}$$
$$- 48\sqrt{2x^2 + 2}\sqrt{2x + 2}(x+1)^2(\sqrt{x}+1)^5.$$

Still, we are unable to decide the nonnegativity of the expression $u_{228}(x)$. Let's apply again, argument given in Note 1. Let's consider

$$h_{228}(x) = \begin{pmatrix} \sqrt{x}(\sqrt{x}-1)^2(1085x^4 + 3169x^3 \\ +22406x^2 + 3169x + 1085) + 193x^6 \\ +807x^{11/2} + 2630x^{9/2} + x^4 + 8657x^{7/2} \\ +8657x^{5/2} + x^2 + 2630x^{3/2} + 807\sqrt{x} + 193 \end{pmatrix}^2$$
$$- \left[48\sqrt{2x^2 + 2}\sqrt{2x + 2}(x+1)^2(\sqrt{x}+1)^5\right]^2.$$

After simplifications,

$$h_{228}(x) = (\sqrt{x}-1)^2 \times w_{228}(x),$$

where

$$w_{228}(x) = \begin{pmatrix} 28033x^{11} + 694218x^{21/2} + 3641647x^{10} - 531700x^{19/2} \\ +19496491x^9 - 9983326x^{17/2} + 131353029x^8 \\ -145745968x^{15/2} + 372451146x^7 - 379208300x^{13/2} \\ +1074205206x^6 - 1163916728x^{11/2} + 1074205206x^5 \\ -379208300x^{9/2} + 372451146x^4 - 145745968x^{7/2} \\ +131353029x^3 - 9983326x^{5/2} + 19496491x^2 \\ -531700x^{3/2} + 3641647x + 694218\sqrt{x} + 28033 \end{pmatrix}.$$

Let's consider

$$k_{228}(t) := w_{228}(t^2) = \begin{pmatrix} 28033t^{22} + 694218t^{21} + 3641647t^{20} \\ -531700t^{19} + 19496491t^{18} - 9983326t^{17} \\ +131353029t^{16} - 145745968t^{15} \\ +372451146t^{14} - 379208300t^{13} \\ +1074205206t^{12} - 1163916728t^{11} \\ +1074205206t^{10} - 379208300t^9 \\ +372451146t^8 - 145745968t^7 \\ +131353029t^6 - 9983326t^5 + 19496491t^4 \\ -531700t^3 + 3641647t^2 + 694218t + 28033 \end{pmatrix}.$$

The polynomial equation $k_{228}(t) = 0$ of 22nd degree admits 22 solutions. Out of them 18 are convex (not written here) and four are reals: -0.5945560541, -0.1110587195, -9.004245724, and -16.81927201. Since we are working with $t > 0$. This means that there are no real positive solutions of the equation $k_{228}(t) = 0$. Thus we conclude that either $k_{228}(t) > 0$ or $k_{228}(t) < 0$, for all $t > 0$. Since $k_{228}(1) = 968884224$, this gives that $k_{228}(t) > 0$ for all $t > 0$, consequently, $w_{228}(x) \geq 0$, for all $x > 0$ giving $v_{228}(x) \geq 0$ and $u_{228}(x)$ nonnegative for all $x > 0$. Finally, we have the required result.

229. For $\frac{12R+7G-7S}{12} \leq \frac{N_1+12N-N_2}{12}$: We have to show that

$$\frac{1}{12}(N_1 + 12N - N_2 - 12R - 7G + 7S) \geq 0.$$

We can write $N_1 + 12N - N_2 - 12R - 7G + 7S = bg_{229}(a/b)$, where $g_{229}(x) = \frac{1}{4}u_{229}(x)/(x+1)$ with

$$u_{229}(x) = (x+1)\left[14\sqrt{2x^2+2} - (\sqrt{x}+1)\sqrt{2x+2}\right]$$
$$- \left(15x^2 + 10x^{3/2} - 2x + 10\sqrt{x} + 15\right).$$

After simplifications,

$$u_{229}(x) = (x+1)\left[14\sqrt{2x^2+2} - (\sqrt{x}+1)\sqrt{2x+2}\right]$$
$$- \left[15x^2 + 9x^{3/2} + \sqrt{x}(\sqrt{x}-1)^2 + 9\sqrt{x} + 15\right].$$

Before proceeding further we shall prove that the expression $14\sqrt{2x^2+2} - (\sqrt{x}+1)\sqrt{2x+2}$ is nonnegative for all $x > 0$. It is true in view of Result 165 – Group 1. Now we shall show that $u_{229}(x) \geq 0$, $\forall x > 0$. Let's consider

$$v_{229}(x) = \left\{(x+1)\left[14\sqrt{2x^2+2} - (\sqrt{x}+1)\sqrt{2x+2}\right]\right\}^2$$
$$- \left[15x^2 + 9x^{3/2} + \sqrt{x}(\sqrt{x}-1)^2 + 9\sqrt{x} + 15\right]^2.$$

After simplifications,

$$v_{229}(x) = \left(\begin{array}{l}(148x^3 + 124x^2 + 124x + 148)(\sqrt{x}-1)^2 \\ +53x(x-1)^2 + 21x^4 + 427x^3 + 427x + 21\end{array}\right)$$
$$- 28\sqrt{2x^2+2}\sqrt{2x+2}(x+1)^2(\sqrt{x}+1).$$

Still, we are unable to decide the nonnegativity of the expression $u_{229}(x)$. Let's apply again, argument given in Note 1. Let's consider

$$h_{229}(x) = \left(\frac{(148x^3 + 124x^2 + 124x + 148)(\sqrt{x} - 1)^2}{+53x(x-1)^2 + 21x^4 + 427x^3 + 427x + 21} \right)^2$$
$$- \left[28\sqrt{2x^2 + 2}\sqrt{2x + 2}(x+1)^2(\sqrt{x} + 1) \right]^2.$$

After simplifications,

$$h_{229}(x) = (\sqrt{x} - 1)^2 \times w_{229}(x),$$

where

$$w_{229}(x) = \begin{pmatrix} 25425x^7 - 55470x^{13/2} + 186611x^6 \\ -131676x^{11/2} + 260177x^5 + 42158x^{9/2} \\ +418667x^4 - 287560x^{7/2} + 418667x^3 \\ +42158x^{5/2} - 260177x^2 - 131676x^{3/2} \\ +186611x - 55470\sqrt{x} + 25425 \end{pmatrix}.$$

Let's consider

$$k_{229}(t) := w_{229}(t^2) = \begin{pmatrix} 25425t^{14} - 55470t^{13} + 186611t^{12} \\ -131676t^{11} + 260177t^{10} + 42158t^9 \\ +418667t^8 - 287560t^7 + 418667t^6 \\ +42158t^5 + 260177t^4 - 131676t^3 \\ +186611t^2 - 55470t + 25425 \end{pmatrix}.$$

The polynomial equation $k_{229}(t) = 0$ of 14th degree admits 14 solutions. All these 14 solutions are complex (not written here), i.e., it doesn't have any real solution. Since we are working with $t > 0$, then there are two possibilities, i.e., the graph of $k_{229}(t)$ is either in 1st quadrant or in 4th quadrant. Since $k_{229}(1) = 1204224$, this means that $k_{229}(t) > 0$, for all $t > 0$. Thus, in view of Note 1 $w_{229}(x) > 0$, for all $x > 0$ implies $v_{229}(x) \geq 0$, for all $x > 0$, hence proving the required result.

230. For $\frac{N_1 + 12N - N_2}{12} \leq N$**:** We have to show that $\frac{1}{12}(N_2 - N_1) \geq 0$. It is true by definition. See inequalities (30).

Combining Results 225 to 230 we get the proof of Proposition 17. $\square$

Proposition 18. *The following inequalities hold:*

$$N \leq \begin{cases} \dfrac{9S+8P_5-8C}{9} \leq P_5 \\ \left.\dfrac{3N_1+2A-2N_2}{3}\right\} \\ \dfrac{4N_2+R-S}{4} \end{cases} \leq \begin{cases} N_2 \\ \dfrac{6C+7H-7A}{6} \leq A \end{cases}$$
$$\dfrac{4N_2+R-S}{4} \leq \dfrac{27S+8P_2-8P_5}{27} \leq A.$$

Proof. The proof is based on the following results.

231. For $N \leq \frac{3N_1+2A-2N_2}{3}$: We have to show that $\frac{1}{3}(3N_1 + 2A - 2N_2 - 3N) \geq 0$. We can write $3N_1 + 2A - 2N_2 - 3N = bg_{231}(a/b)$, where $g_{231}(x) = \frac{1}{4}u_{231}(x)$ with

$$u_{231}(x) = 3x + 2\sqrt{x} + 3 - 2(\sqrt{x}+1)\sqrt{2x+2}.$$

Let's consider

$$v_{231}(x) = (3x + 2\sqrt{x} + 3)^2 - \left[2(\sqrt{x}+1)\sqrt{2x+2}\right]^2.$$

After simplifications,

$$v_{231}(x) = (\sqrt{x}-1)^4.$$

In view of Note 1, $v_{231}(x) \geq 0$ implies $u_{231}(x) \geq 0$, $\forall x > 0$, hence proving the required result.

232. For $N \leq \frac{4N_2+R-S}{4}$: We have to show that $\frac{1}{4}(4N_2 + R - S - 4N) \geq 0$. We can write $4N_2 + R - S - 4N = bg_{232}(a/b)$, where $g_{232}(x) = \frac{1}{6}u_{232}(x)/(x+1)$ with

$$u_{232}(x) = 3(x+1)\left[2(\sqrt{x}+1)\sqrt{2x+2} - \sqrt{2x^2+2}\right] - 4(x+\sqrt{x}+1)^2.$$

Before proceeding further we have to show that the expression $2(\sqrt{x}+1)\sqrt{2x+2} - \sqrt{2x^2+2}$ is nonnegative for all $x > 0$. It is true in view of Result 96 given in Group 1. Now we shall show that $u_{232}(x) \geq 0$, $\forall x > 0$. Let's consider

$$v_{232}(x) = \left\{3(x+1)\left[2(\sqrt{x}+1)\sqrt{2x+2} - \sqrt{2x^2+2}\right]\right\}^2 - \left[4(x+\sqrt{x}+1)^2\right]^2.$$

After simplifications,

$$v_{232}(x) = \begin{pmatrix} 74x^4 + 80x^{7/2} + 164x^3 + 176x^{5/2} \\ +164x^2 + 176x^{3/2} + 164x + 80\sqrt{x} + 74 \end{pmatrix}$$
$$- 36\sqrt{2x^2 + 2}\sqrt{2x + 2}(\sqrt{x} + 1)(x + 1)^2.$$

Still, we are unable to decide the nonnegativity of $v_{232}(x)$, we shall apply again the argument given in Note 1. Let's write

$$w_{232}(x) = \begin{pmatrix} 74x^4 + 80x^{7/2} + 164x^3 + 176x^{5/2} \\ +164x^2 + 176x^{3/2} + 164x + 80\sqrt{x} + 74 \end{pmatrix}^2$$
$$- \left[36\sqrt{2x^2 + 2}\sqrt{2x + 2}(\sqrt{x} + 1)(x + 1)^2 \right]^2.$$

After simplifications,

$$w_{232}(x) = 4(\sqrt{x} - 1)^4 \begin{pmatrix} 73x^6 + 660x^{11/2} + 2094x^5 \\ +4820x^{9/2} + 8379x^4 + 11304x^{7/2} \\ +12732x^3 + 11304x^{5/2} + 8379x^2 \\ +4820x^{3/2} + 2094x + 660\sqrt{x} + 73 \end{pmatrix}.$$

In view of Note 1, $w_{232}(x) \geq 0$ implies $v_{232}(x) \geq 0$ implies $u_{232}(x) \geq 0$, $\forall x > 0$. This proves the required result.

233. For N $\leq \frac{9S + 8P_5 - 8C}{9}$: We have to show that $\frac{1}{9}(9S + 8P_5 - 8C - 9N) \geq 0$. We can write $4N_2 + R - S - 4N = bg_{233}(a/b)$, where

$$g_{233}(x) = \frac{u_{233}(x)}{2(x + 1)(\sqrt{x} + 1)^2}$$

with

$$u_{233}(x) = 9\sqrt{2x^2 + 2}(x + 1)(\sqrt{x} + 1)^2$$
$$- \left(6x^3 + 50x^{5/2} - 2x^2 + 36x^{3/2} - 2x + 50\sqrt{x} + 6 \right).$$

After simplifications,

$$u_{233}(x) = 9\sqrt{2x^2 + 2}(x + 1)(\sqrt{x} + 1)^2$$
$$- \begin{pmatrix} 6x^3 + \sqrt{x}(x + 1)(\sqrt{x} - 1)^2 \\ +49x^{5/2} + 34x^{3/2} + 49\sqrt{x} + 6 \end{pmatrix}.$$

Let's consider

$$v_{233}(x) = \left[9\sqrt{2x^2 + 2}(x+1)(\sqrt{x}+1)^2\right]^2$$
$$- \left(\begin{array}{c}6x^3 + \sqrt{x}(x+1)(\sqrt{x}-1)^2 \\ +49x^{5/2} + 34x^{3/2} + 49\sqrt{x}+6\end{array}\right)^2.$$

This gives

$$v_{233}(x) = (\sqrt{x}-1)^2 \left(\begin{array}{c}89\sqrt{x}(x^3+1) + 63x^4 + 135x^3 + 135x \\ +187\sqrt{x}(\sqrt{x}-1)^2(x+\sqrt{x}+1)^2 \\ +x^2(\sqrt{x}-2)^2 + x(2\sqrt{x}-1)^2 + 63\end{array}\right).$$

In view of Note 1, $v_{233}(x) \geq 0$ implies $u_{233}(x) \geq 0$, $\forall x > 0$. This proves the required result.

234. For $\frac{9S+8P_5-8C}{9} \leq P_5$: We have to show that $\frac{1}{9}(P_5 - 9S + 8C) \geq 0$. We can write $P_5 - 9S + 8C = bg_{234}(a/b)$, where

$$g_{234}(x) = \frac{u_{234}(x)}{2(x+1)(\sqrt{x}+1)^2}$$

with

$$u_{234}(x) = 18x^3 + 32x^{5/2} + 22x^2 + 22x + 32\sqrt{x} + 18$$
$$- 9\sqrt{2x^2 + 2}(x+1)(\sqrt{x}+1)^2.$$

Let's consider

$$v_{234}(x) = \left(18x^3 + 32x^{5/2} + 22x^2 + 22x + 32\sqrt{x} + 18\right)^2$$
$$- \left[9\sqrt{2x^2 + 2}(x+1)(\sqrt{x}+1)^2\right]^2.$$

After simplifications,

$$v_{234}(x) = 2(\sqrt{x}-1)^2 \left(\begin{array}{c}81x^5 + 414x^{9/2} + 1007x^4 + 1332x^{7/2} \\ +1080x^3 + 812x^{5/2} + 1080x^2 \\ +1332x^{3/2} + 1007x + 414\sqrt{x} + 81\end{array}\right).$$

In view of Note 1, $v_{234}(x) \geq 0$ implies $u_{234}(x) \geq 0$, $\forall x > 0$, hence proving the required result.

235. For $\frac{3N_1+2A-2N_2}{3} \leq \frac{6C+7H-7A}{6}$: We have to show that

$$\frac{1}{6}(6C + 7H - 11A - 6N_1 + 4N_2) \geq 0.$$

We can write $6C + 7H - 11A - 6N_1 + 4N_2 = bg_{235}(a/b)$, where $g_{235}(x) = u_{235}(x)/(x+1)$ with

$$u_{235}(x) = \sqrt{2x+2}(x+1)(\sqrt{x}+1) - (x^2 + 3x^{3/2} + 3\sqrt{x} + 1).$$

Let's consider

$$v_{235}(x) = \left[\sqrt{2x+2}(x+1)(\sqrt{x}+1)\right]^2 - (x^2 + 3x^{3/2} + 3\sqrt{x} + 1)^2.$$

After simplifications,

$$v_{235}(x) = (\sqrt{x}-1)^4(x^2 + 2x^{3/2} + x + 2\sqrt{x} + 1).$$

In view of Note 1, $v_{235}(x) \geq 0$ implies $u_{235}(x) \geq 0$, $\forall x > 0$, hence proving the required result.

236. For $\frac{6C+7H-7A}{6} \leq A$: We have to show that $\frac{1}{6}(13A - 6C - 7H) \geq 0$. We can write $13A - 6C - 7H = bg_{236}(a/b)$, where

$$g_{236}(x) = \frac{(x-1)^2}{2(x+1)}.$$

Since, $g_{236}(x) \geq 0$, $\forall x > 0$, hence proving the required result.

237. For $\frac{3N_1+2A-2N_2}{3} \leq N_2$: We have to show that $\frac{1}{3}(5N_2 - 3N_1 - 2A) \geq 0$. We can write $5N_2 - 3N_1 - 2A = bg_{237}(a/b)$, where $g_{237}(x) = \frac{1}{4}u_{237}(x)$ with

$$u_{237}(x) = 5\sqrt{2x+2}(\sqrt{x}+1) - (7x + 6\sqrt{x} + 7).$$

Let's consider

$$v_{237}(x) = \left[5\sqrt{2x+2}(\sqrt{x}+1)\right]^2 - (7x + 6\sqrt{x} + 7)^2.$$

After simplifications,

$$v_{237}(x) = (\sqrt{x}-1)^2(x^2 + 18\sqrt{x} + 1).$$

In view of Note 1, $v_{237}(x) \geq 0$ implies $u_{237}(x) \geq 0$, $\forall x > 0$, hence proving the required result.

238. For $\frac{4N_2+R-S}{4} \leq N_2$: We have to show that $\frac{1}{4}(S - R) \geq 0$. It is true by definition. See inequalities (30).

239. For $\frac{4N_2+R-S}{4} \leq \frac{27S+8P_2-8P_5}{27}$: We have to show that

$$\frac{1}{108}(135S + 32P_2 - 32P_5 - 108N_2 - 27R) \geq 0.$$

We can write $135S + 32P_2 - 32P_5 - 108N_2 - 27R = bg_{239}(a/b)$, where

$$g_{239}(x) = \frac{u_{239}(x)}{2(x+1)(x^2+1)(\sqrt{x}+1)^2}$$

with

$$u_{239}(x) = 27(x+1)(x^2+1)(\sqrt{x}+1)^2$$
$$\times \left[5\sqrt{2x^2+2} - 2(\sqrt{x}+1)\sqrt{2x+2}\right]$$
$$- \left(\begin{array}{l}100 + 72x^{9/2} + 200x^4 - 56x^{7/2} + 172x^3 \\ -112x^{5/2} + 172x^2 - 56x^{3/2} + 200x + 72\sqrt{x}\end{array}\right),$$

i.e.,

$$u_{239}(x) = 27(x+1)(x^2+1)(\sqrt{x}+1)^2$$
$$\times \left[5\sqrt{2x^2+2} - 2(\sqrt{x}+1)\sqrt{2x+2}\right]$$
$$- \left(\begin{array}{l}72x^{9/2} + 172x^4 + 28x(x+1)^2(\sqrt{x}-1)^2 \\ +100x^5 + 88x^3 + 88x^2 + 172x + 72\sqrt{x} + 100\end{array}\right).$$

Before proceeding further we shall prove that the expression $5\sqrt{2x^2+2} - 2(\sqrt{x}+1)\sqrt{2x+2}$ is nonnegative for all $x > 0$. Let's consider

$$m_{239}(x) = \left[5\sqrt{2x^2+2}\right]^2 - \left[2(\sqrt{x}+1)\sqrt{2x+2}\right]^2.$$

After simplifications,

$$m_{239}(x) = 2(21x^2 - 8x - 8x^{3/2} - 8\sqrt{x} + 21)$$
$$= 8(2x + 3\sqrt{x} + 2)(\sqrt{x}-1)^2 + 5x^2 + 8x + 5.$$

In view of Note 1, it proves that the expression $5\sqrt{2x^2+2} - 2(\sqrt{x}+1)\sqrt{2x+2}$ is positive for all $x > 0$. We shall show that $u_{239}(x) \geq 0$, $\forall x > 0$. Let's consider

$$v_{239}(x) = \left[27(x+1)(x^2+1)(\sqrt{x}+1)^2\right]^2$$
$$\times \left[5\sqrt{2x^2+2} - 2(\sqrt{x}+1)\sqrt{2x+2}\right]^2$$
$$- \left(\begin{array}{l}72x^{9/2} + 172x^4 + 28x(x+1)^2(\sqrt{x}-1)^2 \\ +100x^5 + 88x^3 + 88x^2 + 172x + 72\sqrt{x} + 100\end{array}\right)^2.$$

After simplifications,

$$
v_{239}(x) = \begin{pmatrix}
32282x^{10} + 166392x^{19/2} + 351392x^9 \\
+641416x^{17/2} + 932394x^8 + 1454704x^{15/2} \\
+1822688x^7 + 2441616x^{13/2} + 2641916x^6 \\
+3022976x^{11/2} + 2844288x^5 + 3022976x^{9/2} \\
+2641916x^4 + 2441616x^{7/2} + 1822688x^3 \\
+1454704x^{5/2} + 932394x^2 + 641416x^{3/2} \\
+351392x + 166392\sqrt{x} + 32282
\end{pmatrix}
$$
$$
- 14580\sqrt{2x+2}\sqrt{2x^2+2}(x+1)^2(x^2+1)^2(\sqrt{x}+1)^5.
$$

Since we are unable to decide the nonnegativity of $v_{239}(x)$, we shall apply again the argument given in Note 1. Let's consider

$$
w_{239}(x) = \begin{pmatrix}
32282x^{10} + 166392x^{19/2} + 351392x^9 \\
+641416x^{17/2} + 932394x^8 + 1454704x^{15/2} \\
+1822688x^7 + 2441616x^{13/2} + 2641916x^6 \\
+3022976x^{11/2} + 2844288x^5 + 3022976x^{9/2} \\
+2641916x^4 + 2441616x^{7/2} + 1822688x^3 \\
+1454704x^{5/2} + 932394x^2 + 641416x^{3/2} \\
+351392x + 166392\sqrt{x} + 32282
\end{pmatrix}^2
$$
$$
- \left[14580\sqrt{2x+2}\sqrt{2x^2+2}(x+1)^2(x^2+1)^2(\sqrt{x}+1)^5\right]^2.
$$

After simplifications,

$$
w_{239}(x) = 4(\sqrt{x}-1)^4 \times h_{239}(x)
$$

where

$h_{239}(x) =$

$$
\begin{pmatrix}
47955481 + 4684004586x + 47955481x^{18} + 751791196x^{35/2} \\
+4684004586x^{17} + 17866609580x^{33/2} + 49944341613x^{16} \\
+111306437344x^{31/2} + 210683703912x^{15} + 352074620096x^{29/2} \\
+538355245656x^{14} + 768497832032x^{27/2} + 1041525608872x^{13} \\
+1343318978240x^{25/2} + 1657725207728x^{12} + 1957053680128x^{23/2} \\
+2229905018776x^{11} + 2458387179104x^{21/2} + 2641735035506x^{10} \\
+2757781946584x^{19/2} + 2802045773532x^9 + 2757781946584x^{17/2} \\
+2641735035506x^8 + 2458387179104x^{15/2} + 2229905018776x^7 \\
+1957053680128x^{13/2} + 1657725207728x^6 + 1343318978240x^{11/2} \\
+1041525608872x^5 + 768497832032x^{9/2} + 538355245656x^4 \\
+352074620096x^{7/2} + 210683703912x^3 + 111306437344x^{5/2} \\
+49944341613x^2 + 17866609580x^{3/2} + 751791196\sqrt{x}
\end{pmatrix}
$$

In view of Note 1, $w_{239}(x) \geq 0$ implies $v_{239}(x) \geq 0$ implies $u_{239}(x) \geq 0$, $\forall x > 0$, hence proving the required result.

240. For $\frac{27S+8P_2-8P_5}{27} \leq \mathbf{A}$: We have to show that $\frac{1}{27}(27A - 27S - 8P_2 + 8P_5) \geq 0$. We can write $27A - 27S - 8P_2 + 8P_5 = bg_{240}(a/b)$, where

$$g_{240}(x) = \frac{u_{240}(x)}{2(x^2+1)(\sqrt{x}+1)^2}$$

with

$$u_{240}(x) = \begin{pmatrix} 43x^4 + 54x^{7/2} + 70x^3 + 22x^{5/2} \\ +54x^2 + 22x^{3/2} + 70x + 54\sqrt{x} + 43 \end{pmatrix} \\ - 27\sqrt{2x^2 + 2}(x^2+1)(\sqrt{x}+1)^2.$$

Let's consider

$$v_{240}(x) = \begin{pmatrix} 43x^4 + 54x^{7/2} + 70x^3 + 22x^{5/2} \\ +54x^2 + 22x^{3/2} + 70x + 54\sqrt{x} + 43 \end{pmatrix}^2 \\ - \left[27\sqrt{2x^2 + 2}(x^2+1)(\sqrt{x}+1)^2 \right]^2.$$

After simplifications,

$$v_{240}(x) = (\sqrt{x} - 1)^2 \times w_{240}(x),$$

where

$$w_{240}(x) = \begin{pmatrix} 391x^7 - 406x^{13/2} - 1015x^6 + 1996x^{11/2} \\ +11095x^5 + 13502x^{9/2} + 6105x^4 - 1128x^{7/2} \\ +6105x^3 + 13502x^{5/2} + 11095x^2 \\ +1996x^{3/2} - 1015x - 406\sqrt{x} + 391 \end{pmatrix}.$$

We need to show that $v_{240}(x) \geq 0$, $\forall x > 0$. Let's consider $x = t^2$, then we can write

$$k_{240}(t) := w_{240}(t^2) = \begin{pmatrix} 391t^{14} - 406t^{13} - 1015t^{12} + 1996t^{11} \\ +11095t^{10} + 13502t^9 + 6105t^8 - 1128t^7 \\ +6105t^6 + 13502t^5 + 11095t^4 \\ +1996t^3 - 1015t^2 - 406t + 391 \end{pmatrix}.$$

The polynomial equation $k_{240}(t) = 0$ is of 14th degree. It admits 14 solutions. All these 14 solutions are complex (not written here), i.e., it doesn't have any real solution. Since we are working with $t > 0$, then there are two possibilities, i.e., the graph of $k_{240}(t)$ is either in 1st quadrant or in 4th quadrant. Since $k_{240}(1) = 62208$, this means that $k_{240}(t) > 0$, for all $t > 0$. Thus, in view of Note 1 $w_{240}(x) > 0$, for all $x > 0$ implies $v_{240}(x) \geq 0$, for all $x > 0$, hence proving the required result.

Combining Results 231 to 240 we get the proof of Proposition 18. $\quad\Box$

Proposition 19. *The following inequalities hold:*

$$N_2 \leq \frac{42N + C - P_4}{42} \leq \begin{cases} \frac{14C+9P_3-9P_5}{14} \leq \begin{cases} P_5 \\ R \end{cases} \\ \frac{12N_3+C-S}{12} \\ \frac{36R+11G-11C}{36} \end{cases} \leq (a)$$

$$(a) \leq \frac{8A + P_4 - P_5}{8} \leq \begin{cases} \frac{6S+5P_5-5C}{6} \leq P_5 \\ \frac{4P_2+45S-45R}{4} \leq \frac{8C+9H-9A}{8} \\ \frac{18S+5P_2-5P_5}{18} \end{cases} \leq A.$$

Proof. The proof is based on the following results.

241. For $N_2 \leq \frac{42N+C-P_4}{42}$: We have to show that $\frac{1}{42}(42N + C - P_4 - 42N_2) \geq 0$. We can write $42N + C - P_4 - 42N_2 = bg_{241}(a/b)$, where

$$g_{241}(x) = \frac{u_{241}(x)}{2(x+1)(\sqrt{x}+1)^2}$$

with

$$u_{241}(x) = 30x^3 + 88x^{5/2} + 134x^2 + 168x^{3/2} + 134x + 88\sqrt{x} + 30 \\ - 21\sqrt{2x+2}(x+1)(\sqrt{x}+1)^3.$$

Let's consider

$$v_{241}(x) = \begin{pmatrix} 30x^3 + 88x^{5/2} + 134x^2 \\ +168x^{3/2} + 134x + 88\sqrt{x} + 30 \end{pmatrix}^2 \\ - \left[21\sqrt{2x+2}(x+1)(\sqrt{x}+1)^3\right]^2.$$

After simplifications,

$$v_{241}(x) = 2(\sqrt{x}-1)^4 \begin{pmatrix} 9x^4 + 30x^{7/2} + 20x^3 + 10x^{5/2} \\ +30x^2 + 10x^{3/2} + 20x + 30\sqrt{x} + 9 \end{pmatrix}.$$

In view of Note 1, $v_{241}(x) \geq 0$ implies $u_{241}(x) \geq 0$, $\forall x > 0$, hence proving the required result.

242. For $\frac{42N+C-P_4}{42} \leq \frac{12N+C-S}{12}$**:** We have to show that $\frac{1}{84}(5C - 7S + 2P_4) \geq 0$. We can write $5C - 7S + 2P_4 = bg_{242}(a/b)$, where

$$g_{242}(x) = \frac{u_{242}(x)}{2(x+1)(\sqrt{x}+1)^2}$$

with

$$u_{242}(x) = 10x^3 + 20x^{5/2} + 26x^2 + 26x + 20\sqrt{x} + 10$$
$$- 7\sqrt{2x^2 + 2}(x+1)(\sqrt{x}+1)^2.$$

Let's consider

$$v_{242}(x) = \left(10x^3 + 20x^{5/2} + 26x^2 + 26x + 20\sqrt{x} + 10\right)^2$$
$$- \left[7\sqrt{2x^2 + 2}(x+1)(\sqrt{x}+1)^2\right]^2.$$

After simplifications,

$$v_{242}(x) = 2(\sqrt{x} - 1)^4 \left(\begin{array}{l} x^4 + 8x^{7/2} + 94x^3 + 264x^{5/2} \\ +386x^2 + 264x^{3/2} + 94x + 8\sqrt{x} + 1 \end{array} \right).$$

In view of Note 1, $v_{242}(x) \geq 0$ implies $u_{242}(x) \geq 0$, $\forall x > 0$, hence proving the required result.

243. For $\frac{42N+C-P_4}{42} \leq \frac{36R+11G-11C}{36}$**:** We have to show that

$$\frac{1}{252}(252R + 77G - 83C - 252N + 6P_4) \geq 0.$$

We can write $252R + 77G - 83C - 252N + 6P_4 = bg_{243}(a/b)$, where

$$g_{243}(x) = \frac{u_{243}(x)}{(x+1)(\sqrt{x}+1)^2}$$

with

$$u_{243}(x) = x^3 - 5x^{5/2} + 11x^2 - 14x^{3/2} + 11x - 5\sqrt{x} + 1$$
$$= (x - \sqrt{x} + 1)(\sqrt{x} - 1)^4.$$

Since $u_{243}(x) \geq 0$, $\forall x > 0$, hence proving the required result.

244. For $\frac{42N+C-P_4}{42} \leq \frac{14C+9P_3-9P_5}{14}$**:** We have to show that

$$\frac{1}{42}(41C + 27P_3 - 27P_5 - 42N + P_4) \geq 0.$$

We can write $41C + 27P_3 - 27P_5 - 42N + P_4 = bg_{244}(a/b)$, where

$$g_{244}(x) = \frac{u_{244}(x)}{(x+1)(x-\sqrt{x}+1)(\sqrt{x}+1)^2}$$

with

$$u_{244}(x) = \begin{pmatrix} 40x^{7/2} - 119x^3 + 116x^{5/2} - 74x^2 \\ +116x^{3/2} - 119x + 40\sqrt{x} \end{pmatrix}$$
$$= \sqrt{x}(40x + 41\sqrt{x} + 40)(\sqrt{x} - 1)^4.$$

Since $u_{244}(x) \geq 0$, $\forall x > 0$, hence proving the required result.

245. For $\frac{14C+9P_3-9P_5}{14} \leq P_5$**:** We have to show that $\frac{1}{14}(23P_5 - 14C - 9P_3) \geq 0$. We can write $23P_5 - 14C - 9P_3 = bg_{245}(a/b)$, where

$$g_{245}(x) = \frac{u_{245}(x)}{(x+1)(x-\sqrt{x}+1)(\sqrt{x}+1)^2}$$

with

$$u_{245}(x) = \begin{pmatrix} 9x^4 - 37x^{7/2} + 83x^3 - 101x^{5/2} \\ +92x^2 - 101x^{3/2} + 83x - 37\sqrt{x} + 9 \end{pmatrix}.$$

After simplifications,

$$u_{245}(x) = (\sqrt{x} - 1)^2 \times w_{245}(x),$$

where

$$w_{245}(x) = 9x^3 - 19x^{5/2} + 36x^2 - 10x^{3/2} + 36x - 19\sqrt{x} + 9$$
$$= (x+1)\left[5(\sqrt{x} - 1)^4 + 2(x-1)^2 + 2x^2 + x + 2\right]$$
$$+ x^{5/2} + 30x^{3/2} + \sqrt{x}.$$

Since $w_{245}(x) > 0$ giving $u_{245}(x) \geq 0$, $\forall x > 0$, hence proving the required result.

246. For $\frac{14C+9P_3-9P_5}{14} \leq R$**:** We have to show that $\frac{1}{14}(14R - 14C - 9P_3 + 9P_5) \geq 0$. We can write $14R - 14C - 9P_3 + 9P_5 = bg_{246}(a/b)$, where

$$g_{246}(x) = \frac{u_{246}(x)}{3(x+1)(x-\sqrt{x}+1)(\sqrt{x}+1)^2}$$

with

$$u_{246}(x) = \begin{pmatrix} 13x^4 + 109x^3 - 41x^{7/2} - 121x^{5/2} \\ +80x^2 - 121x^{3/2} + 109x - 41\sqrt{x} + 13 \end{pmatrix}.$$

After simplifications,

$$u_{246}(x) = (\sqrt{x} - 1)^2 \begin{pmatrix} 5x^3 + 8(x^2 + 1)(\sqrt{x} - 1)^2 + x^{5/2} \\ +58x^2 + 26x^{3/2} + 58x + \sqrt{x} + 5 \end{pmatrix}.$$

This gives $u_{246}(x) \geq 0$, $\forall x > 0$, hence proving the required result.

247. For $\frac{12N+C-S}{12} \leq \frac{8A+P_4-P_5}{8}$: We have to show that

$$\frac{1}{24}(24A + 3P_4 - 3P_5 - 24N - 2C + 2S) \geq 0.$$

We can write $24A + 3P_4 - 3P_5 - 24N - 2C + 2S = bg_{247}(a/b)$, where

$$g_{247}(x) = \frac{u_{247}(x)}{(x+1)(\sqrt{x}+1)^2}$$

with

$$u_{247}(x) = \sqrt{2x^2 + 2}(x + 1)(\sqrt{x} + 1)^2$$
$$- (x^3 + 4x^{5/2} + 3x^2 + 3x + 4\sqrt{x} + 1).$$

Let's consider

$$v_{247}(x) = \left[\sqrt{2x^2 + 2}(x + 1)(\sqrt{x} + 1)^2\right]^2$$
$$- (x^3 + 4x^{5/2} + 3x^2 + 3x + 4\sqrt{x} + 1)^2.$$

After simplifications,

$$v_{247}(x) = (x - 1)^6.$$

In view of Note 1, $v_{247}(x) \geq 0$ implies $u_{247}(x) \geq 0$, $\forall x > 0$, hence proving the required result.

248. For $\frac{36R+11G-11C}{36} \leq \frac{8A+P_4-P_5}{8}$: We have to show that

$$\frac{1}{72}(72A + 9P_4 - 9P_5 - 72R - 22G + 22C) \geq 0.$$

We can write $72A + 9P_4 - 9P_5 - 72R - 22G + 22C = bg_{248}(a/b)$, where

$$g_{248}(x) = \frac{u_{248}(x)}{(x+1)(\sqrt{x}+1)^2}$$

with

$$u_{248}(x) = x^3 - 2x^{5/2} - x^2 + 4x^{3/2} - x - 2\sqrt{x} + 1$$
$$= (\sqrt{x}+1)^2(\sqrt{x}-1)^4.$$

Since $u_{248}(x) \geq 0$, $\forall x > 0$, hence proving the required result.

249. For $\frac{8A+P_4-P_5}{8} \leq \frac{8C+9H-9A}{8}$: We have to show that

$$\frac{1}{8}(8C + 9H - 17A - P_4 + P_5) \geq 0.$$

We can write $8C + 9H - 17A - P_4 + P_5 = bg_{249}(a/b)$, where

$$g_{249}(x) = \frac{u_{249}(x)}{2(x+1)(\sqrt{x}+1)^2}$$

with

$$u_{249}(x) = x^3 - 2x^{5/2} - x^2 + 4x^{3/2} - x - 2\sqrt{x} + 1$$
$$= (\sqrt{x}-1)^4(\sqrt{x}+1)^2.$$

Since $u_{249}(x) \geq 0$, $\forall x > 0$, hence proving the required result.

250. For $\frac{8A+P_4-P_5}{8} \leq \frac{18S+5P_2-5P_5}{18}$: We have to show that

$$\frac{1}{72}(72S + 20P_2 - 11P_5 - 72A - 9P_4) \geq 0.$$

We can write $72S + 20P_2 - 11P_5 - 72A - 9P_4 = bg_{250}(a/b)$, where

$$g_{250}(x) = \frac{u_{250}(x)}{(x^2+1)(\sqrt{x}+1)^2}$$

with

$$u_{250}(x) = 36\sqrt{2x^2 + 2}(x^2+1)(\sqrt{x}+1)^2$$
$$- \begin{pmatrix} 47x^4 + 72x^{7/2} + 110x^3 + 32x^{5/2} \\ +54x^2 + 32x^{3/2} + 110x + 72\sqrt{x} + 47 \end{pmatrix}.$$

Let's consider

$$v_{250}(x) = \left[36\sqrt{2x^2+2}(x^2+1)(\sqrt{x}+1)^2\right]^2$$
$$- \left(\begin{array}{l} 47x^4 + 72x^{7/2} + 110x^3 + 32x^{5/2} \\ +54x^2 + 32x^{3/2} + 110x + 72\sqrt{x} + 47 \end{array}\right)^2.$$

After simplifications,

$$v_{250}(x) = (\sqrt{x}-1)^4 \left(\begin{array}{l} 383x^6 + 5132x^{11/2} + 18258x^5 \\ +35292x^{9/2} + 40349x^4 + 30824x^{7/2} \\ +22916x^3 + 30824x^{5/2} + 40349x^2 \\ +35292x^{3/2} + 18258x + 5132\sqrt{x} + 383 \end{array}\right).$$

In view of Note 1, $v_{250}(x) \geq 0$ implies $u_{250}(x) \geq 0$, $\forall x > 0$, hence proving the required result.

251. For $\frac{8A+P_4-P_5}{8} \leq \frac{6S+5P_5-5C}{6}$: We have to show that

$$\frac{1}{24}(24S + 23P_5 - 20C - 24A - 3P_4) \geq 0.$$

We can write $24S + 23P_5 - 20C - 24A - 3P_4 = bg_{251}(a/b)$, where

$$g_{251}(x) = \frac{u_{251}(x)}{(x^2+1)(\sqrt{x}+1)^2}$$

with

$$u_{251}(x) = 12\sqrt{2x^2+2}(x+1)(\sqrt{x}+1)^2$$
$$- \left(9x^3 + 64x^{5/2} - x^2 + 48x^{3/2} - x + 64\sqrt{x} + 9\right).$$

After simplifications,

$$u_{251}(x) = 12\sqrt{2x^2+2}(x+1)(\sqrt{x}+1)^2$$
$$- \left(\begin{array}{l} 9x^3 + 63x^{5/2} + \sqrt{x}(x+1)(\sqrt{x}-1)^2 \\ +x^2 + 46x^{3/2} + 63\sqrt{x} + x + 9 \end{array}\right).$$

Let's consider

$$u_{251}(x) = \left[12\sqrt{2x^2+2}(x+1)(\sqrt{x}+1)^2\right]^2$$
$$- \left(\begin{array}{l} 9x^3 + 63x^{5/2} + \sqrt{x}(x+1)(\sqrt{x}-1)^2 \\ +x^2 + 46x^{3/2} + 63\sqrt{x} + x + 9 \end{array}\right)^2.$$

This gives

$$v_{251}(x) = (\sqrt{x} - 1)^4 \begin{pmatrix} 206x^4 + 828\sqrt{x}(\sqrt{x} - 1)^2 (x + \sqrt{x} + 1)^2 \\ + 178x^3 + 178x + 118x(x+1)(\sqrt{x} - 1)^2 \\ + (\sqrt{x} - 1)^2(\sqrt{x} + 1)^2(x+1)^2 + 206 \end{pmatrix}.$$

In view of Note 1, $v_{251}(x) \geq 0$ implies $u_{251}(x) \geq 0$, $\forall x > 0$, hence proving the required result.

252. For $\frac{6S + 5P_5 - 5C}{6} \leq P_5$: We have to show that $\frac{1}{6}(P_5 - 6S + 5C) \geq 0$. We can write $P_5 - 6S + 5C = bg_{252}(a/b)$, where

$$g_{252}(x) = \frac{u_{252}(x)}{(x^2 + 1)(\sqrt{x} + 1)^2}$$

with

$$u_{252}(x) = 6x^3 + 10x^{5/2} + 8x^2 + 8x + 10\sqrt{x} + 6 \\ - 3\sqrt{2x^2 + 2}(x+1)(\sqrt{x} + 1)^2.$$

Let's consider

$$v_{252}(x) = \left(6x^3 + 10x^{5/2} + 8x^2 + 8x + 10\sqrt{x} + 6\right)^2 \\ - \left[3\sqrt{2x^2 + 2}(x+1)(\sqrt{x} + 1)^2\right]^2.$$

After simplifications,

$$v_{252}(x) = 2(\sqrt{x} - 1)^2 \begin{pmatrix} 9x^5 + 42x^{9/2} + 101x^4 + 132x^{7/2} \\ + 108x^3 + 80x^{5/2} + 108x^2 \\ + 132x^{3/2} + 101x + 42\sqrt{x} + 9 \end{pmatrix}.$$

In view of Note 1, $v_{252}(x) \geq 0$ implies $u_{252}(x) \geq 0$, $\forall x > 0$, hence proving the required result.

253. For $\frac{8C + 9H - 9A}{8} \leq \frac{4P_2 + 45S - 45R}{4}$: We have to show that

$$\frac{1}{8}(8P_2 + 90S - 90R - 8C - 9H + 9A) \geq 0.$$

We can write $8P_2 + 90S - 90R - 8C - 9H + 9A = bg_{253}(a/b)$, where

$$g_{253}(x) = \frac{u_{253}(x)}{2(x^2 + 1)(x + 1)}$$

with

$$u_{253}(x) = 90\sqrt{2x^2 + 2}(x+1)(x^2+1)$$
$$- (127x^4 + 122x^3 + 222x^2 + 122x + 127).$$

Let's consider

$$v_{253}(x) = \left[90\sqrt{2x^2 + 2}(x+1)(x^2+1)\right]^2$$
$$- \left(127x^4 + 122x^3 + 222x^2 + 122x + 127\right)^2.$$

After simplifications,

$$v_{253}(x) = (x-1)^4\left(71x^4 + 1696x^3 - 114x^2 + 1696x + 71\right)$$
$$= (x-1)^4\left(71x^4 + 1639x^3 + 57x(x-1)^2 + 1639x + 71\right).$$

In view of Note 1, $v_{253}(x) \geq 0$ implies $u_{253}(x) \geq 0$, $\forall x > 0$, hence proving the required result.

254. For $\frac{4P_2 + 45S - 45R}{4} \leq \mathbf{A}$: We have to show that $\frac{1}{4}(4A - 4P_2 - 45S + 45R) \geq 0$. We can write $4A - 4P_2 - 45S + 45R = bg_{254}(a/b)$, where

$$g_{254}(x) = \frac{u_{254}(x)}{2(x+1)(x^2+1)}$$

with

$$u_{254}(x) = 64x^4 + 60x^3 + 112x^2 + 60x + 64$$
$$- 45\sqrt{2x^2 + 2}(x+1)(x^2+1).$$

Let's consider

$$v_{254}(x) = \left(64x^4 + 60x^3 + 112x^2 + 60x + 64\right)^2$$
$$- \left[45\sqrt{2x^2 + 2}(x+1)(x^2+1)\right]^2.$$

h After simplifications,

$$v_{254}(x) = 2(x-1)^2 \times w_{84}(x)$$

where

$$w_{254}(x) = 23x^6 - 164x^5 + 517x^4 - 392x^3 + 517x^2 - 164x + 23.$$

The polynomial equation $w_{254}(x) = 0$ of 6th degree admits 6 solutions. All of them are convex (not written here). This means that there are no real positive solutions of the equation $w_{254}(x) = 0$. Thus we conclude that either $w_{254}(x) > 0$ or $w_{254}(x) < 0$, for all $x > 0$. Since $w_{254}(1) = 360$, this gives that $w_{254}(x) > 0$ for all $t > 0$, consequently, $u_{254}(x) \geq 0$, for all $x > 0$. This proves the required result.

255. For $\frac{18S + 5P_2 - 5P_5}{18} \leq A$**:** We have to show that $\frac{1}{18}(18A - 18S - 5P_2 + 5P_5) \geq 0$. We can write $18A - 18S - 5P_2 + 5P = bg_{255}(a/b)$, where

$$g_{255}(x) = \frac{u_{255}(x)}{(x^2 + 1)(\sqrt{x} + 1)^2}$$

with

$$u_{255}(x) = \left(\begin{array}{c} 14x^4 + 18x^{7/2} + 23x^3 + 8x^{5/2} \\ +18x^2 + 8x^{3/2} + 23x + 18\sqrt{x} + 14 \end{array} \right) \\ - 9\sqrt{2x^2 + 2}(x^2 + 1)(\sqrt{x} + 1)^2,$$

Let's consider

$$v_{255}(x) = \left(\begin{array}{c} 14x^4 + 18x^{7/2} + 23x^3 + 8x^{5/2} + 18x^2 \\ +8x^{3/2} + 23x + 18\sqrt{x} + 14 \end{array} \right)^2 \\ - \left[9\sqrt{2x^2 + 2}(x^2 + 1)(\sqrt{x} + 1)^2 \right]^2.$$

After simplifications,

$$v_{255}(x) = (x - 1)^2 \times w_{255}(x),$$

where

$$w_{255}(x) = \left(\begin{array}{c} 34x^7 - 76x^{13/2} - 190x^6 + 100x^{11/2} + 1063x^5 \\ +1322x^{9/2} + 489x^4 - 300x^{7/2} + 489x^3 + 1322x^{5/2} \\ +1063x^2 + 100x^{3/2} - 190x - 76\sqrt{x} + 34 \end{array} \right).$$

For simplicity, let's consider

$$k_{255}(t) := w_{255}(t^2) = \left(\begin{array}{c} 34t^{14} - 76t^{13} - 190t^{12} + 100t^{11} \\ +1063t^{10} + 1322t^9 + 489t^8 - 300t^7 \\ +489t^6 + 1322t^5 + 1063t^4 \\ +100t^3 - 190t^2 - 76t + 34 \end{array} \right).$$

The polynomial equation $k_{255}(t) = 0$ of 14th degree admits 14 solutions. All of them are convex (not written here). This means that there are no real positive solutions of the equation $k_{255}(t) = 0$. Thus we conclude that either $k_{255}(t) > 0$ or $k_{255}(t) < 0$, for all $t > 0$. Since $k_{255}(1) = 5184$, this gives that $k_{255}(t) > 0$ for all $t > 0$, consequently, $v_{255}(x) \geq 0$, for all $x > 0$. This proves the required result.

Combining Results 241 to 255 we get the proof of Proposition 19. $\square$

Proposition 20. *The following inequalities hold:*

$$A \leq \frac{5S + P_3 - C}{5} \leq \left\{ \begin{matrix} P_5 \leq C + P_3 - P_4 \\ \left\{ \begin{matrix} \frac{3A+2C-2S}{3} \leq \frac{9P_2+14A-14P_3}{9} \\ S \leq \frac{18R+C-P2}{18} \end{matrix} \right\} \leq \frac{3P_2+5A-5P_3}{3} \end{matrix} \right\} \leq C.$$

Proof. It is based on the following results.

256. For $A \leq \frac{5S+P_3-C}{5}$: We have to show that $\frac{1}{5}(5S + P_3 - C - 5A) \geq 0$. We can write $5S + P_3 - C - 5A = bg_{256}(a/b)$, where

$$g_{256}(x) = \frac{u_{256}(x)}{2(x+1)(x - \sqrt{x} + 1)}$$

with

$$u_{256}(x) = 5\sqrt{2x^2 + 2}(x+1)(x - \sqrt{x} + 1) \\ - (7x^3 - 7x^{5/2} + 15x^2 - 10x^{3/2} + 15x - 7\sqrt{x} + 7).$$

After simplifications,

$$u_{256}(x) = 5\sqrt{2x^2 + 2}(x+1)(x - \sqrt{x} + 1) \\ - \left(\begin{matrix} (7x^2 + 5x + 7)(\sqrt{x} - 1)^2 \\ +7\sqrt{x}(x^2 + 1) + 3x(x+1) \end{matrix} \right).$$

Let's consider

$$v_{256}(x) = \left[5\sqrt{2x^2 + 2}(x+1)(x - \sqrt{x} + 1) \right]^2 \\ - \left(\begin{matrix} (7x^2 + 5x + 7)(\sqrt{x} - 1)^2 \\ +7\sqrt{x}(x^2 + 1) + 3x(x+1) \end{matrix} \right)^2.$$

This gives

$$v_{256}(x) = (\sqrt{x} - 1)^6 (x^3 + 4x^{5/2} + 10x^{3/2} + 4\sqrt{x} + 1).$$

In view of Note 1, $v_{256}(x) \geq 0$ implies $u_{256}(x) \geq 0$, $\forall x > 0$, hence proving the required result.

257. For $\frac{5S+P_3-C}{5} \leq P_5$: We have to show that $\frac{1}{5}(5P_5 - 5S - P_3 + C) \geq 0$. We can write $5P_5 - 5S - P_3 + C = bg_{257}(a/b)$, where

$$g_{257}(x) = \frac{u_{257}(x)}{2(x+1)(\sqrt{x}+1)^2(x - \sqrt{x}+1)}$$

with

$$u_{257}(x) = \begin{pmatrix} 12x^4 - 8x^{7/2} + 38x^3 - 32x^{5/2} \\ +60x^2 - 32x^{3/2} + 38x - 8\sqrt{x} + 12 \end{pmatrix} \\ - 5\sqrt{2x^2 + 2}(x+1)(x - \sqrt{x}+1)(\sqrt{x}+1)^2.$$

After simplifications,

$$u_{257}(x) = 2\begin{pmatrix} 2\left(x^3 + 4x^2 + 4x + 1\right)\left(\sqrt{x} - 1\right)^2 \\ +4x^4 + 9x^3 + 14x^2 + 9x + 4 \end{pmatrix} \\ - 5\sqrt{2x^2 + 2}(x+1)(x - \sqrt{x}+1)(\sqrt{x}+1)^2.$$

Let's consider

$$v_{257}(x) = 4\begin{pmatrix} 2\left(x^3 + 4x^2 + 4x + 1\right)\left(\sqrt{x} - 1\right)^2 \\ +4x^4 + 9x^3 + 14x^2 + 9x + 4 \end{pmatrix}^2 \\ - \left[5\sqrt{2x^2 + 2}(x+1)(x - \sqrt{x}+1)(\sqrt{x}+1)^2\right]^2.$$

This gives

$$v_{257}(x) = 2(\sqrt{x} - 1)^2 \times w_{257}(x),$$

where

$$w_{257}(x) = \begin{pmatrix} (\sqrt{x} - 1)^2\left(26x^6 + 131x^5 + 322x^4\right) \\ +(\sqrt{x} - 1)^2\left(442x^3 + 322x^2 + 131x + 26\right) \\ +21x^7 + 105x^6 + 259x^5 + 415x^4 \\ +415x^3 + 259x^2 + 105x + 21 \end{pmatrix}.$$

In view of Note 1, $v_{257}(x) \geq 0$ implies $u_{257}(x) \geq 0$, $\forall x > 0$, hence proving the required result.

258. For $P_5 \leq C + P_3 - P_4$: We have to show that $C + P_3 - P_4 - P_5 \geq 0$. We can write $C + P_3 - P_4 - P_5 = bg_{258}(a/b)$, where

$$g_{258}(x) = \frac{u_{258}(x)}{(x+1)(\sqrt{x}+1)^2(x-\sqrt{x}+1)}$$

with

$$u_{258}(x) = 2x^{7/2} - 10x^2 - 7x^3 + 10x^{3/2} + 10x^{5/2} - 7x + 2\sqrt{x}$$
$$= \sqrt{x}(\sqrt{x}-1)^4(2x+\sqrt{x}+2).$$

Since $u_{258}(x) \geq 0$, $\forall x > 0$, hence proving the required result.

259. For $\frac{5S+P_3-C}{5} \leq R$: We have to show that $\frac{1}{5}(5R - 5S - P_3 + C) \geq 0$. We can write $5R - 5S - P_3 + C = bg_{259}(a/b)$, where

$$g_{259}(x) = \frac{u_{259}(x)}{6(x+1)(x-\sqrt{x}+1)}$$

with

$$u_{259}(x) = 2\left(13x^3 - 13x^{5/2} + 20x^2 - 10x^{3/2} + 20x - 13\sqrt{x} + 13\right)$$
$$- 15\sqrt{2x^2 + 2}(x+1)(x-\sqrt{x}+1).$$

After simplifications,

$$u_{259}(x) = 2\left(\begin{array}{l}(13x^2 + 5x + 13)(\sqrt{x}-1)^2 \\ +13\sqrt{x}(x^2+1) + 2x(x+1)\end{array}\right)$$
$$- 15\sqrt{2x^2+2}(x+1)(x-\sqrt{x}+1).$$

Let's consider

$$v_{259}(x) = 4\left(\begin{array}{l}(13x^2 + 5x + 13)(\sqrt{x}-1)^2 \\ +13\sqrt{x}(x^2+1) + 2x(x+1)\end{array}\right)^2$$
$$- \left[15\sqrt{2x^2+2}(x+1)(x-\sqrt{x}+1)\right]^2.$$

This gives

$$v_{259}(x) = 2(\sqrt{x}-1)^2\left(\begin{array}{l}113x^5 + 140x^4 + 70x^{7/2} + 277x^3 \\ +277x^2 + 58x^2(\sqrt{x}-1)^2 \\ +70x^{3/2} + 140x + 113\end{array}\right).$$

In view of Note 1, $v_{259}(x) \geq 0$ implies $u_{259}(x) \geq 0$, $\forall x > 0$, hence proving the required result.

260. For R $\leq \frac{3A+2C-2S}{3}$: We have to show that $\frac{1}{3}(3A + 2C - 2S - 3R) \geq 0$. We can write $3A + 2C - 2S - 3R = bg_{260}(a/b)$, where $g_{260}(x) = \frac{1}{2}u_{260}(x)/(x+1)$ with

$$u_{260}(x) = 3x^2 + 2x + 3 - 2\sqrt{2x^2 + 2}(x+1).$$

Let's consider

$$v_{260}(x) = \left(3x^2 + 2x + 3\right)^2 - \left[2\sqrt{2x^2 + 2}(x+1)\right]^2.$$

After simplifications,

$$v_{260}(x) = (x-1)^2.$$

In view of Note 1, $v_{260}(x) \geq 0$ implies $u_{260}(x) \geq 0$, $\forall x > 0$, hence proving the required result.

261. For $\frac{3A+2C-2S}{3} \leq \frac{9P_2+14A-14P_3}{9}$: We have to show that

$$\frac{1}{9}(9P_2 + 5A - 14P_3 - 6C + 6S) \geq 0.$$

We can write $9P_2 + 5A - 14P_3 - 6C + 6S = bg_{261}(a/b)$, where

$$g_{261}(x) = \frac{u_{261}(x)}{2(x+1)(x^2+1)(x-\sqrt{x}+1)}$$

with

$$u_{261}(x) = 6\sqrt{2x^2 + 2}(x+1)\left(x^2 + 1\right)(x - \sqrt{x} + 1)$$
$$- \left(\begin{matrix} 7x^5 - 7x^{9/2} + 7x^4 + 28x^{7/2} - 22x^3 \\ +22x^{5/2} - 22x^2 + 28x^{3/2} + 7x - 7\sqrt{x} + 7 \end{matrix}\right).$$

After simplifications,

$$u_{261}(x) = 6\sqrt{2x^2 + 2}(x+1)\left(x^2 + 1\right)(x - \sqrt{x} + 1)$$
$$- \left(\begin{matrix} \left(7x^4 + 11x^{5/2} + 11x^{3/2} + 7\right)(\sqrt{x} - 1)^2 \\ +7\sqrt{x}(x^4 + 1) + 17x^{3/2}(x^2 + 1) \end{matrix}\right).$$

Let's consider

$$v_{261}(x) = \left[6\sqrt{2x^2 + 2}(x+1)\left(x^2 + 1\right)(x - \sqrt{x} + 1)\right]^2$$

$$-\left(\begin{matrix}(7x^4+11x^{5/2}+11x^{3/2}+7)\,(\sqrt{x}-1)^2\\+7\sqrt{x}(x^4+1)+17x^{3/2}(x^2+1)\end{matrix}\right)^2.$$

After simplifications,

$$v_{261}(x)=(\sqrt{x}-1)^4\left(\begin{matrix}23x^8+46x^{15/2}+259x^7+126x^{13/2}\\+554x^6+578x^{11/2}+813x^5+658x^{9/2}\\+414x^4+658x^{7/2}+813x^3+578x^{5/2}\\+554x^2+126x^{3/2}+259x+46\sqrt{x}+23\end{matrix}\right).$$

In view of Note 1, $v_{261}(x)\geq0$ implies $u_{261}(x)\geq0$, $\forall x>0$, hence proving the required result.

262. For S $\leq\frac{18R+C-P_2}{18}$: We have to show that $\frac{1}{18}(18R+C-P_2-18S)\geq0$. We can write $18R+C-P_2-18S=bg_{262}(a/b)$, where

$$g_{262}(x)=\frac{u_{262}(x)}{(x+1)(x^2+1)}$$

with

$$u_{262}(x)=13x^4+11x^3+24x^2+11x+13$$
$$-9\sqrt{2x^2+2}(x+1)(x^2+1).$$

Let's consider

$$v_{262}(x)=\left(13x^4+11x^3+24x^2+11x+13\right)^2$$
$$-\left[9\sqrt{2x^2+2}(x+1)(x^2+1)\right]^2.$$

After simplifications,

$$v_{262}(x)=(x-1)^4\left(7x^4-10x^3+15x^2-10x+7\right)$$
$$=(x-1)^4\left[2x^4+5(x^2+1)(x-1)^2+5x^2+2\right].$$

In view of Note 1, $v_{262}(x)\geq0$ implies $u_{262}(x)\geq0$, $\forall x>0$, hence proving the required result.

263. For $\frac{9P_2+14A-14P_3}{9}\leq\frac{3P_2+5A-5P_3}{3}$: We have to show that $\frac{1}{9}(A-P_3)\geq0$. It is true by definition. See inequalities (30).

264. For $\frac{18R+C-P_2}{18}\leq\frac{3P_2+5A-5P_3}{3}$: We have to show that

$$\frac{1}{18}(19P_2+30A-30P_3-18R-C)\geq0.$$

We can write $19P_2 + 30A - 30P_3 - 18R - C = bg_{264}(a/b)$, where

$$g_{264}(x) = \frac{u_{264}(x)}{(x+1)(x^2+1)(x-\sqrt{x}+1)}$$

with

$$u_{264}(x) = \begin{pmatrix} 2x^5 - 2x^{9/2} + 9x^4 - 37x^{7/2} + 49x^3 \\ -42x^{5/2} + 49x^2 - 37x^{3/2} - 2\sqrt{x} + 9x + 2 \end{pmatrix}.$$

After simplifications,

$$u_{264}(x) = (\sqrt{x}-1)^4 \begin{pmatrix} 2x^3 + 6x^{5/2} + 21x^2 \\ +19x^{3/2} + 21x + 6\sqrt{x} + 2 \end{pmatrix}.$$

Since $u_{264}(x) \geq 0$, $\forall x > 0$, hence proving the required result.

265. For $\frac{3P_2 + 5A - 5P_3}{3} \leq \mathbf{C}$: We have to show that $\frac{1}{3}(3C - 3P_2 - 5A + 5P_3) \geq 0$. We can write $3C - 3P_2 - 5A + 5P_3 = bg_{265}(a/b)$, where

$$g_{265}(x) = \frac{u_{265}(x)}{2(x+1)(x^2+1)(x-\sqrt{x}+1)}$$

with

$$u_{265}(x) = \begin{pmatrix} x^5 - x^{9/2} - 5x^4 + 16x^{7/2} + 16x^{3/2} \\ -16x^3 + 10x^{5/2} - 16x^2 - 5x - \sqrt{x} + 1 \end{pmatrix}.$$

After simplifications,

$$u_{265}(x) = (\sqrt{x}-1)^4 \begin{pmatrix} x^4 + x^{5/2}(\sqrt{x}-2)^2 + 3x^{5/2} \\ +2x^2 + 3x^{3/2} + \sqrt{x}(2\sqrt{x}-1)^2 + 1 \end{pmatrix}.$$

Since $u_{265}(x) \geq 0$, $\forall x > 0$, hence proving the required result.

266. For $\mathbf{C} + \mathbf{P_3} - \mathbf{P_4} \leq \mathbf{C}$: We have to show that $P_4 - P_3 \geq 0$. It is true by definition. See the inequalities (12). $\square$

4.1 Equality Relations

Based on Results 169–266 given in Section 4, we have equalities among the relations. These are given as follows:

(i) In view of Results 172, 204, and 213, the following equalities hold:

$$P_5 - 3P_4 - C + 3G$$
$$= 6N_1 + 3H - 3A - 6G - 2P_5 + 2C$$
$$= 3H + 2C - 2P_5 - 6N_1 + 3A.$$

It leads us to the following equalities among the means given in (30):

(a) $3G + A + P_5 = H + P_4 + 2N_1 + C$;
(b) $G + 2N_1 + P_5 = H + A + P_4 + C$;
(c) $N_1 = \frac{A+G}{2}$.

(ii) In view of Results 185 and 216, the following equalities hold:

$$A + 4P_4 - 4G - H$$
$$= 4P_4 + 3R - 3A - 4G.$$

It leads us to the following equalities among the means given in (30):

$$A = \frac{H + 3R}{4}.$$

(iii) In view of Results 248 and 249, the following equalities hold:

$$72A + 9P_4 - 9P_5 - 72R - 22G + 22C$$
$$= 2(8C + 9H - 17A - P_4 + P_5).$$

It leads us to the following equalities among the means given in (30):

$$11P_4 + 106A + 6C = 22G + 18H + 11P_5 + 72R.$$

4.2 Unification Results

Above we have proved nine Propositions 12 to 20 with three means each side. These propositions are unified in two main theorems. Before, let's write the measures given in Group II in simplified way:

$$m_1 := P_5 + 3P_3 - 3P_4$$

$$m_7 := \frac{3N_1 + 2P_2 - 2N_2}{3}$$

$$m_2 := C + 3P_3 - 3G$$

$$m_8 := N + 8R - 8S$$

$$m_3 := \frac{23N_1 + 14P_1 - 14N_2}{23}$$

$$m_9 := \frac{13A + 6P_1 - 6P_5}{13}$$

$$m_4 := \frac{18N + 11P_1 - 11N_2}{18}$$

$$m_{10} := A + 4P_4 - 4G$$

$$m_5 := \frac{3P_5 + 2H - 2S}{3}$$

$$m_{11} := \frac{2N_1 + P_3 - A}{2}$$

$$m_6 := \frac{16P_5 + 9P_1 - 9C}{16}$$

$$m_{12} := \frac{4P_5 + 5G - 5S}{4}$$

$$m_{13} := \frac{11P_5 + 30N_2 - 30R}{11}$$

$$m_{14} := \frac{11N + 5P_3 - 5R}{11}$$

$$m_{15} := P_3 + 3S - 3R$$

$$m_{16} := S + 2P_5 - 2C$$

$$m_{17} := \frac{12N + N_1 - N_2}{12}$$

$$m_{18} := \frac{2N_1 + H - A}{2}$$

$$m_{19} := \frac{4C + 7H - 7A}{4}$$

$$m_{20} := \frac{4S + 5H - 5A}{4}$$

$$m_{21} := \frac{3H + C - P_5}{3}$$

$$m_{22} := \frac{3G + P_5 - C}{3}$$

$$m_{23} := P_5 + 2S - 2C$$

$$m_{24} := \frac{27R + 26N_2 - 26C}{27}$$

$$m_{25} := P_4 + 6N_2 - 6N$$

$$m_{26} := \frac{3H + 2C - 2P_5}{3}$$

$$m_{27} := P_2 + 2C - 2P_5$$

$$m_{28} := P_5 + 6N_2 - 6A$$

$$m_{29} := \frac{4P_4 + 3R - 3A}{4}$$

$$m_{30} := C + 2N_1 - 2S$$

$$m_{31} := \frac{2N_1 + S - C}{2}$$

$$m_{32} := C + P_4 - S$$

$$m_{33} := \frac{12R + 7G - 7S}{12}$$

$$m_{34} := \frac{20N_2 + 3G - 3R}{20}$$

$$m_{35} := \frac{3N_1 + 2A - 2N_2}{3}$$

$$m_{36} := \frac{6C + 7H - 7A}{6}$$

$$m_{37} := \frac{27S + 8P_2 - 8P_5}{27}$$

$$m_{38} := \frac{4N_2 + R - S}{4}$$

$$m_{39} := \frac{9S + 8P_5 - 8C}{9}$$

$$m_{40} := \frac{8C + 9H - 9A}{8}$$

$$m_{41} := \frac{8A + P_4 - P_5}{8}$$

$$m_{42} := \frac{14C + 9P_3 - 9P_5}{14}$$

$$m_{43} := \frac{18S + 5P_2 - 5N_5}{18}$$

$$m_{44} := \frac{4P_2 + 45S - 45R}{4}$$

$$m_{45} := \frac{12N + C - S}{12}$$

$$m_{46} := \frac{6S + 5P_5 - 5C}{6}$$

$$m_{47} := \frac{36R + 11G - 11C}{36}$$

$$m_{48} := \frac{5S + P_3 - C}{5}$$

$$m_{49} := \frac{3A + 2C - 2S}{3}$$

$$m_{50} := C + P_3 - P_4$$

$$m_{51} := \frac{18R + C - P_2}{18}$$

$$m_{52} := \frac{4P_2 + C - P_1}{4}$$

$$m_{53} := \frac{5H + C - P_3}{5}$$

$$m_{54} := \frac{3P_2 + 5A - 5P_3}{3} \qquad\qquad m_{56} := \frac{42N + C - P_4}{3}$$

$$m_{55} := \frac{9P_2 + 14A - 14P_3}{9}$$

Using the above notations, the results appearing in Propositions 12–20 are summarized in theorem below.

Theorem 6. *The following inequalities hold:*

(i) $\mathbf{P_2} \le \begin{cases} m_3 \le m_4 \le \mathbf{N_2} \\ m_2 \le m_1 \le \mathbf{P_5} \\ m_6 \le \mathbf{A}. \end{cases}$

(ii) $\mathbf{P_3} \le m_8 \le m_7 \le m_9 \le \mathbf{N_1}.$

(iii) $\mathbf{H} \le \begin{cases} m_{14} \le m_{15} \le \begin{cases} \mathbf{N} \\ m_5 \le \begin{cases} \mathbf{P_5} \\ \mathbf{R} \end{cases} \end{cases} \\ \left. \begin{matrix} m_{12} \\ m_{13} \end{matrix} \right\} \le \begin{cases} m_{16} \le \mathbf{P_5} \\ \mathbf{N_1} \end{cases} \\ m_{10} \le \begin{cases} m_{16} \le \mathbf{P_5} \\ \mathbf{A} \end{cases} \\ m_{11} \le \begin{cases} m_{15} \le \mathbf{N} \\ \mathbf{G} \end{cases} \\ m_{52} \le \mathbf{N_1}. \end{cases}$

(iv) $\mathbf{P_4} \le \begin{cases} m_{22} \le m_{18} \le m_{21} \le \begin{cases} m_{43} \le m_{19} \le \mathbf{N_1} \\ \mathbf{G} \le \begin{cases} m_{26} \le m_{27} \le \mathbf{A} \\ m_{26} \le m_{30} \\ m_{31} \le m_{53} \\ m_{25} \\ m_{29} \end{cases} \le \mathbf{N_1} \end{cases} \\ m_{20} \le \begin{cases} m_{23} \le \mathbf{A} \\ \mathbf{N_1}. \end{cases} \end{cases}$

(v) $\mathbf{N_1} \le m_{34} \le \left. \begin{matrix} m_{32} \\ m_{33} \end{matrix} \right\} \le m_{17} \le \mathbf{N} \le \begin{cases} m_{39} \le \mathbf{P_5} \\ \left. \begin{matrix} m_{35} \\ m_{38} \end{matrix} \right\} \le \begin{cases} m_{36} \le \mathbf{A} \\ \mathbf{N_2} \end{cases} \\ m_{38} \le m_{37} \le \mathbf{A}. \end{cases}$

$$\text{(vi)} \quad \mathbf{N_2} \le m_{56} \le \begin{cases} \begin{Bmatrix} m_{45} \\ m_{47} \end{Bmatrix} \le m_{41} \le \begin{Bmatrix} m_{40} \le m_{44} \\ m_{43} \\ m_{46} \le \mathbf{P_5} \end{Bmatrix} \le \mathbf{A} \\ m_{42} \le \begin{cases} \mathbf{P_5} \\ \mathbf{R.} \end{cases} \end{cases}$$

$$\text{(vii)} \quad \mathbf{A} \le m_{48} \le \begin{cases} \mathbf{R} \le \begin{Bmatrix} m_{49} \le m_{55} \\ \mathbf{S} \le m_{51} \end{Bmatrix} \le m_{54} \\ \mathbf{P_5} \le m_{50} \end{cases} \le \mathbf{C}.$$

The best possible continued inequalities derived from Propositions 13–20 are summarized in the theorem below.

Theorem 7. *The following inequalities give refinement over the inequalities (10):*

$$\mathbf{H} \le \begin{Bmatrix} \mathbf{P_4} \le m_{31} \le m_{26} \le m_{29} \\ m_{16} \end{Bmatrix} \le \mathbf{G} \le \begin{Bmatrix} m_{35} \le m_{40} \\ m_{41} \le m_{65} \\ m_{34} \\ m_{39} \end{Bmatrix} < \mathbf{N_1}, \quad (44)$$

$$\mathbf{N_1} \le m_{44} \le \begin{Bmatrix} m_{42} \\ m_{43} \end{Bmatrix} \le m_{22} \le \mathbf{N} \le \begin{Bmatrix} m_{45} \\ m_{48} \end{Bmatrix} \le \mathbf{N_2}, \quad (45)$$

$$\mathbf{N_2} \le m_{68} \le \begin{Bmatrix} m_{55} \\ m_{57} \end{Bmatrix} \le m_{51} \le \begin{Bmatrix} m_{50} \le m_{54} \\ m_{53} \end{Bmatrix} \le \mathbf{A}, \quad (46)$$

and

$$\mathbf{A} \le m_{58} \le \begin{cases} \mathbf{P_5} \le m_{62} \\ \mathbf{R} \le \begin{Bmatrix} m_{59} \le m_{67} \\ \mathbf{S} \le m_{63} \end{Bmatrix} \le m_{66} \end{cases} \le \mathbf{C}. \quad (47)$$

5. GENERAL THEOREM

This section gives a general theorem combining the results given in Sections 3 and 4. Instead combining Theorems 5 and 7, we shall combine Theorems 6 and 8.

Theorem 8. *The following inequalities hold:*

$$\text{(i)} \quad \mathbf{P_1} \le \mathbf{P_2} \le \frac{P_1 + 2P_3}{3} \le \mathbf{P_3} \le \frac{H + P_2}{2} \le \frac{P_1 + 3H}{4} \le \mathbf{H}.$$

(ii) $\mathbf{H} \leq \dfrac{2P_4 + P_3}{3} \leq \dfrac{4P_4 + P_2}{5} \leq \dfrac{8P_4 + P_1}{9} \leq \mathbf{P_4}.$

(iii) $\mathbf{P_4} \leq \dfrac{3G + P_5 - C}{3} \leq \dfrac{2N_1 + H - A}{2} \leq \dfrac{3H + C - P_5}{3} \leq \dfrac{3G + P_3}{4}$

$$\leq \dfrac{P_1 + 9G}{10} \leq \mathbf{G}.$$

(iv) $\mathbf{G} \leq \left\{ \begin{array}{l} \dfrac{4P_4 + S}{5} \leq \dfrac{C + 6P_4}{7} \\[2mm] \dfrac{4P_4 + 3R - 3A}{4} \\[2mm] \dfrac{2N_1 + S - C}{2} \\[2mm] P_4 + 6N_2 - 6N \\[2mm] \dfrac{3H + 2C - 2P_5}{3} \leq \left\{ \begin{array}{l} \dfrac{A + 2P_4}{3} \\[2mm] C + 2N_1 - 2S \end{array} \right. \end{array} \right\} \leq \dfrac{10P_4 + 3R}{13} \leq \dfrac{A + 2P_4}{3}$

$$\leq \left\{ \begin{array}{l} \dfrac{6N_1 + P_2}{7} \leq \dfrac{P_1 + 10N_1}{11} \\[2mm] \left. \begin{array}{l} \dfrac{S + 3G}{4} \\[2mm] \dfrac{C + 5G}{6} \end{array} \right\} \leq \dfrac{7G + 3R}{10} \end{array} \right\} \leq \dfrac{G + 2N_2}{3} \leq \mathbf{N_1}.$$

(v) $\mathbf{N_1} \leq \dfrac{20N_2 + 3G - 3R}{20} \leq \left\{ \begin{array}{l} \dfrac{4N_2 + P_4}{5} \leq \dfrac{9N + H}{10} \\[2mm] C + P_4 - S \\[2mm] \dfrac{12R + 7G - 7S}{12} \end{array} \right\}$

$$\leq \left\{ \begin{array}{l} \dfrac{P_3 + 15N}{16} \leq \dfrac{P_2 + 21N}{22} \leq \dfrac{P_1 + 33N}{34} \leq \dfrac{N_1 + 12N - N_2}{12} \\[2mm] \left. \begin{array}{l} \dfrac{C + 14N_1}{15} \\[2mm] \dfrac{S + 8N_1}{9} \\[2mm] \dfrac{8P_4 + 7S}{15} \end{array} \right\} \leq \dfrac{P_4 + 14N_2}{15} \end{array} \right\} \leq \mathbf{N}.$$

(vi) $\mathbf{N} \leq \dfrac{H + 20N_2}{21} \leq \left\{ \begin{array}{l} \dfrac{3R + 11N_1}{14} \\[2mm] \dfrac{P_3 + 32N_2}{33} \leq \dfrac{4N_2 + R - S}{4} \\[2mm] \dfrac{3N_1 + 2A - 2N_2}{3} \end{array} \right\} \leq \dfrac{P_1 + 68N_2}{69} \leq \mathbf{N_2}.$

(vii) $\mathbf{N_2} \leq \dfrac{42N + C - P_4}{42} \leq \left\{ \begin{array}{l} \dfrac{12N + C - S}{12} \\[2mm] \dfrac{36R + 11G - 11C}{36} \end{array} \right\} \leq \dfrac{8A + P_4 - P_5}{8} \leq \dfrac{P_5 + 9N}{10}$

$$\leq \dfrac{7A + H}{8} \leq \left\{ \begin{array}{l} \dfrac{4P_2 + 45S - 45R}{4} \\[2mm] \dfrac{P_3 + 11A}{12} \leq \left\{ \begin{array}{l} \dfrac{P_2 + 15A}{16} \leq \left\{ \begin{array}{l} \dfrac{P_1 + 23A}{24} \\[2mm] \dfrac{2S + H}{3} \end{array} \right. \\[4mm] \dfrac{9R + 4P_4}{13} \end{array} \right. \\[4mm] \dfrac{2N + R}{3} \end{array} \right\} \leq \mathbf{A}.$$

(viii) $\mathbf{A} \le \dfrac{5S + P_3 - C}{5} \le \left\{ \begin{array}{c} \dfrac{P_5 + 2N_2}{3} \le \dfrac{6R + P_2}{7} \\[2mm] \dfrac{P_3 + 3S}{4} \le \dfrac{9R + 2P_3}{11} \le \left\{ \begin{array}{c} \dfrac{6R + P_2}{7} \\[2mm] \dfrac{4S + P_2}{5} \end{array} \right\} \end{array} \right\} \le \dfrac{P_1 + 9R}{10}$

$\le \left\{ \begin{array}{c} \dfrac{3H + 5C}{8} \le \dfrac{21R + P_3}{22} \le \left\{ \begin{array}{l} \mathbf{P_5} \le C + P_3 - P_4 \le \mathbf{C} \\[2mm] \dfrac{A + 2S}{3} \le \dfrac{11S + P_3}{12} \le \mathbf{R} \end{array} \right. \\[4mm] \dfrac{4N + 3C}{7} \\[2mm] \dfrac{16N_2 + 11C}{27} \end{array} \right\} \le \mathbf{R}.$

(ix) $\mathbf{R} \le \left\{ \begin{array}{c} \dfrac{3A + 2C - 2S}{3} \\[2mm] \dfrac{P_1 + 20S}{21} \end{array} \right\} \le \dfrac{9P_2 + 14A - 14P_3}{9} \\[2mm] \begin{array}{c} \dfrac{P_1 + 20S}{21} \le \mathbf{S} \le \left\{ \begin{array}{c} \dfrac{18R + C - P_2}{18} \\[2mm] \dfrac{5C + 2P_4}{7} \\[2mm] \dfrac{5C + 4N_2}{9} \end{array} \right\} \end{array} \right\} \le \left\{ \begin{array}{c} \dfrac{3P_2 + 5A - 5P_3}{3} \\[2mm] \dfrac{7C + P_1}{8} \end{array} \right\} \le \mathbf{C}.$

Proof. In view of Propositions 1–20, it is sufficient to show only those results not shown before.

267. For $\dfrac{3\mathbf{H} + \mathbf{C} - \mathbf{P_5}}{3} \le \dfrac{3\mathbf{G} + \mathbf{P_3}}{4}$**:** We have to show that

$$\frac{1}{12}(9G + 3P_3 + 4P_5 - 12H - 4C) \ge 0.$$

We can write $9G + 3P_3 + 4P_5 - 12H - 4C = bg_{267}(a/b)$, where

$$g_{267}(x) = \frac{u_{267}(x)}{(x+1)(\sqrt{x}+1)^2(x - \sqrt{x} + 1)}$$

with

$$u_{267}(x) = x^{7/2} - 25x^{5/2} + 4x^3 + 40x^2 - 25x^{3/2} + 4x + \sqrt{x}$$
$$= \sqrt{x}(\sqrt{x} - 1)^2(x + 8\sqrt{x} + 1).$$

Since $u_{267}(x) \ge 0$, $\forall x > 0$, hence proving the required result.

268. For $\dfrac{4\mathbf{P_4} + 3\mathbf{R} - 3\mathbf{A}}{4} \le \dfrac{10\mathbf{P_4} + 3\mathbf{R}}{13}$**:** We have to show that $\frac{3}{52}(13A - 4P_4 - 9R) \ge 0$. It is already proved in Result 70.

269. For $\dfrac{2\mathbf{N_1} + \mathbf{S} - \mathbf{C}}{2} \le \dfrac{10\mathbf{P_4} + 3\mathbf{R}}{13}$**:** We have to show that

$$\frac{1}{26}(20P_4 + 6R + 13C - 26N_1 - 13S) \ge 0.$$

We can write $20P_4 + 6R + 13C - 26N_1 - 13S = bg_{269}(a/b)$, where

$$g_{269}(x) = \frac{u_{269}(x)}{2(x+1)(\sqrt{x}+1)^2}$$

with

$$u_{269}(x) = 21x^3 + 16x^{5/2} + 67x^2 + 44x(\sqrt{x}-1)^2 + 67x + 16\sqrt{x} + 21$$
$$- 13\sqrt{2x^2 + 2}(x+1)(\sqrt{x}+1)^2.$$

Let's consider

$$v_{269}(x) = \left(\begin{array}{c} 21x^3 + 16x^{5/2} + 67x^2 \\ +44x(\sqrt{x}-1)^2 + 67x + 16\sqrt{x} + 21 \end{array} \right)^2$$
$$- \left[13\sqrt{2x^2+2}(x+1)(\sqrt{x}+1)^2 \right]^2.$$

After simplifications,

$$v_{269}(x) = (\sqrt{x}-1)^4 \left(\begin{array}{c} 103x^4 - 268x^{7/2} + 524x^3 - 84x^{5/2} \\ +4442x^2 - 84x^{3/2} + 524x - 268\sqrt{x} + 103 \end{array} \right).$$

Still, we need to show that the second expression of $v_{261}(x)$ is nonnegative. Let's consider

$$h_{269}(x) = \left(103x^4 + 524x^3 + 4442x^2 + 524x + 103 \right)^2$$
$$- \left(268x^{7/2} + 84x^{5/2} + 84x^{3/2} + 268\sqrt{x} \right)^2.$$

After simplifications,

$$h_{269}(x) = \left(\begin{array}{c} 10609x^8 + 36120x^7 + 1144604x^6 \\ +4711080x^5 + 0143974x^4 + 4711080x^3 \\ +1144604x^2 + 36120x + 10609 \end{array} \right).$$

In view of Note 1, $h_{269}(x) \geq 0$ giving $v_{269}(x) \geq 0$ giving $u_{269}(x) \geq 0$ $\forall x > 0$, hence proving the required result.

270. For $P_4 + 6N_2 - 6N \leq \frac{10P_4 + 3R}{13}$: We have to show that

$$\frac{1}{13}(78N + 3R - 78N_2 - 3P_4) \geq 0.$$

We can write $78N + 3R - 78N_2 - 3P_4 = bg_{270}(a/b)$, where

$$g_{270}(x) = \frac{u_{270}(x)}{2(x+1)(\sqrt{x}+1)^2}$$

with

$$u_{270}(x) = \left(\begin{array}{l} 56x^3 + 164x^{5/2} + 244x^2 \\ +320x^{3/2} + 244x + 164\sqrt{x} + 56 \end{array} \right)$$
$$- 39\sqrt{2x+2}(x+1)(\sqrt{x}+1)^3.$$

Let's consider

$$v_{270}(x) = \left(\begin{array}{l} 56x^3 + 164x^{5/2} + 244x^2 \\ +320x^{3/2} + 244x + 164\sqrt{x} + 56 \end{array} \right)^2$$
$$- \left[39\sqrt{2x+2}(x+1)(\sqrt{x}+1)^3 \right]^2.$$

After simplifications,

$$v_{270}(x) = 2(\sqrt{x}-1)^4 \left(\begin{array}{l} 47x^4 + 246x^{7/2} + 436x^3 \\ +594x^{5/2} + 786x^2 + 594x^{3/2} \\ +436x + 246\sqrt{x} + 47 \end{array} \right).$$

In view of Note 1, $v_{270}(x) \geq 0$ giving $u_{270}(x) \geq 0 \; \forall x > 0$, hence proving the required result.

271. For $\frac{3H+2C-2P_5}{3} \leq \frac{A+2P_4}{3}$**:** We have to show that

$$\frac{1}{3}(A + 2P_4 + 2P_5 - 3H - 2C) \geq 0.$$

We can write $A + 2P_4 + 2P_5 - 3H - 2C = bg_{271}(a/b)$, where

$$g_{271}(x) = \frac{u_{271}(x)}{2(x+1)(\sqrt{x}+1)^2}$$

with

$$u_{271}(x) = x^3 - 6x^{5/2} + 15x^2 - 20x^{3/2} + 15x - 6\sqrt{x} + 1 = (\sqrt{x}-1)^6.$$

Since $u_{271}(x) \geq 0, \; \forall x > 0$, hence proving the required result.

272. For $C + 2N_1 - 2S \leq \frac{6N_1 + P_2}{7}$: We have to show that $\frac{1}{7}(P_2 + 14S - 8N_1 - 7C) \geq 0$. We can write $P_2 + 14S - 8N_1 - 7C = bg_{272}(a/b)$, where

$$g_{272}(x) = \frac{u_{272}(x)}{(x+1)(x^2+1)}$$

with

$$u_{272}(x) = 7\sqrt{2x^2 + 2}(x+1)(x^2+1)$$
$$- \left(\begin{array}{l} 9x^4 + 4x^{7/2} + 3x^3 + 4x^{5/2} \\ +16x^2 + 4x^{3/2} + 3x + 4\sqrt{x} + 9 \end{array} \right).$$

Let's consider

$$v_{272}(x) = \left[7\sqrt{2x^2 + 2}(x+1)(x^2+1) \right]^2$$
$$- \left(\begin{array}{l} 9x^4 + 4x^{7/2} + 3x^3 + 4x^{5/2} \\ +16x^2 + 4x^{3/2} + 3x + 4\sqrt{x} + 9 \end{array} \right)^2.$$

After simplifications,

$$v_{272}(x) = (\sqrt{x} - 1)^4 \left(\begin{array}{l} 15x^6 + 2(x^5 + 1)(\sqrt{x} - 1)^2 + 6x^5 \\ +28x^{9/2} + 94x^4 + 20x^{7/2} + 10x^3 \\ +20x^{5/2} + 94x^2 + 28x^{3/2} + 6x + 15 \end{array} \right).$$

In view of Note 1, $v_{272}(x) \geq 0$ giving $u_{272}(x) \geq 0 \; \forall x > 0$, hence proving the required result.

273. For $C + 2N_1 - 2S \leq \frac{S + 3G}{4}$: We have to show that $\frac{1}{12}(27S + 9G - 12C - 24N_1) \geq 0$. We can write $27S + 9G - 12C - 24N_1 = bg_{273}(a/b)$, where $g_{273}(x) = \frac{3}{2}u_{273}(x)/(x+1)$ with

$$u_{273}(x) = 9\sqrt{2x^2 + 2}(x+1) - 2(6x^2 + x^{3/2} + 4x + \sqrt{x} + 6).$$

Let's consider

$$v_{273}(x) = \left[9\sqrt{2x^2 + 2}(x+1) \right]^2 - 4(6x^2 + x^{3/2} + 4x + \sqrt{x} + 6)^2.$$

After simplifications,

$$v_{273}(x) = 2(\sqrt{x} - 1)^2 \left(\begin{array}{l} 3(x^2 + 1)(\sqrt{x} - 1)^2 + 6x^3 \\ +40x^2 + 52x^{3/2} + 40x + 6 \end{array} \right).$$

In view of Note 1, $v_{273}(x) \geq 0$ giving $u_{273}(x) \geq 0 \; \forall x > 0$, hence proving the required result.

274. For $C + 2N_1 - 2S \leq \frac{C+5G}{6}$: We have to show that $\frac{1}{6}(5G + 12S - 5C - 12N_1) \geq 0$. We can write $5G + 12S - 5C - 12N_1 = bg_{274}(a/b)$, where $g_{274}(x) = u_{274}(x)/(x+1)$ with

$$u_{274}(x) = 6\sqrt{2x^2 + 2}(x+1) - \left(8x^2 + x^{3/2} + 6x + \sqrt{x} + 8\right).$$

Let's consider

$$v_{274}(x) = \left[6\sqrt{2x^2 + 2}(x+1)\right]^2 - \left(8x^2 + x^{3/2} + 6x + \sqrt{x} + 8\right)^2.$$

After simplifications,

$$v_{274}(x) = (\sqrt{x} - 1)^2\left(8x^3 + 39x^2 + 50x^{3/2} + 39x + 8\right).$$

In view of Note 1, $v_{274}(x) \geq 0$ giving $u_{274}(x) \geq 0 \; \forall x > 0$, hence proving the required result.

275. For $\frac{20N_2 + 3G - 3R}{20} \leq \frac{4N_2 + P_4}{6}$: We have to show that

$$\frac{1}{20}(4P_4 + 3R - 4N_2 - 3G) \geq 0.$$

We can write $4P_4 + 3R - 4N_2 - 3G = bg_{275}(a/b)$, where

$$g_{275}(x) = \frac{u_{275}(x)}{(x^3 + 1)(\sqrt{x} + 1)^2}$$

with

$$u_{275}(x) = 2x^3 + x^{5/2} + 13x^2 + 13x + x(\sqrt{x} - 1)^2 + \sqrt{x} + 2$$
$$- \sqrt{2x + 2}(x + 1)(\sqrt{x} + 1)^3.$$

Let's consider

$$v_{275}(x) = \left[2x^3 + x^{5/2} + 13x^2 + 13x + x(\sqrt{x} - 1)^2 + \sqrt{x} + 2\right]^2$$
$$- \left[\sqrt{2x + 2}(x + 1)(\sqrt{x} + 1)^3\right]^2.$$

After simplifications,

$$v_{275}(x) = (\sqrt{x} - 1)^4 \left(\begin{array}{c} 6x(x + 1)(\sqrt{x} - 1)^2 \\ + 2x^4 + 3x^3 + 6x^2 + 3x + 2 \end{array}\right).$$

In view of Note 1, $v_{275}(x) \geq 0$ giving $u_{275}(x) \geq 0 \; \forall x > 0$, hence proving the required result.

276. For $C + P_4 - S \leq \frac{P_3 + 15N}{16}$: We have to show that

$$\frac{1}{16}(P_3 + 15N + 16S - 16C - 16P_4) \geq 0.$$

We can write $P_3 + 15N + 16S - 16C - 16P_4 = bg_{276}(a/b)$, where

$$g_{276}(x) = \frac{u_{276}(x)}{(x+1)(\sqrt{x}+1)^2(x - \sqrt{x}+1)}$$

with

$$u_{276}(x) = 8\sqrt{2x^2 + 2}(x+1)(\sqrt{x}+1)^2(x - \sqrt{x}+1)$$
$$- \left(\begin{matrix} 11x^4 + 6x^{7/2} + 35x(x+1)(\sqrt{x}-1)^2 \\ +13x^3 + 68x^2 + 13x + 6\sqrt{x} + 11 \end{matrix} \right).$$

Let's consider

$$u_{276}(x) = \left[8\sqrt{2x^2 + 2}(x+1)(\sqrt{x}+1)^2(x - \sqrt{x}+1) \right]^2$$
$$- \left(\begin{matrix} 11x^4 + 6x^{7/2} + 35x(x+1)(\sqrt{x}-1)^2 \\ +13x^3 + 68x^2 + 13x + 6\sqrt{x} + 11 \end{matrix} \right)^2.$$

After simplifications,

$$v_{276}(x) = (\sqrt{x}-1)^6 \left(\begin{matrix} 7x^5 + 166x^{9/2} + 183x^4 + 417x^{7/2} \\ +63x^{3/2}(x+1)(\sqrt{x}-1)^2 + 1014x^{5/2} \\ +417x^{3/2} + 183x + 166\sqrt{x} + 7 \end{matrix} \right).$$

In view of Note 1, $v_{276}(x) \geq 0$ giving $u_{276}(x) \geq 0 \; \forall x > 0$, hence proving the required result.

277. For $C + P_4 - S \leq \frac{C + 14N_1}{15}$: We have to show that

$$\frac{1}{15}(15S + 14N_1 - 15P_4 - 14C) \geq 0.$$

We can write $15S + 14N_1 - 15P_4 - 14C = bg_{277}(a/b)$, where

$$g_{277}(x) = \frac{u_{277}(x)}{2(x+1)(\sqrt{x}+1)^2}$$

with

$$u_{277}(x) = 15\sqrt{2x^2 + 2}(x+1)(\sqrt{x}+1)^2$$
$$- \begin{pmatrix} 21x^3 + 28x^{5/2} + 28x(\sqrt{x}-1)^2 \\ +71x^2 + 71x + 28\sqrt{x} + 21 \end{pmatrix}.$$

Let's consider

$$v_{277}(x) = \left[15\sqrt{2x^2 + 2}(x+1)(\sqrt{x}+1)^2\right]^2$$
$$- \begin{pmatrix} 21x^3 + 28x^{5/2} + 28x(\sqrt{x}-1)^2 \\ +71x^2 + 71x + 28\sqrt{x} + 21 \end{pmatrix}^2.$$

After simplifications,

$$v_{277}(x) = (\sqrt{x}-1)^2 \begin{pmatrix} \sqrt{x}(x+1)\left(34x^2 + 537x + 34\right)(\sqrt{x}-1)^2 \\ +9x^5 + 608x^{9/2} + x^4 + 827x^{7/2} \\ +6710x^{5/2} + 827x^{3/2} + x + 608\sqrt{x} + 9 \end{pmatrix}.$$

In view of Note 1, $v_{277}(x) \geq 0$ giving $u_{277}(x) \geq 0 \; \forall x > 0$, hence proving the required result.

278. For $C + P_4 - S \leq \frac{S+8N_1}{9}$: We have to show that

$$\frac{1}{9}(10S + 8N_1 - 9C - 9P_4) \geq 0.$$

We can write $10S + 8N_1 - 9C - 9P_4 = bg_{278}(a/b)$, where

$$g_{278}(x) = \frac{u_{278}(x)}{(x+1)(\sqrt{x}+1)^2}$$

with

$$u_{278}(x) = 15\sqrt{2x^2 + 2}(x+1)(\sqrt{x}+1)^2$$
$$- \begin{pmatrix} 21x^3 + 28x^{5/2} + 28x(\sqrt{x}-1)^2 \\ +71x^2 + 71x + 28\sqrt{x} + 21 \end{pmatrix}.$$

Let's consider

$$v_{278}(x) = \left[15\sqrt{2x^2 + 2}(x+1)(\sqrt{x}+1)^2\right]^2$$
$$- \begin{pmatrix} 21x^3 + 28x^{5/2} + 28x(\sqrt{x}-1)^2 \\ +71x^2 + 71x + 28\sqrt{x} + 21 \end{pmatrix}^2.$$

After simplifications,

$$v_{278}(x) = (\sqrt{x} - 1)^2 \left(\begin{array}{l} \sqrt{x}(x+1)\left(6x^2 + 31x + 6\right)(\sqrt{x} - 1)^2 \\ +x^5 + 56x^{9/2} + x^4 + 690x^{5/2} \\ +77x^{7/2} + 77x^{3/2} + x + 56\sqrt{x} + 1 \end{array} \right).$$

In view of Note 1, $v_{278}(x) \geq 0$ giving $u_{278}(x) \geq 0$ $\forall x > 0$, hence proving the required result.

279. For $C + P_4 - S \leq \frac{8P_4 + 7S}{15}$: We have to show that $\frac{1}{15}(22S - 7P_4 - 15C) \geq 0$. We can write $22S - 7P_4 - 15C = bg_{279}(a/b)$, where

$$g_{279}(x) = \frac{u_{279}(x)}{(x+1)(\sqrt{x} + 1)^2}$$

with

$$u_{279}(x) = 11\sqrt{2x^2 + 2}(x+1)(\sqrt{x} + 1)^2$$
$$- \left(15x^3 + 30x^{5/2} + 43x^2 + 43x + 30\sqrt{x} + 15\right).$$

Let's consider

$$v_{279}(x) = \left[11\sqrt{2x^2 + 2}(x+1)(\sqrt{x} + 1)^2\right]^2$$
$$- \left(15x^3 + 30x^{5/2} + 43x^2 + 43x + 30\sqrt{x} + 15\right)^2.$$

After simplifications,

$$v_{279}(x) = (\sqrt{x} - 1)^2 \left(\begin{array}{l} 17x^5 + 68x^{9/2} + 34\sqrt{x}(x^3 + 1)(\sqrt{x} - 1)^2 \\ +x^4 + 54x^{7/2} + 734x^3 + 1772x^{5/2} \\ +734x^2 + 54x^{3/2} + x + 68\sqrt{x} + 17 \end{array} \right).$$

In view of Note 1, $v_{279}(x) \geq 0$ giving $u_{279}(x) \geq 0$ $\forall x > 0$, hence proving the required result.

280. For $\frac{12R + 7G - 7S}{12} \leq \frac{P_3 + 15N}{16}$: We have to show that

$$\frac{1}{48}(3P_3 + 45N + 28S - 48R - 28G) \geq 0.$$

We can write $3P_3 + 45N + 28S - 48R - 28G = bg_{280}(a/b)$, where

$$g_{280}(x) = \frac{u_{280}(x)}{(x+1)(x - \sqrt{x} + 1)}$$

with

$$u_{280}(x) = 14\sqrt{2x^2 + 2}(x+1)(x - \sqrt{x} + 1)$$
$$- \left(\frac{2(x^2+1)(\sqrt{x}-1)^2 + 15x^3}{+x^2 + 24x^{3/2} + x + 15} \right).$$

Let's consider

$$v_{280}(x) = \left[14\sqrt{2x^2 + 2}(x+1)(x - \sqrt{x} + 1) \right]^2$$
$$- \left(\frac{2(x^2+1)(\sqrt{x}-1)^2 + 15x^3}{+x^2 + 24x^{3/2} + x + 15} \right)^2.$$

After simplifications,

$$v_{280}(x) = (\sqrt{x} - 1)^6 \left(\frac{15(x^2+1)(\sqrt{x}-1)^2 + 88x^3}{+102x^2 + 68x^{3/2} + 102x + 88} \right).$$

In view of Note 1, $v_{280}(x) \geq 0$ giving $u_{280}(x) \geq 0 \ \forall x > 0$, hence proving the required result.

281. For $\frac{12R+7G-7S}{12} \leq \frac{C+14N_1}{15}$: We have to show that

$$\frac{1}{60}(4C + 56N_1 + 35S - 60R - 35G) \geq 0.$$

We can write $4C + 56N_1 + 35S - 60R - 35G = bg_{281}(a/b)$, where $g_{281}(x) = \frac{1}{2}u_{281}(x)/(x+1)$ with

$$u_{281}(x) = 35\sqrt{2x^2 + 2}(x+1)$$
$$- \left(44x^2 + 14x^{3/2} + 24x + 14\sqrt{x} + 44 \right).$$

Let's consider

$$v_{281}(x) = \left[35\sqrt{2x^2 + 2}(x+1) \right]^2$$
$$- \left(44x^2 + 14x^{3/2} + 24x + 14\sqrt{x} + 44 \right)^2.$$

After simplifications,

$$v_{281}(x) = (\sqrt{x} - 1)^2 \left(\frac{206x^3 + 51(x^2+1)(\sqrt{x}-1)^2}{+784x^2 + 820x^{(3/2)} + 784x + 206} \right).$$

In view of Note 1, $v_{281}(x) \geq 0$ giving $u_{281}(x) \geq 0 \ \forall x > 0$, hence proving the required result.

282. For $\frac{12R+7G-7S}{12} \leq \frac{S+8N_1}{9}$: We have to show that

$$\frac{1}{36}(25S + 32N_1 - 36R - 21G) \geq 0.$$

We can write $25S + 32N_1 - 36R - 21G = bg_{282}(a/b)$, where $g_{282}(x) = \frac{1}{2}u_{282}(x)/(x+1)$ with

$$u_{282}(x) = 25\sqrt{2x^2 + 2}(x+1)$$
$$- (32x^2 + 10x^{3/2} + 16x + 10\sqrt{x} + 32).$$

Let's consider

$$u_{282}(x) = \left[25\sqrt{2x^2 + 2}(x+1)\right]^2$$
$$- (32x^2 + 10x^{3/2} + 16x + 10\sqrt{x} + 32)^2.$$

After simplifications,

$$v_{282}(x) = 2(\sqrt{x} - 1)^2 \left(\begin{array}{c} 47(x^2 + 1)(\sqrt{x} - 1)^2 + 66x^3 \\ +340x^2 + 388x^{3/2} + 340x + 66 \end{array}\right).$$

In view of Note 1, $v_{282}(x) \geq 0$ giving $u_{282}(x) \geq 0 \ \forall x > 0$, hence proving the required result.

283. For $\frac{12R+7G-7S}{12} \leq \frac{8P_4+7S}{15}$: We have to show that

$$\frac{1}{60}(32P_4 + 63S - 60R - 35G) \geq 0.$$

We can write $32P_4 + 63S - 60R - 35G = bg_{283}(a/b)$, where

$$g_{283}(x) = \frac{u_{283}(x)}{2(x+1)(\sqrt{x}+1)^2}$$

with

$$u_{283}(x) = 63\sqrt{2x^2 + 2}(x+1)(\sqrt{x}+1)^2$$
$$- \left(\begin{array}{c} 80x^3 + 230x^{5/2} + 44x^2 \\ +300x^{3/2} + 44x + 230\sqrt{x} + 80 \end{array}\right).$$

Let's consider

$$v_{283}(x) = \left[63\sqrt{2x^2 + 2}(x+1)(\sqrt{x}+1)^2\right]^2$$
$$- \left(\begin{array}{c} 80x^3 + 230x^{5/2} + 44x^2 \\ +300x^{3/2} + 44x + 230\sqrt{x} + 80 \end{array}\right)^2 .$$

After simplifications,

$$v_{283}(x) = 2(\sqrt{x}-1)^2 \times w_{283}(x),$$

where

$$w_{283}(x) = \left(\begin{array}{c} 769x^5 - 986x^{9/2} - 959x^4 + 12576x^{7/2} \\ +12158x^3 + 33524x^{5/2} + 12158x^2 \\ +12576x^{3/2} - 959x - 986\sqrt{x} + 769 \end{array}\right).$$

Still, we need to show that the second expression of is $w_{283}(x)$ nonnegative. Let's write

$$k_{283}(t) := w_{283}\left(t^2\right) = \left(\begin{array}{c} 769t^{10} - 986t^9 - 959t^8 + 12576t^7 \\ +12158t^6 + 33524t^5 + 12158t^4 \\ +12576t^3 - 959t^2 - 986t + 769 \end{array}\right).$$

The polynomial equation $k_{283}(t) = 0$ is of 10th degree. It admits 10 solutions. Out of them 8 solutions are complex (not written here) and two are real: -0.4276330193 and -2.338453662. Both these solutions are negative, i.e., it doesn't have any real positive solutions. Since we are working with $t > 0$, then there are two possibilities, one whole the graph of $k_{283}(t)$ is either in first quadrant or in 4th quadrant. Since $k_{283}(1) = 80640$, this means that $k_{283}(t) > 0$, for all $t > 0$. Thus we have $w_{283}(x) > 0$, for all $x > 0$ giving $v_{283}(x) \geq 0$, for all $x > 0$. In view of Note 1, $v_{283}(x) \geq 0$ giving $u_{283}(x) \geq 0 \forall x > 0$, hence proving the required result.

284. For $\frac{P_1 + 33N}{34} \leq \frac{12N + N_1 - N_2}{12}$: We have to show that

$$\frac{1}{204}(17N_1 + 6N - 17N_2 - 6P_1) \geq 0.$$

We can write $17N_1 + 6N - 17N_2 - 6P_1 = bg_{284}(a/b)$, where $g_{284}(x) = \frac{1}{4}u_{284}(x)/(x^3 + 1)$ with

$$u_{284}(x) = 25x^4 + 42x^{7/2} + x^3 + x + 42\sqrt{x} + 25$$

$$- 17\sqrt{2x+2}(x^3+1)(\sqrt{x}+1).$$

Let's consider

$$v_{284}(x) = \left(25x^4 + 42x^{7/2} + x^3 + x + 42\sqrt{x} + 25\right)^2$$
$$- \left[17\sqrt{2x+2}(x^3+1)(\sqrt{x}+1)\right]^2.$$

After simplifications,

$$v_{284}(x) = (\sqrt{x}-1)^2 \begin{pmatrix} 47x^7 + 1038x^{13/2} + 2687x^6 + 3264x^{11/2} \\ +3264x^5 + 3264x^{9/2} + 2158x^4 + 924x^{7/2} \\ +2158x^3 + 3264x^{5/2} + 3264x^2 \\ +3264x^{3/2} + 2687x + 1038\sqrt{x} + 47 \end{pmatrix}.$$

In view of Note 1, $v_{284}(x) \geq 0$ giving $u_{284}(x) \geq 0 \; \forall x > 0$, hence proving the required result.

285. For $\frac{H+20N_2}{21} \leq \frac{3N_1+2A-2N_2}{3}$: We have to show that

$$\frac{1}{21}(21N_1 + 14A - 34N_2 - H) \geq 0.$$

We can write $21N_1 + 14A - 34N_2 - H = bg_{285}(a/b)$, where $g_{285}(x) = \frac{1}{4}u_{285}(x)/(x+1)$ with

$$u_{285}(x) = 49x^2 + 42x^{3/2} + 90x + 42\sqrt{x} + 49$$
$$- 34\sqrt{2x+2}(x+1)(\sqrt{x}+1).$$

Let's consider

$$v_{285}(x) = \left(49x^2 + 42x^{3/2} + 90x + 42\sqrt{x} + 49\right)^2$$
$$- \left[34\sqrt{2x+2}(x+1)(\sqrt{x}+1)\right]^2.$$

After simplifications,

$$v_{285}(x) = (\sqrt{x}-1)^2\left[76(x+1)(\sqrt{x}-1)^2 + 13x^2 + 42x + 13\right].$$

In view of Note 1, $v_{285}(x) \geq 0$ giving $u_{285}(x) \geq 0 \; \forall x > 0$, hence proving the required result.

286. For $\frac{P_3+32N_2}{33} \leq \frac{4N_2+R-S}{4}$: We have to show that

$$\frac{1}{132}(4N_2 + 33R - 33S - 4P_3) \geq 0.$$

We can write $4N_2 + 33R - 33S - 4P_3 = bg_{286}(a/b)$, where

$$g_{286}(x) = \frac{u_{286}(x)}{2(x+1)(x - \sqrt{x} + 1)}$$

with

$$u_{286}(x) = 22(x^2 + x + 1)(\sqrt{x} - 1)^2 + 22x^3 + 36x^2 + 36x + 22$$
$$- (x+1)(x - \sqrt{x} + 1)[33\sqrt{2x^2 + 2} - 2(\sqrt{x} + 1)\sqrt{2x + 2}].$$

Before proceeding further we shall prove that the expression $33\sqrt{2x^2 + 2} - 2(\sqrt{x} + 1)\sqrt{2x + 2}$ is nonnegative for all $x > 0$. It is true in view of Result 165 – Group 1. Now we shall show that $u_{286}(x) \geq 0$, $\forall x > 0$. Let's consider

$$v_{286}(x) = \left[22(x^2 + x + 1)(\sqrt{x} - 1)^2 + 22x^3 + 36x^2 + 36x + 22\right]^2$$
$$\left[(x+1)(x - \sqrt{x} + 1)\right]^2$$
$$\times \left[33\sqrt{2x^2 + 2} - 2(\sqrt{x} + 1)\sqrt{2x + 2}\right]^2.$$

After simplifications,

$$v_{286}(x) = 132\sqrt{2x^2 + 2}\sqrt{2x + 2}(x^3 + 1)^2(\sqrt{x} + 1)$$
$$- 2\begin{pmatrix}(121x^5 + 535x^4 + 171x^{5/2} + 535x + 121)(\sqrt{x} - 1)^2 \\ +4x^6 + 313x^5 + 622x^4 + 117x^{7/2} \\ +117x^{5/2} + 622x^2 + 313x + 4\end{pmatrix}.$$

Still, we are unable to decide the nonnegativity of the expression $u_{228}(x)$. Let's apply again, argument given in Note 1. Let's consider

$$h_{286}(x) = \left[132\sqrt{2x^2 + 2}\sqrt{2x + 2}(x^3 + 1)^2(\sqrt{x} + 1)\right]^2$$
$$- 4\begin{pmatrix}(121x^5 + 535x^4 + 171x^{5/2} + 535x + 121)(\sqrt{x} - 1)^2 \\ +4x^6 + 313x^5 + 622x^4 + 117x^{7/2} \\ +117x^{5/2} + 622x^2 + 313x + 4\end{pmatrix}^2.$$

After simplifications,

$$h_{286}(x) = (\sqrt{x} - 1)^4 \times w_{286}(x),$$

where

$$w_{286}(x) = \begin{pmatrix} \sqrt{x}\begin{pmatrix} 20412x^8 + 67265x^7 \\ +180486x^4 + 67265x + 20412 \end{pmatrix}(\sqrt{x}-1)^2 \\ +1799x^{10} + 12436x^{19/2} + 121391x^{17/2} \\ +226927x^{15/2} + 221490x^7 + 53928x^{13/2} \\ +250039x^6 + 551318x^{11/2} + 551318x^{9/2} \\ +250039x^4 + 53928x^{7/2} + 221490x^3 \\ +226927x^{5/2} + 121391x^{3/2} + 12436\sqrt{x} + 1799 \end{pmatrix}.$$

In view of Note 1, $w_{286}(x) > 0$ giving $v_{286}(x) \geq 0$ giving $u_{286}(x) \geq 0 \ \forall x > 0$, hence proving the required result.

287. For $\frac{3N_1+2A-2N_2}{3} \leq \frac{P_1+68N_2}{69}$: We have to show that

$$\frac{1}{69}(P_1 + 114N_2 - 69N_1 - 46A) \geq 0.$$

We can write $P_1 + 114N_2 - 69N_1 - 46A = bg_{287}(a/b)$, where $g_{287}(x) = \frac{1}{4}u_{287}(x)/(x^3+1)$ with

$$u_{287}(x) = 114\sqrt{2x+2}(x^3+1)(\sqrt{x}+1) \\ - (161x^4 + 138x^{7/2} + 157x^3 + 157x + 138\sqrt{x} + 161).$$

Let's consider

$$v_{287}(x) = \left[114\sqrt{2x+2}(x^3+1)(\sqrt{x}+1)\right]^2 \\ - (161x^4 + 138x^{7/2} + 157x^3 + 157x + 138\sqrt{x} + 161)^2.$$

After simplifications,

$$v_{287}(x) = (\sqrt{x}-1)^4\begin{pmatrix} 71x^6 + 7832x^{11/2} + 13288x^5 \\ +15096x^{9/2} + 13256x^4 + 7768x^{7/2} \\ +62x^3 + 7768x^{5/2} + 13256x^2 \\ 15096x^{3/2} + 13288x + 7832\sqrt{x} + 71 \end{pmatrix}.$$

In view of Note 1, $v_{287}(x) \geq 0$ giving $u_{287}(x) \geq 0 \ \forall x > 0$, hence proving the required result.

288. For $\frac{4N_2+R-S}{4} \leq \frac{P_1+68N_2}{69}$: We have to show that

$$\frac{1}{276}(4P_1 + 69S - 4N_2 - 69R) \geq 0.$$

We can write $4P_1 + 69S - 4N_2 - 69R = bg_{288}(a/b)$, where $g_{288}(x) = \frac{1}{2}u_{288}(x)/(x^3 + 1)$ with

$$u_{288}(x) = (x^3 + 1)\left[69\sqrt{2x^2 + 2} - 2(\sqrt{x} + 1)\sqrt{2x + 2}\right]$$
$$- \left[4(x^2 + 1)(x - 1)^2 + 88x^4 + 84x^2 + 88\right].$$

Before proceeding further we shall prove that the expression $69\sqrt{2x^2 + 2} - 2(\sqrt{x} + 1)\sqrt{2x + 2}$ is nonnegative for all $x > 0$. It is true in view of Result 165 − Group 1. Now we shall show that $u_{288}(x) \geq 0$, $\forall x > 0$. Let's consider

$$v_{288}(x) = \left\{(x^3 + 1)\left[69\sqrt{2x^2 + 2} - 2(\sqrt{x} + 1)\sqrt{2x + 2}\right]\right\}^2$$
$$- \left[4(x^2 + 1)(x - 1)^2 + 88x^4 + 84x^2 + 88\right]^2.$$

After simplifications,

$$v_{288}(x) = \begin{pmatrix} 1066x^8 + 16x^{15/2} + 1488x^7 + 16x^{13/2} - 7462x^6 \\ +22004x^5 + 32x^{9/2} - 25488x^4 + 32x^{7/2} + 16x^{3/2} \\ +22004x^3 - 7462x^2 + 1488x + 16\sqrt{x} + 1066 \end{pmatrix}$$
$$- 276\sqrt{2x^2 + 2}\sqrt{2x + 2}(x^3 + 1)^2(\sqrt{x} + 1).$$

Before proceeding further, we need to show that the first expression of $v_{288}(x)$ is nonnegative. Let's consider

$$h_{288}(t) = \begin{pmatrix} 1066t^{16} + 16t^{15} + 1488t^{14} + 16t^{13} - 7462t^{12} \\ +22004t^{10} + 32t^9 - 25488t^8 + 32t^7 + 22004t^6 \\ -7462t^4 + 16t^3 + 1488t^2 + 16t + 1066 \end{pmatrix}.$$

The polynomial equation $h_{288}(t) = 0$ is of 16th degree. It admits 16 solutions. All the solutions are complex (not written here), i.e., it doesn't have any real positive solutions. Since we are working with $t > 0$, then there are two possibilities, one whole the graph of $h_{288}(t)$ is either in first quadrant or in 4th quadrant. Since $h_{288}(1) = 8832$, this means that $h_{288}(t) > 0$, for all $t > 0$. This proves that the first expression of $v_{288}(x)$ is nonnegative. Let's prove now the nonnegativity of $v_{288}(x)$. Let's consider

$$w_{288}(x) = \begin{pmatrix} 1066x^8 + 16x^{15/2} + 1488x^7 + 16x^{13/2} - 7462x^6 \\ +22004x^5 + 32x^{9/2} - 25488x^4 + 32x^{7/2} + 16x^{3/2} \\ +22004x^3 - 7462x^2 + 1488x + 16\sqrt{x} + 1066 \end{pmatrix}^2$$
$$- \left[276\sqrt{2x^2 + 2}\sqrt{2x + 2}(x^3 + 1)^2(\sqrt{x} + 1)\right]^2.$$

After simplifications,

$$w_{288}(x) = 4(\sqrt{x} - 1)^4 \times k_{288}(x),$$

where

$$k_{288}(x) = \begin{pmatrix} 207913 + 2144650x + 687828\sqrt{x} + 5028772x^4 \\ +5028772x^{10} + 52160960x^{19/2} - 39912560x^{17/2} \\ -39912560x^{11/2} + 207913x^{14} + 687828x^{27/2} \\ +2144650x^{13} + 5151364x^{25/2} + 6705021x^{12} \\ +3602528x^{23/2} - 1639796x^{11} - 7134000x^{21/2} \\ +61634244x^9 + 3602528x^{5/2} - 98373382x^8 \\ +39191864x^{15/2} + 157230308x^7 + 39191864x^{13/2} \\ -98373382x^6 + 61634244x^5 + 52160960x^{9/2} \\ -7134000x^{7/2} - 1639796x^3 + 5151364x^{3/2} + 6705021x^2 \end{pmatrix}.$$

Following the same procedure of h_{288}, we can show that k_{288} is nonnegative. In view of Note 1, $w_{288}(x) > 0$ giving $v_{288}(x) \geq 0$ giving $u_{288}(x) \geq 0 \; \forall x > 0$, hence proving the required result.

289. For $\frac{5S + P_3 - C}{5} \leq \frac{P_5 + 2N_2}{3}$: We have to show that

$$\frac{1}{15}(5P_5 + 10N_2 + 3C - 15S - 3P_3) \geq 0.$$

We can write $5P_5 + 10N_2 + 3C - 15S - 3P_3 = bg_{289}(a/b)$, where

$$g_{289}(x) = \frac{u_{289}(x)}{2(x+1)(\sqrt{x}+1)^2(x - \sqrt{x} + 1)}$$

with

$$u_{289}(x) = \begin{pmatrix} 2(x+1)(x^2 + 8x + 1)(\sqrt{x} - 1)^2 \\ +14x^4 + 14x^3 + 24x^2 + 14x + 14 \end{pmatrix}$$
$$- 5(x+1)(\sqrt{x}+1)^2(x - \sqrt{x} + 1)$$
$$\times \left[3\sqrt{2x^2 + 2} - (\sqrt{x}+1)\sqrt{2x+2}\right].$$

Before proceeding further we shall prove that the expression $3\sqrt{2x^2 + 2} - (\sqrt{x}+1)\sqrt{2x+2}$ is nonnegative for all $x > 0$. Let's consider

$$p_{289}(x) = \left[3\sqrt{2x^2 + 2}\right]^2 - [(\sqrt{x}+1)\sqrt{2x+2}]^2.$$

After simplifications,

$$p_{289}(x) = 4\left(4x^2 - x^{3/2} - x - \sqrt{x} + 4\right)$$
$$= 10\left(x^2 + 1\right) + 5(x-1)^2 + (\sqrt{x} - 1)^4.$$

In view of Note 1, it proves that the expression $3\sqrt{2x^2 + 2} - (\sqrt{x} + 1)\sqrt{2x+2}$ is positive for all $x > 0$. Now we shall show that $u_{289}(x) \geq 0$, $\forall x > 0$. Let's consider

$$v_{289}(x) = \left(\begin{matrix} 2(x+1)(x^2 + 8x + 1)(\sqrt{x} - 1)^2 \\ +14x^4 + 14x^3 + 24x^2 + 14x + 14 \end{matrix}\right)^2$$
$$- \left[5(x+1)(\sqrt{x}+1)^2(x - \sqrt{x} + 1)\right]^2$$
$$\times \left[3\sqrt{2x^2 + 2} - (\sqrt{x}+1)\sqrt{2x+2}\right]^2.$$

After simplifications,

$$v_{289}(x) = 150\sqrt{2x+2}\sqrt{2x^2+2}(x+1)^2(\sqrt{x}+1)^5(x - \sqrt{x} + 1)^2$$
$$- 4\left(\begin{matrix} 61x^8 + 307x^{15/2} + 174x^7 + 1256x^{13/2} \\ +434x^6 + 2670x^{11/2} + 462x^5 + 4167x^{9/2} \\ +138x^4 + 4167x^{7/2} + 462x^3 + 2670x^{5/2} \\ +434x^2 + 256x^{3/2} + 174x + 307\sqrt{x} + 61 \end{matrix}\right).$$

Since we are unable to decide the nonnegativity of $v_{289}(x)$, we shall apply again the argument given in Note 1. Let's consider

$$v_{289}(x) = \left[150\sqrt{2x+2}\sqrt{2x^2+2}(x+1)^2(\sqrt{x}+1)^5(x - \sqrt{x} + 1)^2\right]^2$$
$$- 16\left(\begin{matrix} 61x^8 + 307x^{15/2} + 174x^7 + 1256x^{13/2} \\ +434x^6 + 2670x^{11/2} + 462x^5 + 4167x^{9/2} \\ +138x^4 + 4167x^{7/2} + 462x^3 + 2670x^{5/2} \\ +434x^2 + 256x^{3/2} + 174x + 307\sqrt{x} + 61 \end{matrix}\right)^2.$$

After simplifications,

$$w_{289}(x) = 16(\sqrt{x} - 1)^4 \times h_{289}(x)$$

where

$$h_{289}(x) = \begin{pmatrix} 1904 + 1247x - 690106x^8 + 1807668x^{15/2} + 409644x^{7/2} \\ -63651x^3 - 690106x^6 + 1411704x^{11/2} - 421227x^5 \\ +137328x^{5/2} - 13765x^2 + 32814x^{3/2} - 835330x^7 \\ +1807668x^{13/2} + 1247x^{13} + 32814x^{25/2} - 13765x^{12} \\ +137328x^{23/2} - 63651x^{11} + 409644x^{21/2} - 421227x^9 \\ +876930x^{9/2} - 196737x^4 + 3912\sqrt{x} - 196737x^{10} \\ +876930x^{19/2} + 1411704x^{17/2} + 1904x^{14} + 3912x^{27/2} \end{pmatrix}.$$

Still, we need to show that $h_{289}(x)$ is nonnegative. Let's consider

$$k_{289}(t) := h_{289}(t^2)$$

$$= \begin{pmatrix} 1904t^{28} + 3912t^{27} + 1247t^{26} + 32814t^{25} - 13765t^{24} \\ +137328t^{23} - 63651t^{22} + 409644t^{21} - 196737t^{20} \\ +876930t^{19} - 421227t^{18} + 1411704t^{17} - 690106t^{16} \\ +1807668t^{15} - 835330t^{14} + 1807668t^{13} - 690106t^{12} \\ +1411704t^{11} - 421227t^{10} + 876930t^9 - 196737t^8 \\ +409644t^7 - 63651t^6 + 137328t^5 - 13765t^4 \\ +32814t^3 + 1247t^2 + 3912t + 1904 \end{pmatrix}.$$

The polynomial equation $k_{289}(t) = 0$ is of 28th degree. It admits 28 solutions. Out of them 26 solutions are complex (not written here) and two are real: -0.2665254262 and -3.751987247. Both these solutions are negative, i.e., it doesn't have any real positive solutions. Since we are working with $t > 0$, then there are two possibilities, one whole the graph of $k_{289}(t)$ is either in first quadrant or in 4th quadrant. Since $k_{289}(1) = 5760000$, this means that $k_{289}(t) > 0$, for all $t > 0$. Thus we have $w_{289}(x) > 0$, for all $x > 0$ giving $v_{289}(x) \geq 0$, for all $x > 0$. In view of Note 1, $v_{289}(x) \geq 0$ giving $u_{289}(x) \geq 0 \forall x > 0$, hence proving the required result.

290. For $\frac{5S+P_3-C}{5} \leq \frac{P_3+3S}{4}$**:** We have to show that $\frac{1}{20}(P_3+4C-5S) \geq 0$. We can write $P_3 + 4C - 5S = bg_{290}(a/b)$, where

$$g_{290}(x) = \frac{u_{290}(x)}{2(x+1)(x-\sqrt{x}+1)}$$

with

$$u_{290}(x) = 4(x^2+1)(\sqrt{x}-1)^2 + 4x^3 + 6x^2 + 6x + 4$$
$$- 5\sqrt{2x^2+2}(x+1)(x-\sqrt{x}+1).$$

Let's consider

$$v_{290}(x) = \left[4(x^2 + 1)(\sqrt{x} - 1)^2 + 4x^3 + 6x^2 + 6x + 4\right]^2$$
$$- \left[5\sqrt{2x^2 + 2}(x + 1)(x - \sqrt{x} + 1)\right]^2.$$

After simplifications,

$$v_{290}(x) = 2(\sqrt{x} - 1)^4 \begin{pmatrix} 7x^4 + 14x^{7/2} + x^3 + 18x^{5/2} \\ +20x^2 + 18x^{3/2} + x + 14\sqrt{x} + 7 \end{pmatrix}.$$

In view of Note 1, $v_{290}(x) \geq 0$ giving $u_{290}(x) \geq 0 \ \forall x > 0$, hence proving the required result.

291. For $\frac{P_1 + 20S}{21} \leq \frac{9P_2 + 14A - 14P_3}{9}$: We have to show that

$$\frac{1}{63}(63P_2 + 98A - 98P_3 - 3P_1 - 60S) \geq 0.$$

We can write $63P_2 + 98A - 98P_3 - 3P_1 - 60S = bg_{291}(a/h)$, where

$$g_{291}(x) = \frac{u_{291}(x)}{(x^3 + 1)(x^2 + 1)(x - \sqrt{x} + 1)}$$

with

$$u_{291}(x) = \begin{pmatrix} \begin{pmatrix} 25x^6 + 55x^5 + 56x^4 \\ +46x^3 + 56x^2 + 55x + 25 \end{pmatrix}(\sqrt{x} - 1)^2 \\ +10(x^5 + 1)(x - 1)^2 + 14x^7 + x^{13/2} + x^{11/2} \\ +100x^5 + 4x^4 + 4x^3 + 100x^2 + x^{3/2} + \sqrt{x} + 14 \end{pmatrix}$$
$$- 30\sqrt{2x^2 + 2}(x^3 + 1)(x^2 + 1)(x - \sqrt{x} + 1).$$

Let's consider

$$v_{291}(x) = \begin{pmatrix} \begin{pmatrix} 25x^6 + 55x^5 + 56x^4 \\ +46x^3 + 56x^2 + 55x + 25 \end{pmatrix}(\sqrt{x} - 1)^2 \\ +10\left(x^5 + 1\right)(x - 1)^2 + 14x^7 + x^{13/2} + x^{11/2} \\ +100x^5 + 4x^4 + 4x^3 + 100x^2 + x^{3/2} + \sqrt{x} + 14 \end{pmatrix}^2$$
$$- \left[30\sqrt{2x^2 + 2}(x^3 + 1)(x^2 + 1)(x - \sqrt{x} + 1)\right]^2.$$

After simplifications,

$$v_{291}(x) = (\sqrt{x} - 1)^4 \times w_{291}(x),$$

where

$$w_{291}(x) = \begin{pmatrix} x\begin{pmatrix} 719x^9 + 184x^8 + 2280x^6 + 118x^5 \\ 118x^4 + 2280x^3 + 184x + 719 \end{pmatrix}(\sqrt{x}-1)^2 \\ +601x^{12} + 1202x^{23/2} + 3364x^{11} + 1794x^{10} \\ +12292x^9 + 1316x^{17/2} + 7740x^8 + 6236x^7 \\ +23550x^6 + 6236x^5 + 7740x^4 + 1316x^{7/2} \\ +12292x^3 + 1794x^2 + 3364x + 1202\sqrt{x} + 601 \end{pmatrix}.$$

In view of Note 1, $v_{291}(x) \geq 0$ giving $u_{291}(x) \geq 0 \ \forall x > 0$, hence proving the required result.

292. For $\frac{5C+2P_4}{7} \leq \frac{3P_2+5A-5P_3}{3}$: We have to show that

$$\frac{1}{21}(21P_2 + 35A - 35P_3 - 15C - 6P_4) \geq 0.$$

We can write $21P_2 + 35A - 35P_3 - 15C - 6P_4 = bg_{292}(a/b)$, where

$$g_{292}(x) = \frac{u_{292}(x)}{(\sqrt{x}+1)^2(x^2+1)(x+1)(x-\sqrt{x}+1)}$$

with

$$u_{292}(x) = (\sqrt{x}-1)^4\begin{pmatrix} 5x^4 + 25x^{7/2} + 64x^3 + 151x^{5/2} \\ +178x^2 + 151x^{3/2} + 64x + 25\sqrt{x} + 5 \end{pmatrix}.$$

Since $u_{292}(x) \geq 0 \forall x > 0$, hence proving the required result.

293. For $\frac{5C+4N_2}{9} \leq \frac{3P_2+5A-5P_3}{3}$: We have to show that

$$\frac{1}{9}(9P_2 + 15A - 15P_3 - 5C - 4N_2) \geq 0.$$

We can write $9P_2 + 15A - 15P_3 - 5C - 4N_2 = bg_{293}(a/b)$, where

$$g_{293}(x) = \frac{u_{293}(x)}{(x+1)(x^2+1)(x-\sqrt{x}+1)}$$

with

$$u_{293}(x) = \begin{pmatrix} 5x^5 - 5x^{9/2} + 23x^4 - 48x^{7/2} + 64x^3 \\ -46x^{5/2} + 64x^2 - 48x^{3/2} + 23x - 5\sqrt{x} + 5 \end{pmatrix} \\ - 2\sqrt{2x+2}(x+1)(x^2+1)(x-\sqrt{x}+1).$$

After simplifications,

$$u_{293}(x) = \begin{pmatrix} (3x^4 + 24x^3 + 23x^2 + 24x + 3)(\sqrt{x} - 1)^2 \\ + 2(x^3 + 1)(x - 1)^2 + x^{9/2} + 15x^2(x + 1) + \sqrt{x} \end{pmatrix} \\ - 2\sqrt{2x + 2}(x + 1)(x^2 + 1)(x - \sqrt{x} + 1).$$

Let's consider

$$u_{293}(x) = \begin{pmatrix} (3x^4 + 24x^3 + 23x^2 + 24x + 3)(\sqrt{x} - 1)^2 \\ + 2(x^3 + 1)(x - 1)^2 + x^{9/2} + 15x^2(x + 1) + \sqrt{x} \end{pmatrix}^2 \\ - [2\sqrt{2x + 2}(x + 1)(x^2 + 1)(x - \sqrt{x} + 1)]^2.$$

After simplifications,

$$v_{293}(x) = (\sqrt{x} - 1)^4 \times w_{293}(x),$$

where

$$w_{293}(x) = \begin{pmatrix} 17x^8 + 18x^{15/2} + 201x^7 + 38x^{13/2} \\ + 431x^6 + 795x^5 + 1496x^4 + 795x^3 \\ + 431x^2 + 38x^{3/2} + 201x + 18\sqrt{x} + 17 \\ + x^2 (179x^3 + 169x^2 + 169x + 179) (\sqrt{x} - 1)^2 \end{pmatrix}.$$

In view of Note 1, $v_{293}(x) \geq 0$ giving $u_{293}(x) \geq 0$ $\forall x > 0$, hence proving the required result.

294. For $\frac{9P_2 + 14A - 14P_3}{9} \leq \frac{7C + P_1}{8}$**:** We have to show that

$$\frac{1}{72}(63C + 9P_1 + 112P_3 - 72P_2 - 112A) \geq 0.$$

We can write $63C + 9P_1 + 112P_3 - 72P_2 - 112A = bg_{294}(a/b)$, where

$$g_{294}(x) = \frac{u_{294}(x)}{(x^2 + 1)(x^3 + 1)(x - \sqrt{x} + 1)}$$

with

$$u_{294}(x) = (\sqrt{x} - 1)^2 \times w_{294}(x),$$

and

$$w_{294}(x) = \begin{pmatrix} 7x^6 + 7x^{11/2} - 56x^5 + 63x^{9/)} + 61x^4 - 2x^{7/2} \\ -112x^3 - 2x^{5/2} + 61x^2 + 63x^{3/2} - 56x + 7\sqrt{x} + 7 \end{pmatrix}.$$

Still, we need to show that $w_{294}(x)$ is nonnegative. Let's consider

$$k_{294}(t) := w_{294}(t^2) = \begin{pmatrix} 7t^{12} + 7t^{11} - 56t^{10} + 63t^9 + 61t^8 - 2t^7 \\ -112t^6 - 2t^5 + 61t^4 + 63t^3 - 56t^2 + 7t + 7 \end{pmatrix}.$$

The polynomial equation $k_{294}(t) = 0$ is of 12th degree. It admits 12 solutions. Out of them 10 solutions are complex (not written here) and two are real: -0.2715364716 and -3.682746536. Both these solutions are negative, i.e., it doesn't have any real positive solutions. Since we are working with $t > 0$, then there are two possibilities, one whole the graph of $k_{294}(t)$ is either in first quadrant or in 4th quadrant. Since $k_{294}(1) = 48$, this means that $k_{294}(t) > 0$, for all $t > 0$. Thus we have $w_{294}(x) > 0$, for all $x > 0$ giving $u_{294}(x) \geq 0$, for all $x > 0$. This proves the required result.

295. For $\frac{18R+C-P_2}{18} \leq \frac{7C+P_1}{8}$: We have to show that

$$\frac{1}{72}(59C + 9P_1 + 4P_2 - 72R) \geq 0.$$

We can write $59C + 9P_1 + 4P_2 - 72R = bg_{295}(a/b)$, where

$$g_{295}(x) = \frac{[10(x^2 + 1) + (x - 1)^2](x - 1)^4}{(x^2 + 1)(x^3 + 1)}.$$

Since $u_{295}(x) \geq 0 \forall x > 0$, hence proving the required result. ☐

5.1 Equality Relations

There are some measure having equality among each other in Groups I and II. According to notations given in Sections 3.2 and 4.2, the following equalities hold:

(i) The measures n_{28} and m_{11} are equal. This gives

$$\frac{P_3 + G}{2} = \frac{2N_1 + P_3 - A}{2} \Rightarrow N_1 = \frac{A + G}{2}.$$

(ii) The measures n_{36} and m_{18} are equal. This gives

$$\frac{2N_1 + H - A}{2} = \frac{G + H}{2} \Rightarrow N_1 = \frac{A + G}{2}.$$

(iii) The measures n_{41} and m_{19} are equal. This gives

$$\frac{4C + 7H - 7A}{4} = \frac{A + 3H}{4} \Rightarrow A = \frac{C + H}{2}.$$

(iv) The measures $\mathbf{n_{76}}$ and $\mathbf{m_{40}}$ are equal. This gives

$$\frac{8C + 9H - 9A}{8} = \frac{7A + H}{8} \Rightarrow A = \frac{C + H}{2}.$$

(v) The measures $\mathbf{n_{82}}$ and $\mathbf{m_{47}}$ are equal. This gives

$$\frac{36R + 11G - 11C}{36} = \frac{11N + R}{12} \Rightarrow 33R + 11G = 33N + 11C.$$

(vi) The measures $\mathbf{n_{81}}$ and $\mathbf{m_{41}}$ are equal. This gives

$$\frac{8A + P_4 - P_5}{8} = \frac{A + 3N}{4} \Rightarrow 6A + P_4 = P_5 + 6N.$$

Remark 6. In some cases, there are same results, for example in (i) and (ii), (iii) and (iv). For more equality relations see Sections 3.1 and 4.1.

6. FINAL REMARKS

This paper brings improvement over the author's previous work (Taneja, 2017) published in same series. The results appearing in Theorem 2 lead us to two groups of inequalities given in Sections 2.5 and 2.6. One group is with *two means* and another group is with *three means*. Based on these two groups we have improved the inequalities (30). These improvements are given in two separate sections. One section is with Propositions 1 to 11 and another with Propositions 12 to 20. Later these results are summarized in Theorems 4–7. In each case continued inequalities are given. Final section gives a joint theorem combining some of the results appearing in Theorems 5 and 7. In future, Theorems 4 and 6 can be combined to bring unified inequalities. This shall be done elsewhere. The idea of difference divergence measures and difference means was started by the author in 2005 (Taneja, 2005a, 2005b) and later studied many by authors, such as Chu, Wang, and Gong (2011), Jain and Chhabra (2016), Li and Zheng (2013), Ohlan (2015), Shi et al. (2010), Tomar and Ohlan (2014), Wu and Qi (2012), Wu, Qi, and Shi (2014), etc.

REFERENCES

Burbea, J., & Rao, C. R. (1982). On the convexity of some divergence measures based on entropy functions. *IEEE Transactions on Information Theory, IT-28*, 489–495.

Chen, C.-P. (2008). Asymptotic representations for Stolarsky, Gini and the generalized Muirhead means. *RGMIA Research Report Collection, 11*(4), 1–13.

Chu, Y., Wang, M., & Gong, W. (2011). Two sharp double inequalities for Seiffert mean. *Journal of Inequalities and Applications, 44*, 1–7.

Czinder, P., & Pales, Z. (2005). Local monotonicity properties of two-variable Gini means and the comparison theorem revisited. *Journal of Mathematical Analysis and Applications, 301*, 427–438.

Dragomir, S. S., Sunde, J., & Buse, C. (2000). New inequalities for Jeffreys divergence measure. *Tamsui Oxford University Journal of Mathematical Sciences, 16*, 295–309.

Eves, H. (2003). Means appearing in geometrical figures. *Mathematics Magazine, 76*, 292–294.

Gini, C. (1938). Di una formula compressiva delle medie. *Metron, 13*, 3–22.

Hellinger, E. (1909). Neue Begründung der Theorie der quadratischen Formen von unendlichen vielen Veränderlichen. *Journal für die Reine und Angewandte Mathematik, 136*, 210–271.

Jain, K. C., & Chhabra, P. (2016). New series of information divergence measures and their properties. *Applied Mathematics and Information Sciences, 10*(4), 1433–1446.

Jain, K. C., & Srivastava, A. (2007). On symmetric information divergence measures of Csiszar's f-divergence class. *Journal of Applied Mathematics, Statistics and Informatics, 3*, 85–102.

Jeffreys, H. (1946). An invariant form for the prior probability in estimation problems. *Proceedings of the Royal Society of London. Series A, 186*, 453–461.

Kullback, S., & Leibler, R. A. (1951). On information and sufficiency. *The Annals of Mathematical Statistics, 22*, 79–86.

Kumar, P., & Johnson, A. (2005). On a symmetric divergence measure and information inequalities. *Journal of Inequalities in Pure and Applied Mathematics, 6*, 1–13.

LeCam, L. (1986). *Asymptotic methods in statistical decision theory*. New York: Springer.

Lehmer, D. H. (1971). On the compounding of certain means. *Journal of Mathematical Analysis and Applications, 36*, 183–200.

Li, W.-H., & Zheng, M.-M. (2013). Some inequalities for bounding Toader mean. *Journal of Function Spaces and Applications, 2013*, 394194.

Ohlan, A. (2015). A new generalized fuzzy divergence measure and applications. *Fuzzy Information and Engineering, 7*, 507–523.

Sánder, J. (2004). A note on Gini-mean. *General Mathematics, 12*(4), 17–21.

Shi, H. N., Zhang, J., & Li, D. (2010). Schur-geometric convexity for difference of means. *Applied Mathematics E-Notes, 10*, 275–284.

Simic, S. (2009a). A simple proof of monotonicity for Stolarsky and Gini means. *Kragujevac Journal of Mathematics, 32*, 75–79.

Simic, S. (2009b). On certain new inequalities in information theory. *Acta Mathematica Hungarica, 124*(4), 353–361.

Taneja, I. J. (1995). New developments in generalized information measures. In P. W. Hawkes (Series Ed.), *Advances in imaging and electron physics: Vol. 91* (pp. 37–135).

Taneja, I. J. (2005a). On symmetric and non-symmetric divergence measures and their generalizations. In P. W. Hawkes (Series Ed.), *Advances in imaging and electron physics: Vol. 138* (pp. 177–250).

Taneja, I. J. (2005b). Refinement inequalities among symmetric divergence measures. *The Australian Journal of Mathematical Analysis and Applications, 2*, 8.

Taneja, I. J. (2006a). Bounds on triangular discrimination, harmonic mean and symmetric chi-square divergences. *Journal of Concrete and Applicable Mathematics, 4*, 91–111.

Taneja, I. J. (2006b). Refinement of inequalities among means. *Journal of Combinatorics, Information and Systems Sciences*, *31*, 357–378.

Taneja, I. J. (2012). Sequences of inequalities among differences of Gini means and divergence measures. *Journal of Applied Mathematics, Statistics and Informatics*, *8*(2), 49–65.

Taneja, I. J. (2013a). Nested inequalities among divergence measures. *Applied Mathematics and Information Sciences*, *7*(1), 49–72.

Taneja, I. J. (2013b). Generalized symmetric divergence measures and the probability of error. *Communications in Statistics. Theory and Methods*, *42*, 1654–1672.

Taneja, I. J. (2013c). Seven means, generalized triangular discrimination, and generating divergence measures. *Information*, *4*, 198–239. https://doi.org/10.3390/info4020198.

Taneja, I. J. (2014). Refinement of Gini-mean inequalities and their connections with divergence measures. *JUET Research Journal of Science and Technology*, *1*(1), 27–60.

Taneja, I. J. (2017). Information measures, mean differences, and inequalities. In P. W. Hawkes (Series Ed.), *Advances in imaging and electron physics: Vol. 201* (pp. 137–260).

Taneja, I. J., & Kumar, P. (2004). Relative information of type s, Csiszar's f-divergence, and information inequalities. *Information Sciences*, *166*, 105–125.

Tomar, V. P., & Ohlan, A. (2014). Sequence of inequalities among fuzzy mean difference divergence measures and their applications. *Springer Open Plus*, *3*, 623.

Wu, Y., & Qi, F. (2012). Schur-harmonic convexity for differences of some means. *International Mathematical Journal of Analysis and its Applications*, *32*(4), 263–270.

Wu, Y., Qi, F., & Shi, H. (2014). Schur-harmonic convexity for differences of some means. *Journal of Mathematical Inequalities*, *8*(2), 321–330.

CHAPTER THREE

The Optical Transfer Theory of the Electron Microscope: Fundamental Principles and Applications[*]

Karl-Joseph Hanszen
Department of Physics, University of Arizona, Tucson, AZ, United States
Present address: Physikalisch-Technische Bundesanstalt, Braunschweig, Germany

Contents

[*] Reprinted from Advances in Optical and Electron Microscopy 4 (1971) 1–84. This work has been supported by Public Health Service Grant No. GM 11852-5.

Advances in Imaging and Electron Physics, Volume 207
ISSN 1076-5670
https://doi.org/10.1016/bs.aiep.2018.04.001

251

1. THE PROBLEM OF IMAGE FORMATION IN THE ELECTRON MICROSCOPE

In transmission electron microscopy, the object, the thickness of which is small compared with its diameter, is penetrated by a pencil of electrons that suffer various types of interactions with the atoms of the object. As a result,

(a) Electrons are scattered out of their initial path.

(b) The scattering occurs with or without energy losses.

(c) A total loss of kinetic energy is possible (*electron absorption*).

In the ideal case, an electron microscope would map each point[1] of the object plane in a one-to-one correspondence onto an element of the image plane. Such an ideal mapping is impossible because of geometrical aberrations and the diffraction error.

In the object plane, an intrinsic diversity of interactions occurs; in the image plane, however, the recording of only one physical parameter is possible, that is the local distribution of the electron current density (e.g. as brightness distribution on a screen or as density distribution in a photographic emulsion). Even if we could eliminate all aberrations from the imaging system, it would be impossible to record the multiplicity of interactions mentioned above on a screen or on a plate. Therefore we could not retrieve all the information about the object structure that is carried by the ray pencil leaving the object. It is certainly possible to predict unequivocally the current distribution in the image if we know the atomic structure of the object and the imaging properties of the microscope. To determine the object structure from the image is impossible, as a rule.

Nevertheless the electron microscopist wants to determine an unknown or partially known object structure from the electron micrograph. This he

---

[1] For reasons of brevity we shall henceforth write "object points" instead of "elements of the object plane".

can only do if the object and the imaging system fulfill certain conditions. The following article deals with these cases.

In order to obtain unambiguous images, the interaction parameters have to be reduced to one. Therefore, either (1) we restrict ourselves to sufficiently thin objects and to electron pencils of sufficiently high energy that the energy losses are of minor importance, or (2) we use a filter lens to remove the decelerated electrons.

Since in the first case there is no absorption, the only interaction is scattering without energy losses. Consequently the interaction can be treated wave-mechanically as Kirchhoff diffraction at the atomic potential distribution in the object, in which case the distribution behaves as a phase structure. Hence the object structure can be derived unambiguously from the image if the electron optical system permits an unequivocal imaging.

In the second case, all electrons eliminated by the filter lens can be treated as absorbed electrons in image formation (Boersch, 1947, 1948). In addition to the information about the phase structure, the mononergetic electron pencil used for image formation now provides information about amplitude structure in the object. In order to obtain unambiguous information, the imaging of one of these structures must be prevented. Fortunately we can suppress the phase structure while improving the imaging of the amplitude structure by means of a sufficiently wide illumination aperture. Therefore, an essential requirement for an unequivocal image interpretation is also fulfilled.

Previous studies about electron microscopical resolution have mainly dealt with atomic interactions, while the question of image formation has been of secondary importance. In those studies, approximate distributions of particular model objects were used, and by so doing, the contrast properties and the possibility of resolving atomic distances on the basis of the point resolution theory were discussed. The conclusions drawn in this way are valid only for that particular object and that distance. The imaging properties of the microscope, however, depend strictly on the object under investigation. It is impossible, therefore, to explain the imaging properties and the resolution of the electron microscope merely from the result of those investigations. Being aware of this, we seek to discover objects the imaging properties of which are equal, in the sense that they depend only on the parameters of the imaging system. It will be shown later, that usually in electron microscopy these objects are the weak phase and weak amplitude objects. The imaging properties of these important objects can be described by very simple equations. The characteristic features of the

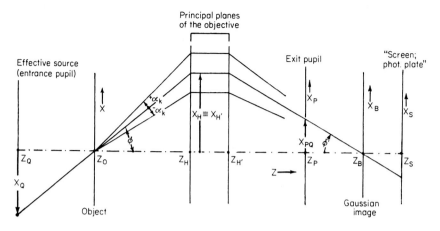

Figure 1 Explanation of the symbols used. φ is considered to be so small, that the corresponding ray on the image side strikes the axis in the Gaussian image plane.

image formation of other objects and the problems arising will also be discussed.

It is the aim of the writer to present the electron microscopical imaging theory in a form which is clear and complete within the limits indicated and comprehensible to the user of the instrument.

2. SYMBOLS AND DEFINITIONS; NUMERICAL DATA

2.1 Symbols and Definitions (See Fig. 1)

The coordinate along the optic axis will be denoted by z, the coordinates in the perpendicular directions by x and y, with the x-coordinate lying in the plane of the drawing. The z-coordinate of the effective electron source (entrance pupil) is denoted by z_Q, the coordinate of the object by z_O, the coordinates of the principal planes of the objective, z_H and $z_{H'}$, that of the exit pupil (briefly called "pupil") by z_P, of the Gaussian image plane by z_B and of the recording plane by z_S. The x and y coordinates in these planes are marked with the same index as the z-coordinate; in the object plane, the index O is omitted for simplicity.

We consider the cardinal elements to be independent of the magnification. Therefore, in a strict sense, our equations are only valid for Newtonian imaging fields. When, as commonly happens, the object is located within the field, the real cardinal elements are to be used (Glaser, 1956, especially p. 194ff).[2]

We now examine a ray, emerging from a single point of the electron source located off the axis. This ray strikes the axis in the object plane at a small angle φ and in the image plane at a small angle φ', so that φ, $\varphi' \ll \frac{\pi}{2}$. For $\varphi \to 0$ we have

$$\varphi = \frac{x_Q}{z_Q - z_O} = \frac{x_H}{z_H - z_O}; \qquad \varphi' = \frac{x_{H'}}{z_{H'} - z_B} = \frac{x_H}{z_{H'} - z_B}. \qquad (1)$$

The signs are so determined that the z-coordinate increases downstream along the beam. The angles are positive if a ray, leaving its point of inter-section with the axis, is entering the first quadrant of the coordinate system and they are negative if it enters the fourth quadrant. By means of φ and φ', the magnification M' is also determined. If we denote the object size by x and the Gaussian image size by x_B, then we have[3]

$$M' =_{df} \frac{x_B}{x} = \frac{\varphi}{\varphi'}. \qquad (2)$$

The magnification is negative, when the image is inverted.

In order to avoid a repeated use of the magnification factor, we use the reduced image coordinate

$$x' =_{df} \frac{-x_B}{M'}. \qquad (3)$$

The front and rear focal lengths of the objective, assumed to be a single lens, are denoted by f and f', and the coordinates of the focal planes by z_F and $z_{F'}$. Then we have:

$$f = -f' \quad (f \text{ is negative}). \qquad (4)$$

For the object distance $(z_Q - z_H)$ and the image distance $(z_B - z_{H'})$ we have in the case of Gaussian imaging

$$z_O - z_H = f(1 - 1/M'); \qquad z_B - z_{H'} = f'(1 - M'). \qquad (5)$$

---

[2] It must be pointed out, however, that in this case the position of the pupil cannot be obtained by constructing the asymptotes of the rays in the image space. Here, we are confronted with a similar problem to that discussed by Hanszen and Lauer (1965) (see especially Fig. 3 in that article).

[3] $=_{df}$ means "is defined to be equal to".

The image of the electron source located in the plane of the exit pupil has a magnification M'_Q. This quantity can be expressed by

$$
\left.
\begin{aligned}
M'_Q &=_{df} \frac{z_P - z_{H'}}{z_Q - z_H}; \\
z_Q - z_H &= f(1 - 1/M'_Q); \\
z_P - z_{H'} &= f'(1 - M'_Q).
\end{aligned}
\right\}
\tag{6}
$$

For $z_Q \to -\infty$, z_P coincides with $z_{F'}$.

We consider the object as composed of sinusoidal structures having the period lengths ϵ_k. A ray falling upon the object will be diffracted into the angles $\pm\alpha_k$. For small φ and α_k, we have

$$
\alpha_k = \pm\lambda/\epsilon_k; \quad \lambda = \text{electron wavelength.} \tag{7}
$$

In the pupil plane, we make use of the reduced coordinates

$$
S =_{df} -\frac{x_P M'}{\lambda f(M' - M'_Q)} = +\frac{x_P M'}{\lambda f'(M' - M'_Q)}. \tag{8}
$$

In that plane, the undiffracted ray has the natural coordinate x_{PQ} and the reduced coordinate

$$
Q =_{df} -\frac{x_{PQ} M'}{\lambda f(M' - M'_Q)} = \frac{\varphi}{\lambda}. \tag{9}
$$

In the same plane, the diffracted rays have the distances $(x_{Pk} - x_{PQ})$ with respect to the undiffracted beam. The corresponding reduced coordinates are

$$
R_k =_{df} S_k - Q = -\frac{(x_{Pk} - x_{PQ})}{\lambda f} \cdot \frac{M'}{(M' - M'_Q)} = \frac{\alpha_k}{\lambda}. \tag{10}
$$

The physical meaning of the reduced coordinate R_k can be understood by comparing (10) and (7). $|R_k|$ is identical with $1/\epsilon_k$. Also the distance x_{PQ} between the undiffracted ray and the axis is measured on this scale. Likewise, (9) and (8) can be explained.

In the important case $(z_Q - z_H) \to -\infty$, i.e. $M'_Q = 0$, we have

$$\left.\begin{array}{l} S(M'_Q = 0) = -\dfrac{x_P}{\lambda f}; \\[3mm] Q(M'_Q = 0) = -\dfrac{x_{PQ}}{\lambda f}; \\[3mm] R_k(M'_Q = 0) = -\dfrac{x_{Pk} - x_{PQ}}{\lambda f}. \end{array}\right\} \tag{11}$$

We need other reduced coordinates besides x', Q, R, and S. They will be introduced and discussed later (see Section 5).

We also need notations for spherical aberration and defocusing. The spherical aberration can be expressed either by the radius δ_s of the aberration disk in the image plane (transverse aberration)

$$\delta_s =_{df} M' C_s \alpha^3 = M'^4 C_s \alpha'^3; \tag{12}$$

if $C_s = C_s(M') = $ spherical aberration constant and

$$\alpha' = \alpha/M', \tag{13}$$

or by the longitudinal aberration

$$\Delta z_s =_{df} -M'^4 C_s d'^2 \approx M'^4 C_s \left(\dfrac{-x_P}{z_B - z_P}\right)^2. \tag{14}$$

Here α and α' are the half aperture angles of the ray pencils on the object and image side, respectively; their values are determined, by the object aperture stop.

In the electron microscope, defocusing is produced by deviations ΔI of the objective current I and ΔU of the accelerating voltage U from the nominal values. The voltage deviation leads to the wavelength alteration (in non-relativistic approximation)

$$\Delta \lambda \approx -\dfrac{\lambda}{2U} \Delta U. \tag{15}$$

It is possible to express both deviations by the chromatic aberration of the cardinal elements Δf, Δz_H and $\Delta z_{H'}$ in terms of the chromatic aberration constant C_{ch}; see e.g. Hanszen (1966a). In microscopical optics, i.e. in the case of $|M'| \gg 1$, the defocusing dependence of $\Delta z_{H'}$ can be neglected.

The defocusing by ΔI and ΔU can be compensated by a displacement Δz of the object:

$$\Delta z =_{df} \left(1 - \frac{2}{M'}\right)\Delta f + \Delta z_H =_{df} C_{ch}\left(\frac{\Delta U}{U} - 2\frac{\Delta I}{I}\right). \qquad (16)$$

We regard Δz as the characteristic defocusing quantity and call it simply "defocusing". The corresponding displacement Δz_d on the image side is

$$\Delta z_d = z_B - z_S =_{df} \Delta z M'^2. \qquad (17)$$

All other quantities are explained, later in this article. New definitions are always denoted by the sign $=_{df}$.

2.2 Numerical Values

In order to be independent of the characteristics of a particular electron microscope, the important results are given in dimensionless form. For illustration and in order to help the user of the instrument, several numerical examples will be given. These are based on the following set of data:

electron wave length: $\lambda = 3.7 \times 10^{-2}$ Å (corresponding to 100 keV electrons);

objective focal length: $f' = -f = -2.7$ mm;

magnifications: $M' = -27$; $M'_Q = -\frac{1}{46}$; $\frac{M'}{M'-M'_Q} = -1.001$;

spherical aberration constant $C_s = 4$ mm;

chromatic aberration constant $C_{ch} = 2.1$ mm;

object field diameter: $2x_e = 1000$ Å;

distance between object and electron source: $(z_O - z_Q) = 100$ mm;

beam aperture: $2\beta = 2x_e/(z_O - z_Q) \approx 10^{-6}$.

We call an objective having the above characteristics a "normal objective".

3. THE ILLUMINATION

3.1 Fundamentals

The entrance pupil acts as the effective illumination source. It can be either the cross-over of the electron beam formed by the field of the electron gun, the image of this cross-over produced by the condenser or especially, if the cross-over is imaged onto the object plane, the cross-section of the beam in the plane of the condenser stop. Usually there is a finite distance $(z_Q - z_O)$ between the entrance pupil and the object.

The pupil cross-section may have an intricate shape such as a circular disc, centered or not centered with respect to the axis, an annulus, a zone plate, etc. In Fig. 1, the radiation emerging from a single point of the effective source in the direction of the center of the object plane is characterized by the ray drawn. The total radiation, arriving at each object point, can be calculated by appropriate integration over the radiating area of the source.

We now assume that the electron source is located at $z_Q = -\infty$, and identify the ray mentioned in Fig. 1 with the direction of propagation of a plane wave, having an oblique incident angle φ with the axis. In this simplified description, a "finite size of the source" means that the rays entering the object point form a finite angular interval $\varphi - \Delta\varphi < \varphi < \varphi + \Delta\varphi$. The illumination conditions can be determined by integration over φ, and they are equal in each object point. This is not the case when the source is located at a finite distance from the object. The resulting differences, however, lead only to higher order terms (Born & Wolf, 1964, esp. p. 522ff), and can thus be neglected. A detailed discussion of the relation between the position of the exit pupil and that of the source and related problems will be given in Section 5. The investigation of the influence of the illumination aperture has recently been completed. In Section 10 a few examples of a finite aperture are given. Otherwise, our considerations are concerned exclusively with the imaging properties at vanishingly small illumination aperture, i.e. with coherent illumination.

3.2 Coherent, Partially Coherent, and Incoherent Illumination

The maximum size of the illumination aperture that can be used for *coherent* illumination is given by the spatial coherence condition:

$$\sin(\Delta\varphi) \approx \Delta\varphi \ll \lambda/4x_e, \qquad (18)$$

in which $\Delta\varphi$ is the semi-angular illumination aperture and $2x_e$ the object field diameter.

Example. In high resolution electron microscopy, structures between 5 Å and 2 Å are of main interest. If we desire an image-field diameter of only 500 image points, corresponding to the value $2x_e = 1000$ Å stated in Section 2.2, the coherence imposes the severe condition $\Delta\varphi \ll 10^{-5}$: which has rarely ever been fulfilled in practice. In fact, the condition (18) can be weakened. For details see the recent studies by Hanszen and Trepte (1971c) on the influence of a finite illumination aperture on the imaging and the summary in Section 10.1.

As long as the illumination aperture is small compared to the objective aperture, we call the illumination *partially coherent*. As is shown by light-optical transfer theory, we may call an illumination *incoherent* when the size of the illumination aperture is at least equal to that of the objective aperture.

3.3 The Equation for the Incident Wave

In the sense of wave mechanics, we can express the *wave function* ψ_Q of a wave incident on the object plane under the angle φ at a fixed time by

$$\psi_Q(x) = \Psi \exp(-2\pi i\varphi x/\lambda) = \Psi \exp(-2\pi iQx); \quad \Psi = \text{real}, \quad (19)$$

where Ψ is the amplitude of this wave. The exponent gives the phase of the wave at the moment of its entering the object plane. We denote the *intensity* of the incident wave by

$$I_Q(x) =_{df} \psi_Q(x)\psi_Q^*(x). \quad (20)$$

It is proportional to the current density in the object, see e.g. Glaser (1956), Eq. (50.14). If the wave proceeds in the direction of the axis ($\varphi = 0$), the phase is always constant throughout the object plane.

4. THE ELECTRON MICROSCOPICAL OBJECT

4.1 Definition of the Object Transparency

The amplitude and the phase are locally changed by the interactions of the incoming wave with the object. As has been pointed out before, the wavelength is assumed to be constant. Our investigations are limited to objects causing changes in amplitude and phase which can be described by a factor

$$F(x, y) = A(x, y) \exp[i\Phi(x, y)]; \quad A < 1 \text{ (real)}; \quad \Phi \text{ real} \quad (21)$$

multiplying the incoming wave. The amplitude modification of the wave is described by A, the phase modification by Φ. We call $A(x, y)$ the amplitude distribution and $\Phi(x, y)$ the phase distribution in the object. The wave function $\psi(x, y)$ and the intensity $I(x, y)$ as the wave emerges from the object are

$$\psi(x, y) = \psi_Q(x) \cdot F(x, y); \quad (22)$$

$$I(x, y) = \psi(x, y)\psi^*(x, y) = A^2(x, y)I_Q(x) =_{df} F_I(x, y) \cdot I_Q(x). \qquad (23)$$

F is called the *object transparency for the wave function*, and $F_I = |F|^2 = A^2$ the *object transparency for the intensity*.

4.2 Amplitude Object and Phase Object

When the object transparency for the wave function is real, i.e. when

$$F(x, y) =_{df} F_A(x, y) = A(x, y), \qquad (24)$$

we call the object an *amplitude object*. When the absolute value of the transparency is constant, and only its phase varies, i.e.

$$F(x, y) =_{df} F_\Phi(x, y) = \exp[i\Phi(x, y)] \qquad (25)$$

we call the object a *phase object*. We now explain these special cases in detail.

A pure phase object is an idealization. Proceeding from a thick specimen to a thin one, one observes that $A(x, y)$ approaches unity faster than $\Phi(x, y)$ approaches zero in the whole object field. The amplitude modulation $A(x, y)$ therefore becomes imperceptible when the phase modulation is still observable. For this reason, thin objects are, to a good approximation, pure phase objects. When the stricter condition $A(x, y) = 1$; $\Phi \ll 1$ is fulfilled, we are dealing with *weak phase objects*. Many objects do indeed behave as weak phase objects.

A pure amplitude object is also an idealization. Only thick, unsupported, opaque objects could be in a close approximation pure amplitude objects. In this case, the image would appear as a silhouette, i.e. $\langle A \rangle = \langle 0; 1 \rangle$. Therefore, these objects would always be strong amplitude objects. Pure amplitude objects that are weak do not exist under natural conditions.

4.3 Fourier Representation of the Object Transparency

In order to simplify the notation, we limit our considerations to one-dimensional objects. It is known from light optics (see Hopkins, 1961, especially p. 485, and Hauser, 1962, especially p. 138), that this limitation is possible in transfer theory without the loss of any general features. Minor misinterpretations which may occur when extending our considerations to two-dimensional objects will be dealt with in Section 5.13.

There are two significant possibilities for dividing the object transparency into independent contributions:

(1) Decomposition of the transparency into "object points", using the delta function:

$$F_O(x) = \int_{-\infty}^{+\infty} F_O(x_i)\delta(x - x_i)\, dx_i \tag{26}$$

in which $F_O(x_i)$ is the weight-function for the δ-sources.

(2) Disintegration of the real and imaginary parts of the transparency into sinusoidal gratings with the period lengths (grating constants) $\epsilon = 1/R$, i.e. into a Fourier integral. This disintegration is appropriate for describing image formation by the wave concept.

The period lengths and therefore also their reciprocals R, which are called spatial frequencies, are positive definite quantities. Because of the axial symmetry of the electron lenses, it is appropriate to expand in terms of cosines.

First, the position of the elementary gratings with respect to the axis intersection $x = 0$ has to be established. This is done by means of the term $\xi(R)$—the so-called lateral phase—in the argument of the cosine, see Eq. (28). The lateral phase is related to the coordinate Δx of the first maximum on the side with negative coordinates by the expression:

$$\Delta x(R) =_{df} -\xi(R)/2\pi R; \quad 0 \leqq \xi(R) < 2\pi \tag{27}$$

Δx is the *lateral displacement*.

To begin with, we give the formal mathematics of the Fourier expansion. The physical interpretation of these results is given in Sections 4.4 to 4.12.

The Fourier integral for the object transparency of the wave function is

$$F(x) =_{df} \int_0^{\infty} \left\{ \tilde{a}_{re}(R) \cos\left[2\pi Rx + \xi_{re}(R)\right] \right.$$
$$\left. + i\tilde{a}_{im}(R) \cos\left[2\pi Rx + \xi_{im}(R)\right] \right\} dR. \tag{28}$$

Here, the coefficients $\tilde{a}_{re}$ and $\tilde{a}_{im}$ of the expansion, as well as ξ_{re} and ξ_{im}, are real quantities. In order to normalize, we put:

$$\tilde{a}_{re}(R = 0) = 2 \times 2x_e; \quad \text{where } 2x_e = \text{object field, cf. (42) and (85);}$$
$$\tilde{a}_{im}(R = 0) = 0; \qquad \xi_{re}(R = 0) = 0. \tag{29}$$

It is convenient to write Eq. (28) in complex notation:

$$F(x) = \int_0^{\infty} \left\{ \frac{\tilde{a}_{re}(R)}{2} \left(\exp\{i[2\pi Rx + \xi_{re}(R)]\} \right. \right.$$

$$+\exp\{-i[2\pi Rx + \xi_{re}(R)]\}\big)$$
$$+i\frac{\tilde{a}_{im}(R)}{2}\big(\exp\{i[2\pi Rx + \xi_{im}(R)]\}$$
$$+\exp\{-i[2\pi Rx + \xi_{im}(R)]\}\big)\bigg\}dR. \tag{30}$$

Formally admitting negative values of R, we obtain

$$F(x) = 1 + \int_{-\infty}^{+\infty} \frac{\tilde{a}_{re}(R \neq 0)}{2} \exp\{i[2\pi Rx + \xi_{re}]\}dR$$
$$+ i\int_{-\infty}^{+\infty} \frac{\tilde{a}_{im}(R \neq 0)}{2} \exp\{i[2\pi Rx + \xi_{im}]\}dR. \tag{31}$$

This is valid if

$$\tilde{a}_{re}(R) = \tilde{a}_{re}(-R); \qquad \tilde{a}_{im}(R) = \tilde{a}_{im}(-R);$$
$$\text{i.e. } \tilde{a}_{re}, \tilde{a}_{im} \text{ are even functions;} \tag{32}$$

$$\xi_{re}(R) = -\xi_{re}(-R); \qquad \xi_{im}(R) = -\xi_{im}(-R);$$
$$\text{i.e. } \xi_{re}, \xi_{im} \text{ are odd functions.} \tag{33}$$

When the notations of Eq. (1.1) in Table 1 are used (31) becomes

$$F(x) = 1 + \int_{-\infty}^{+\infty} \tilde{F}(R \neq 0) \exp(2\pi iRx)dR. \tag{34}$$

In a similar manner, the Fourier integral for the transparency of the intensity, as defined in (23), can be written:

$$F_I(x) =_{df} 1 + \int_{0}^{+\infty} \tilde{A}_1(R \neq 0)\cos[2\pi Rx + \xi_1(R)]dR$$
$$= \int_{-\infty}^{+\infty} |\tilde{F}_I(R \neq 0)| \exp[i\xi_I(R)]\exp(2\pi iRx)dR. \tag{35}$$

Here, we have for the real functions $|\tilde{F}_I(R)|$ and $\xi_I(R)$:

$$|\tilde{F}_I(R)| = |\tilde{F}_I(-R)|; \qquad \xi_I(R) = -\xi_I(-R). \tag{36}$$

Eq. (35) has this simple form, because $F_I(x)$ is real.

Table 1 Relations between the Fourier coefficients

General relations between the complex Fourier coefficients in Eqs. (28)–(36):

(1.1)
$$\tilde{F}(R \neq 0) =_{df} |\tilde{F}| \exp(i\theta) =_{df} \frac{\tilde{a}_{re}}{2}\exp(i\xi_{re}) + i\frac{\tilde{a}_{im}}{2}\exp(i\xi_{im});$$

$$\tilde{F}(R=0) =_{df} \tilde{a}_{re}(R=0)/2 = 2x_e;$$

$$|\tilde{F}(R\neq 0)|^2 = \frac{1}{4}(\tilde{a}_{re}\cos\xi_{re} - \tilde{a}_{im}\sin\xi_{im})^2 + \frac{1}{4}(\tilde{a}_{re}\sin\xi_{re} + \tilde{a}_{im}\cos\xi_{im})^2;$$

$$\cos\theta(R\neq 0) = \frac{1}{2}(\tilde{a}_{re}\cos\xi_{re} - \tilde{a}_{im}\sin\xi_{im})/|\tilde{F}(R\neq 0)|$$

Both $\tilde{F}(R)$ and $\theta(R)$ have no symmetry with respect to $R=0$

	Coefficients of the mathematical Fourier expansion (28):	Relations replacing (1.1) when *weak objects* are present; see (57)–(61):	Coefficients resulting from the empirical division of the object in phase and amplitude components; see (57)–(61):
(1.2)	$\tilde{a}_{re}(\pm R)$	$= 2|\tilde{F}_A(\pm R)|$	$= \tilde{A}(+R)$
(1.3)	$\tilde{a}_{im}(\pm R)$	$= 2|\tilde{F}_\Phi(\pm R)|$	$= \tilde{\Phi}(+R)$
(1.4)	$\xi_{re}(+R) = -\xi_{re}(-R)$	$= \theta_A(+R) = -\theta_A(-R)$	$= \xi_A(+R)$
(1.5)	$\xi_{im}(+R) = -\xi_{im}(-R)$	$= \theta_\Phi(+R) = -\theta_\Phi(-R)$	$= \xi_\Phi(+R)$
(1.6)			$\tilde{A}(R=0) = 2 \times 2x_e;\ \tilde{\Phi}(R=0) = 0$ (Normalization, see (29))

4.4 Diffraction at the Object

As diffraction theory shows (Lohmann & Wegener, 1955), the coordinates $\pm R$ in the Fourier space, introduced in (31) to (34), correspond to the angles $\alpha = \pm \lambda R$ at which a plane wave is diffracted at a plane elementary grating having a spatial frequency R; see (7). Knowing this, the formal extension of R toward negative values becomes clear. Thus, if we interpret $R = \alpha/\lambda$ as *reduced angular coordinate*, the diffraction spectrum behind the object is given by $\tilde{F}(R)$ in (34) for $-\infty < R < +\infty$. If, however, we consider R as a spatial frequency, the frequency representation of both the real part and the imaginary part of the transparency is given by (28) for $0 < R < +\infty$. Here the Fourier coefficients are subject to the symmetries (32) and (33).

Example. An amplitude object, the transparency of which is modulated only by the spatial frequency R_1.

Fourier representation according to (28):

$$F(x) = 1 + \tilde{a}_1 \cos(2\pi R_1 x - \pi/2) = 1 + \tilde{a}_1 \sin 2\pi R_1 x. \tag{37}$$

Fourier representation according to (34)

$$F(x) = 1 - i\frac{\tilde{a}_1}{2} \exp(2\pi i R_1 x) + i\frac{\tilde{a}_1}{2} \exp(-2\pi i R_1 x). \tag{38}$$

Because of this, the wave field behind the object consists only of the following partial waves:

1. The undiffracted wave with the relative amplitude 1.
2. Two waves, diffracted into the angles $\alpha = \pm R_1 \lambda$, having the relative amplitude $\tilde{\alpha}_1/2$ and the phases $\pm\pi/2$ with respect to the undiffracted wave.

We now know the Fourier description of the object transparency and its relation to the diffraction spectrum, but the applicability of the concept "object transparency" is still uncertain. Furthermore we do not know how the amplitude and phase components of the object can be read from the transparency. Information on these details will be given in the next sections.

4.5 The Influence of the Object Thickness (a Purely Geometrical Consideration)

Eq. (34) is equivalent to physical diffraction only for objects of vanishing thickness. When thick objects are present, their z-dimension can be divided into thickness-elements Δz, and the diffraction of each partial wave

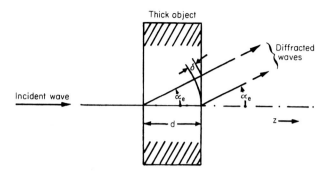

Figure 2 Optical path difference δ between two diffracted waves, originating from front and rear of a thick object.

at each consecutive element must be considered, for details and further references see Jeschke and Niedrig (1970). Within the wave-field emerging from the object, there will be partial waves in each direction which originate from different elements Δz. These waves may weaken one another by interference.

We call an object thin, if the phase differences of the waves diffracted at the front and at the rear of the object by the highest spatial frequency are much smaller than $\lambda/2$, so that there is no extinction for any of the diffracted waves. If d is the object thickness, the optical path-difference between the two waves is, according to Fig. 2:

$$\delta = d\alpha_e^2/2 \tag{39}$$

According to (7) and (10), the resulting thickness limitation is

$$d \ll \frac{1}{\lambda R_e^2} = \frac{\epsilon_e^2}{\lambda} \tag{40}$$

Example. If we are interested in image structures of the size of 2 Å, with the numerical values given in Section 2.2, we obtain the thickness limitation $d \ll 100$ Å.

4.6 Inadequacy of the Concept "Transparency" for Thick Objects

In (21), the transparency was introduced, as a function dependent on position (x, y) only, and in particular, independent of the direction of the incident wave. If the illumination is at an angle to the axis, the wave field behind the object is also inclined without other change in the diffracted waves (Morgenstern, 1965), as is shown in Fig. 3A and B. This is undoubtedly true for diffraction in weak objects. Nor is there any objection to applying (21) to thick, amorphous objects, if the permissible thickness as a function of the angle of incidence is carefully considered according to (40).

Thick crystalline objects raise serious difficulties, however, because diffracted waves appear only at discrete incident angles (Bragg angles); see Fig. 3C and D. Thus the transparency should be described by a function of both the object coordinate and the angular position of the lattice planes, with respect to the incident wave. In order to avoid complications of this type, we restrict our considerations to objects, the thickness d of which is small compared to the extinction thickness d_{ex}:

$$d_{ex} \approx \lambda \frac{U}{\Delta U_h}; \tag{41}$$

U = acceleration voltage;
ΔU_h = Fourier coefficient of the crystal potential in the direction of the penetrating beam;
$\lambda = \lambda(U)$.

Example. According to Komoda (1964), the extinction thickness, valid for diffraction at the (111)-planes of a gold crystal is 161 Å. According to transfer theory, therefore, the thickness restrictions for crystalline objects do not seem to be more severe than those for amorphous ones.

4.7 The Object Modulation

The undiffracted wave $R = 0$ contains the information about the average intensity (background). The diffracted waves $R \neq 0$ carry the essential information about the object structure, i.e. about the amplitude and the phase components of each spatial frequency, and the lateral phases of these frequencies. Some new concepts will now be introduced, enabling us to describe the optical transfer of the amplitude and phase components quantitatively in terms of the spatial frequency.

In the object, both an amplitude distribution $A(x)$ and a phase distribution $\Phi(x)$ are present, as a rule. It is now interesting to find out which spatial frequencies are present in the amplitude and which ones in the phase distribution. In other words, we are looking for the Fourier integrals

$$A(x) = \frac{\tilde{A}(R=0)}{2 \times 2x_e} + \int_0^\infty \tilde{A}(R \neq 0) \cos[2\pi R x + \xi_A(R)] dR;$$

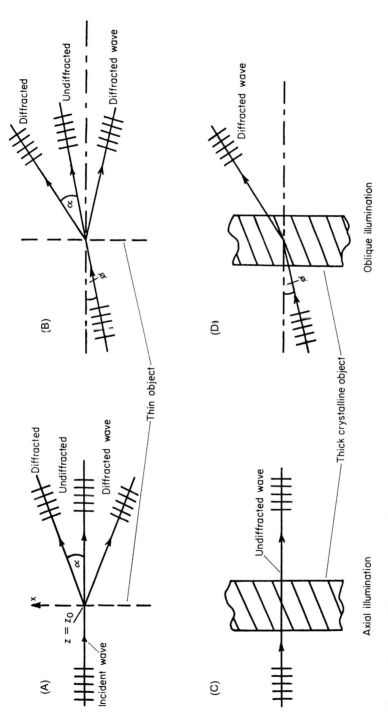

Figure 3 Electron diffraction at a crystalline object, with incident waves of different inclination to the axis. (A) and (B): thin object; (C) and (D): thick object (Morgenstern, 1965).

$$\tilde{A}(R) = \text{real}; \quad \tilde{A}(R=0) = 2\int_{-\infty}^{+\infty} A(x)\,dx \tag{42}$$

$$\Phi(x) = \int_0^{\infty} \tilde{\Phi}(R\neq 0)\cos\left[2\pi Rx + \xi_\Phi(R)\right]dR; \qquad \tilde{\Phi}(R) = \text{real}. \tag{43}$$

For completeness we give a similar expression for the intensity transparency:

$$F_I(x) = A^2(x) = \frac{\tilde{A}_I(R=0)}{2\times 2x_e} + \int_0^{\infty} \tilde{A}_I(R\neq 0)\cos\left[2\pi Rx + \xi_I(R)\right]dR;$$

$$\tilde{A}_I(R) = \text{real}; \quad \tilde{A}_I(R=0) = 2\int_{-\infty}^{+\infty} F_I(x)\,dx. \tag{44}$$

Because the x-independent term has a special meaning, it has been taken outside the integral. In the phase expansion, (43), it was put equal to zero. The strength of each spatial frequency $R > 0$ with respect to the background and the existing lateral phases will be described by the following quantities:

Modulation of the absolute value of the wave function in the object:

$$K_A(R) =_{df} \frac{\tilde{A}(R)}{\tilde{A}(R=0)}\exp\left[i\xi_A(R)\right]; \tag{45}$$

Phase modulation in the object:

$$K_\Phi(R) =_{df} \frac{\tilde{\Phi}(R)}{1}\exp\left[i\xi_\Phi(R)\right]; \tag{46}$$

Intensity modulation in the object:

$$K_I(R) =_{df} \frac{\tilde{A}_I(R)}{\tilde{A}_I(R=0)}\exp\left[i\xi_I(R)\right]. \tag{47}$$

The phase distribution is given in an angular scale. Since there is no constant term in (43), the reference quantity in (46) can be prescribed arbitrarily. We use the radian as our angular unit. For simplicity we normalize:

$$\frac{\tilde{A}(R=0)}{2\times 2x_e} = 1; \qquad \frac{\tilde{A}_I(R=0)}{2\times 2x_e} = 1. \tag{48}$$

It is possible to describe all objects by two of the above given equations, i.e. by (45) and (46), or by (47) and (46). For describing a pure amplitude object we need (45) or (47) only, and for a pure phase object, only (46) is necessary.

Eqs. (45) to (48) give a description of the object based on the empirical separation into amplitude and phase components, while the description in Eqs. (28) to (36) is actually related to the symmetry conditions in the diffraction pattern. The relationship between these two descriptions will be given in the next section.

4.8 The Strong Phase Object

The separation of the object transparency into amplitude and phase components according to (45)–(46) was a purely phenomenological one, and is not directly related to the object diffraction, described by (34). In the special case of a pure amplitude object, (42) is identical with the real part of (28). In the case of pure phase objects, however, there are differences between (43) and the imaginary part of (28), as explained by the following example[4]:

Example. A pure phase object, the phase of which is modulated by one spatial frequency only (see Hanszen, Morgenstern, & Rosenbruch, 1963). The transparency for the wave function of this object is given by

$$F(x) = \exp(i\tilde{a}\cos 2\pi R_1 x); \quad \text{with } \tilde{a} = 2\pi\,\Delta nd/\lambda, \tag{49}$$

where (in non-relativistic approximation)

$$\Delta n = \frac{1}{2}\frac{\Delta U_m}{U} \tag{50}$$

is the deviation of the refractive index (due to the "mean inner potential" ΔU_m) and d the object thickness. The Fourier representation of $F(x)$ is

$$F(x) = \sum_{k=-\infty}^{+\infty} \tilde{F}_k \exp(2\pi i k R_1 x), \tag{51}$$

where the Fourier coefficients

$$\tilde{F}_k = \exp(ik\pi/2).J_k(\tilde{a}) \tag{52}$$

---

[4] Similar conditions exist in the Fourier expansion of the object intensity $I(x)$. Each spatial frequency in the intensity transparency has a spectrum with numerous diffraction orders of the wave function $|\psi| = \sqrt{I(x)}$. In the case of a coherently illuminated *amplitude object*, it is thus not expedient to expand the transparency of the intensity according to Fourier, although it is expedient for other reasons to use $I(x)$, which can be directly observed.

contain the Bessel functions $J_k(\tilde{a})$. Although there is only one spatial frequency in the object, there is a sequence of diffraction orders k behind the object on each side.

The unequivocal correlation between *one* spatial frequency and *two* diffracted waves which was seen in the case of an amplitude object (see (37) and (38)) is therefore, not generally valid for phase objects. Yet this correlation is decisive for the possibility of establishing a transfer theory. Therefore, it is impossible to cover all phase objects by the theory. On the contrary, we have to restrict our theory to objects whose higher diffraction orders are negligible. That holds true, indeed, for *weak objects*: For $\tilde{a} \to 0$, the first terms in the expansion of the Bessel functions are[5]:

$$\left.\begin{aligned}
J_0(\tilde{a}) &= 1 - \frac{\tilde{a}^2}{4} + \cdots ; \\
J_1(\tilde{a}) &= \frac{\tilde{a}}{2} - \cdots ; \\
J_2(\tilde{a}) &= \frac{\tilde{a}^2}{8} - \cdots ; \\
J_3(\tilde{a}) &= 0 \ldots .
\end{aligned}\right\} \tag{53}$$

Example. An object may be called weak, if

$$J_2(\tilde{a})/J_1(\tilde{a}) \leqq 0.2; \quad \text{i.e.} \quad \tilde{a} \leqq 0.8. \tag{54}$$

According to (49) and (50) the object thickness is restricted to

$$d \leqq \frac{0.8\lambda(U) \cdot U}{\pi \Delta U_m}; \quad \Delta U_m = \text{mean inner potential.} \tag{55}$$

Except for the slightly different meaning of ΔU, this value differs from that given by (41) only by a factor ≈ 0.25. The limitations of thickness as presented in the previous section are therefore correct as to the order of magnitude. For example, the maximum permitted thickness of carbon specimens ($\Delta U_m \approx 6$ V) is 150 Å.

It must be pointed out that the above estimate is based on the numerical value of the inner potential, i.e. the average of the local potential distribution in matter. In the imaging of atoms the real potential, which may differ locally much more from the anode potential than the inner potential, must

---

[5] Details about the spectra of strong phase objects can be found in the article of Nagendra Nath (1939), esp. in Fig. 10.

be taken into account. For such cases the above theory may not be sufficient. For treating problems of this kind see Hauser (1962).

4.9 The Weak Phase Object

A weak phase object is defined by $\Phi(x, y) \ll 1$. Then, (25) can be evaluated as a series. Thus we obtain for the one dimensional object

$$F_\Phi(x) = 1 + i\Phi(x) - \tfrac{1}{2}\Phi^2(x) + \cdots \tag{56}$$

We truncate this series after the linear term and expand $F_\Phi(x)$ according to (34) with $\tilde{F}(R) \equiv i\tilde{F}_\Phi(R) = i|\tilde{F}_\Phi(R)| \exp[i\theta_\Phi(R)]$, where $\tilde{F}(R)$ is defined in Table 1, Eq. (1.1), and obtain

$$F_\Phi(x) = 1 + i \int_{-\infty}^{+\infty} |\tilde{F}_\Phi(R \neq 0)| \exp\{i[2\pi Rx + \theta_\Phi(R)]\} dR; \tag{57}$$

or, we expand $\Phi(x)$ in (56) according to (43) and obtain

$$F_\Phi(x) = 1 + i \int_0^{+\infty} \tilde{\Phi}(R \neq 0) \cos[2\pi Rx - \xi_\Phi(R)] dR. \tag{58}$$

Since the integral is real, $\tilde{F}_\Phi(R) = \tilde{F}_\Phi^*(-R)$. Comparing the coefficients of both equations leads to Eqs. (1.3), (1.5) in Table 1, which show the connection between the coefficients $\tilde{F}_\Phi$; $\tilde{\Phi}$ and θ_Φ; ξ_ϕ.

In this special case the integral (58) is identical with the imaginary part of (28) if the coefficients are identified as shown in (29) and (1.5, Table 1). This means that the phase modulation *of a weak phase object* affects only the imaginary part of the transparency for the wave function.

These facts are illustrated in Fig. 4. In Fig. 4A, a strong phase object can be seen. Both the imaginary part (containing the odd Bessel functions) and the real part (containing the even Bessel functions) are modulated. In this case, the general Fourier transform (28) is to be used. Fig. 4B shows a weak phase object. The cylindrical surface, containing the whole set of the $F_\Phi(x)$ values, can be approximated by the tangential plane. This means that (28) contracts to (58).

4.10 The Strong and the Weak Amplitude Object

The expansion of (24) according to (34) and (42); (48), i.e.

$$F_A(x) = 1 + \int_{-\infty}^{+\infty} |\tilde{F}_A(R \neq 0)| \exp\{i[2\pi Rx + \theta_A(R)]\} dR;$$

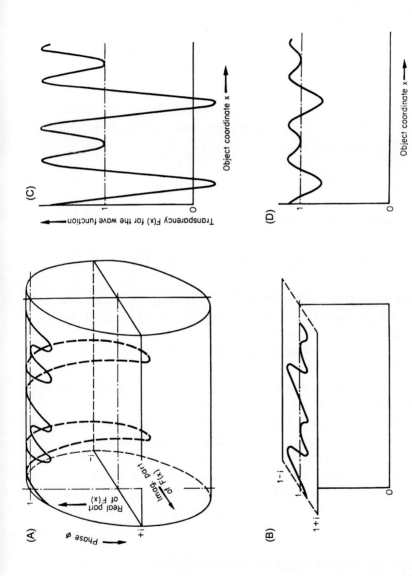

Figure 4 The object transparency $F(x)$ for the wave function. The object has only two spatial frequencies $\neq 0$. (A) strong, (B) weak phase object; (C) strong, (D) weak amplitude object (Hanszen & Morgenstern, 1965).

with

$$\tilde{F}(R) \equiv \tilde{F}_A(R) = \left|\tilde{F}_A(R)\right| \exp\left[i\theta_A(R)\right] = \tilde{F}_A^*(-R), \tag{59}$$

$$F_A(x) = 1 + \int_0^{+\infty} \tilde{A}(R \neq 0) \cos\left[2\pi Rx + \xi_A(R)\right] dR \tag{60}$$

does not depend on the object strength. Comparing both equations, we obtain a connection between the coefficients $\tilde{F}_A$; $\tilde{A}$ and ξ_A; θ_A, given in Eqs. (1.2) and (1.4) of Table 1. In this case, (60) is identical with the real part of (28), if the coefficients are identified as in (29) and (1.4, Table 1). Therefore the amplitude modulation of any pure amplitude object has a real transparency.

Because strong amplitude objects present other difficulties, as shown in Section 5.4, we restrict our further considerations to weak amplitude objects. They are defined by $\tilde{A}(R \neq 0) \ll 1$. Examples of strong and weak amplitude objects are given in Fig. 4C and 4D.

4.11 The Weak Object with Amplitude and Phase Components

For an object having both a weak amplitude and a weak phase modulation, we can write

$$\begin{aligned}
F(x) = F_A(x)F_\Phi(x) &\approx 1 + \int_{-\infty}^{+\infty} \left\{ \left|\tilde{F}_A(R \neq 0)\right| \exp\left(i[2\pi Rx + \theta_A]\right) \right. \\
&\quad + i\left|\tilde{F}_\Phi(R \neq 0)\right| \exp\left(i[2\pi Rx + \theta_\Phi]\right) \right\} dR \\
&\approx 1 + \int_0^\infty \left\{ \tilde{A}(R \neq 0) \cos[2\pi Rx + \xi_A] \right. \\
&\quad + i\tilde{\Phi}(R \neq 0) \cos[2\pi Rx + \xi_\Phi] \right\} dR,
\end{aligned} \tag{61}$$

where all coefficients are $\ll 1$. Due to (29), (61) is equal to (28), when the corresponding coefficients are compared as shown in Table 1.

From this we learn that the information about amplitude and phase modulation in a weak object is contained in the diffraction pattern in such a manner that each spatial frequency is unequivocally related to two diffraction angles of equal width but opposite sign. Unfortunately only $|\tilde{F}|$ and θ, but not the interesting quantities $\tilde{A}$, $\tilde{\Phi}$, ξ_A, and ξ_Φ can be directly extracted from the diffraction pattern. In order to obtain these quantities, $|\tilde{F}|$ and θ must be divided into even and odd functions of R.

We are only interested in knowing whether it is possible in principle to solve this problem. To avoid complicated formulae, however, our further considerations are limited to simple examples of practical importance. For a more general treatise see Menzel (1958, 1960) and Hauser (1962). Eq. (61) together with Table 1 is in fact the basis for the transfer theory of weak objects.

In order to develop the theory, an expression for the intensity transparency of a weak object is needed. From (44) and (61) we obtain the required approximation:

$$F_I(x) = \left\{ 1 + \int_0^{+\infty} \tilde{A}(R \neq 0) \cos[2\pi Rx + \xi_A] dR \right\}^2$$

$$\approx 1 + 2 \int_0^{+\infty} \tilde{A}(R \neq 0) \cos[2\pi Rx + \xi_A] dR. \tag{62}$$

In contrast to the remark in the footnote of Section 4.8, each spatial frequency in the intensity transparency F_I of a weak object now has only one corresponding frequency in the transparency of the wave function F. The Fourier coefficients of F_I are twice as large as those of F.

4.12 A Particular Relationship Between Modulation and Diffraction of a Weak Object

In (45) and (46) the concepts "amplitude and phase modulation" were introduced *ad hoc*. Knowing (61), we can now give the following relation between the transparency for the wave function $F(x)$ and the modulations $K_A(R)$ and $K_\Phi(R)$ in the case of a weak object, and only in that case:

$$F(x) = 1 + \tfrac{1}{2} \int_{-\infty}^{+\infty} \left[K_A(R \neq 0) + iK_\Phi(R \neq 0) \right] \exp(2\pi iRx) dR;$$

$$\text{with } K_A(-R) =_{df} K_A^*(+R); \quad K_\Phi(-R) =_{df} K_\Phi^*(+R). \tag{63}$$

Similarly, for the transparency of the intensity, we obtain:

$$F(x) = 1 + \int_{-\infty}^{+\infty} K_A(R \neq 0) \exp(2\pi iRx) dR. \tag{64}$$

Comparing (64) with (44) and (47), we find the following relation between the modulation of the amplitude of the wave function and the modulation of the intensity

$$K_I(R) = 2K_A(R). \tag{65}$$

In weak objects, the amplitude or phase modulation of a spatial frequency and the amplitude of the waves diffracted by these modulations, are strictly proportional. Therefore, the concept "weak object" has not only the meaning "the modulation of all spatial frequencies are small compared with 1", but also "the amplitude of the wave functions diffracted in any direction is small compared with the amplitude of the undiffracted wave". Thus, the transparency of a weak object can be obtained from the diffraction pattern.

The electron optical lens system alters the diffraction pattern in an undesired manner. Therefore, the image transparency is not identical with the object transparency. With the knowledge obtained in this section it is possible to evaluate the electron optical transfer properties directly from the difference between the so-called object-related and image-related diffraction patterns.

5. THE FORMATION OF THE OPTICAL IMAGE

5.1 Fundamental Problems

In every plane which is traversed by the wave field behind the object, all the information about the interaction between the beam electrons and the atoms in the object is present. The information, however, is generally not separated according to the interactions at the different object points. In principle, the imaging system should decode this information in such a way that the current density in each image point gives information about the interaction in exactly one object point. Some reasons why the electron microscope is not capable of doing so have already been mentioned in the introduction.

Having now a complete knowledge about the weak object in (63)–(65), we can be sure that, even for such objects, it is generally impossible to determine the structure of the object from the image unequivocally. This is so because it is impossible, in principle, to draw unequivocal conclusions from the intensity modulation in a single micrograph about the intensity modulation of the amplitude *and* phase structure in the object. Therefore, we cannot profitably treat weak objects which have phase and amplitude components at the same time. It is better to limit our studies to pure phase or pure amplitude objects.

As previously stated, objects having a thickness of about 100 Å or less behave in high-voltage electron microscopy as weak phase objects without an amplitude component; for an experimental proof see Section 6.5.

The Perfect Image. It is easy to indicate the conditions for perfectly imaging a pure amplitude object. In an ideal case, all electrons emerging from an object point should be focused into the corresponding image point (aberration-free image). Then in a reduced scale (3), the wave function $\psi'(x')$ in the image plane would be equal to the wave function $\psi(x)$ in the object plane; the same would be true for the related intensities $I'(x')$ and $I(x)$.

Under these conditions, however, the image of a phase object would be devoid of information, because with $A(x) = \text{const}$, (23) leads to $I(x)$ and $I'(x') = \text{const}$. In order to map the phase distribution $\Phi(x)$ onto the intensity distribution $I'(x')$, special measures are needed (see Sections 5.10 to 5.12).

The Real Image. A true point-to-point imaging is impossible. Diffraction at the pupil boundaries and the geometrical aberrations make the images of object points spread into aberration disks which overlap in a very complex manner. The relations between $\psi(x)$ and $\psi'(x')$ as well as between $I(x)$ and $I'(x')$ can be expressed by convolution integrals.

In this section, we only explain the processes taking place under coherent illumination.[6] Starting with an expression for the aberration disk of an object point (the so-called point-image), we present the imaging process in terms of spatial frequencies. Fig. 5 illustrates the meaning of the following equations.

The connection between the wave function $\psi(x)$ at a place x in the object plane and the wave function $\psi'(x')$ at a place x' in the image plane, produced by an imaging system having aberrations can be expressed by

$$d\psi'(x') = k(x, x')\psi(x)dx. \tag{66}$$

Here, $k(x, x')$ is called the *spread function* or *point image*. The contributions of all wave functions emerging from individual object points superimpose linearly at the image point x' in question. Therefore, the wave function in the image of an extended object is the integral

$$\psi'(x') = \int_{-\infty}^{+\infty} k(x, x')\psi(x)dx. \tag{67}$$

In Section 3.2, we limited our study to a small paraxial object area. The image of this area is impaired by defocusing and spherical aberration only.[7]

---

[6] For fundamentals see Born and Wolf (1964), p. 480ff; for descriptions especially written from an electron microscopical point of view, see Hanszen, Morgenstern, and Rosenbruch (1963) and Lenz (1965b), p. 274f.

[7] If no mention is made, the axial astigmatism is considered to be corrected.

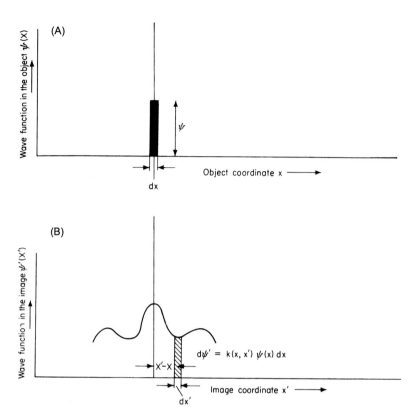

Figure 5 Image formation of coherently illuminated objects. (A) Radiating object point; (B) corresponding aberration disk in the image (the fact that ψ' is complex cannot be shown in this simplified figure).

The shape of the point image does not depend on the position of the object point. An image formation having these properties is called *invariant, stationary,* or *isoplanatic* (e.g. see Linfoot, 1964; Röhler, 1967, especially p. 16). Under the above conditions, k depends only on the distance $(x' - x)$ between the image point in question and the Gaussian image point:

$$k(x, x') = k(x' - x).\tag{68}$$

By means of (67)–(68) the wave function is known at each image point. Since $\psi(x)$ is a function which, according to (22)–(26), depends only on the direction of the illuminating wave and the properties of the object and since the spread function $k(x' - x)$ is introduced as an instrument parameter, we are on the way to a solution of the imaging problem in the manner described in Section 1.

5.2 Introduction of Transfer Functions

We replace the convolution (67)–(68) by a simpler but mathematically equivalent relation. For this purpose, we present not only the object and the image wave function, but also the spread function as a Fourier integral:

$$\psi(x) = \int_{-\infty}^{+\infty} \tilde{\psi}(R)\exp(2\pi iRx)dR; \tag{69a}$$

$$\psi'(x') = \int_{-\infty}^{+\infty} \tilde{\psi}'(R)\exp\big(2\pi iRx'\big)dR; \tag{69b}$$

$$k(x'-x) = \int_{-\infty}^{+\infty} \tilde{k}(R)\exp\big[2\pi iR(x'-x)\big]dR. \tag{69c}$$

Using the Fourier inverse

$$(\tilde{\psi}R) = \int_{-\infty}^{+\infty} \psi(x)\exp(-2\pi iRx)dx \tag{70}$$

and those of (69b) and (69c) we obtain

$$\tilde{\psi}'(R) = \tilde{k}(R)\tilde{\psi}(R); \quad \tilde{k}(R) = \text{complex}. \tag{71}$$

For reasons explained in Section 5.6, (71) is called the *filter-equation*, and $\tilde{k}(R)$ the *transfer-function for the wave function*. The intricate convolution (67); (68) of the wave function in position space can be replaced by a simple multiplication of its Fourier transform with the transfer-function $\tilde{k}(R)$.

When Eq. (19) describing the illumination wave, and (22) and (34) describing the object transparency and its Fourier representation, and the Fourier inverse of (34) are introduced into (70), we obtain

$$\tilde{\psi}(R) = \Psi\tilde{F}(Q+R) \tag{72}$$

where $\tilde{\psi}(R)dR$ is the wave function, diffracted by the object in the direction $\alpha = \lambda R$. Therefore, the filter equation takes the form

$$\tilde{\psi}'(R) = \Psi\tilde{F}(Q+R)\tilde{k}(R). \tag{73}$$

Here, $\tilde{\psi}'(R)dR$ is the wave function in the image, entering from the direction $\alpha' = \lambda R/M'$. We call α the *image-side diffraction-angle*. If we define the "image transparency for the wave function" by

$$F'(x') =_{df} \int_{-\infty}^{+\infty} \tilde{F}'(Q+R)\exp\big[2\pi i(Q+R)x\big]d(Q+R), \tag{74}$$

and if we use its Fourier inverse and similar equations as (19) and (22) for the image side, then, with respect to (10), we have

$$\tilde{F}'(Q+R) = \tilde{k}(R)\tilde{F}(Q+R); \quad \text{or} \quad \tilde{F}'(S) = \tilde{k}(R)\tilde{F}(S), \tag{75}$$

respectively.

By this result, the filter equation is related to the diffraction already known. We learn from (75) that the influence of the imaging system on the diffracted wave function can be described by a complex factor depending only on instrument parameters and the diffraction angle.

5.3 Linear Transfer

It was possible to describe the imaging process in a simple form (71)–(75), because the wave functions superpose linearly. Image formation which obeys these equations is called linear imaging, and it has the following properties. When a wave function $\psi_3(x)$ in the object is composed of the wave function $\psi_1(x)$ and $\psi_2(x)$ of two other objects according to

$$\psi_3(x) = C_1\psi_1(x) + C_2\psi_2(x), \tag{76}$$

the corresponding wave functions $\psi_1'(x')$; $\psi_2'(x')$, $\psi_3'(x')$ in the image are related by

$$\psi_3'(x') = C_1'\psi_1'(x') + C_2'\psi_2'(x'). \tag{77}$$

Therefore, in the case of linear transfer, we can determine the optical transfer properties as follows. First we determine the transfer properties of objects having only one spatial frequency R_k. After having done this for objects with every possible R_k, we also know with certainty the transfer properties for all possible objects since, according to (76) and (77) the object and image properties of any compound object may be obtained by linear superposition of the corresponding properties of the elementary objects. In other words, linear transfer is characterized by the fact that the only object parameter present in the transfer function is the direction R of diffraction.

5.4 The Imaging of Amplitude and Phase of the Object Wave Function and of the Object Intensity into the Image Intensity

Instead of the wave function $\psi'(x')$, the intensity

$$I'(x') = \psi'(x')\psi'^*(x') \tag{78}$$

is usually recorded in the image plane. Therefore, at the end of each imaging process, there is generally a quadratic relation which destroys the entire system of linear transfer for coherent illumination. Another quadratic relation, that is (23), is involved if one refers, as is usual, to the intensity distribution in the object instead of to the wave-function distribution. Also in the case of amplitude objects, to avoid difficulties, we must now restrict ourselves to *weak* objects. It has been shown in proceeding from (61) to (62), that the relation between the object transparencies for the wave function on the one hand and for the intensity on the other can be approximated by a linear equation. The same is true for the image quantities. Therefore, the imaging process of weak objects can be described to a good approximation by a sequence of linear equations.

The transfer function $\tilde{k}(R)$ indicates how the image-side wave function of the diffraction pattern arises from the corresponding object-side wave function. As will be shown later on, $\tilde{k}(R)$ has in general no symmetry properties with respect to $R = 0$. Therefore, $\tilde{\psi}'(R)$; $\tilde{F}'(R)$ and $\tilde{F}'_I(R)$ can be expressed by sets of coefficients, which are related in a complicated way to the set of coefficients for $\tilde{F}(R)$ and $\tilde{\psi}(R)$ respectively, given in Table 1. These coefficients, however, carry the desired information about the modulation properties of the object and the image.

Since only a few functions $\tilde{k}(R)$ are of practical importance, we do not intend to evaluate general equations which are valid for any $\tilde{k}(R)$. Instead of this, we try in the next section to study the physics of the imaging process in order to become acquainted with the physical significance of $\tilde{k}(R)$ and to learn which of the functions are of practical interest. The influence of these functions on the image properties will then be studied by means of simple *model objects*.

The limitation to the theory on weak objects has the effect that, contrary to the discussion in Section 5.3, only weak objects are usable as model objects, i.e. objects with a weak modulation (45)–(47). Such a model object is, for example:

$$F(x) = 1 + \tilde{A}_k \cos[2\pi R_k x + \xi_{A_k}] + i\tilde{\Phi}_k \cos[2\pi R_k x + \xi_{\Phi_k}];$$
$$\tilde{A}_k; \tilde{\Phi}_k \ll 1 \tag{79}$$
$$F_I(x) = 1 + 2\tilde{A}_k \cos[2\pi R_k x + \xi_{A_k}] + \cdots . \tag{80}$$

This object contains *two* frequencies $R = 0$ and $R = R_k$.

We try to determine the imaging properties of objects like (79)–(80) containing any possible R_k. Then, because of the linearity, we not only

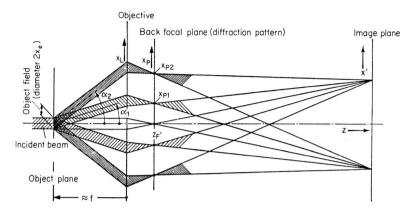

Figure 6 Image formation in the electron microscope. The object is illuminated by a beam parallel to the axis. Each beam, diffracted by the spatial frequencies of the object is focused on a certain point in the pupil plane. Each image point receives waves from all radiating points of the pupil plane.

know the imaging properties of the electron microscope for all model objects, but also for all possible weak objects.

5.5 The Physics of the Imaging Process

The physics of the imaging process is shown in Fig. 6. The object within a field of radius x_e has only the spatial frequencies $R = 0$, R_1, and R_2. Therefore, diffracted beams appear only at the angles $\alpha_1 = \pm\lambda R_1$ and $\alpha_2 = \pm\lambda R_2$. The undiffracted beam is focused in the back focal point $(z_{F'}; x = 0)$, the diffracted beams in the points $(z_{F'}; \pm x_{P_1})$ and $(z_{F'}; \pm x_{P_2})$.

All points of the $z_{F'} =$ plane (plane of the exit pupil or diffraction plane) can be understood as point sources radiating coherently but with different phases. They radiate coherently because they are illuminated by a coherent beam, and they have different phases because the paths of the diffracted beams are not equal in length.

In the image plane, each beam interferes with the others. Therefore, every point in the image plane receives waves from every point in the diffraction plane. The intensity in each area of the image plane can be obtained by squaring the sum of the wave functions. The physical reason for neglecting the higher terms in (62) in the case of a weak object is the following. Only the interferences of the weak diffracted beams with the strong undiffracted beam are taken into account, but not the mutual interferences of the diffracted beams with one another.

We shall now study the physical conditions in the diffraction plane. As we have seen in Fig. 6, the information about the diffraction angles resides in the diffraction plane in such a way, that to each point of this plane there is a specific diffraction angle. Moreover, in the case of a weak object, *one* spatial frequency of the object belongs uniquely to *two* points of the diffraction plane at equal distance from but on opposite sides of the axis. Amplitude and phase at both points carry uniquely the information about amplitude and phase modulation of this frequency and the lateral phases of these components.

Accordingly we can say that the set of spatial frequencies, contained in the object, is imaged on the set of points in the diffraction plane in such a way that for each spatial frequency there are two spectral points. Correspondingly, the diffraction pattern is, in other words, the spatial frequency spectrum of the object. Strictly speaking the electron microscope up to the diffraction plane is a spectroscopic instrument for spatial frequencies.

We now abandon plane wave illumination and move the electron source to a finite distance from the object. The source should be small enough to fulfill the coherence condition. The position of the diffraction plane, which is no longer identical with the back focal plane, is now given by the last equation of (6). Eq. (10) denotes the position x_{Pk} of the spectral point, produced by the spatial frequency R_k on the side with positive coordinates. From this equation we learn further that the reduced pupil coordinate

$$R = -\frac{x_P - x_Q}{\lambda f} \frac{M'}{M' - M'_Q} \tag{81}$$

equals the spatial frequency when the intersection x_Q of the undiffracted beam with this plane (the zero order of the spectrum) is the zero point. Expressions for the reduced coordinate Q of the undiffracted wave and the reduced coordinate $S = Q + R$, measured from the axis, have already been given in (8)–(9).

5.6 Interpretation of the Filter Equation

Having seen that R can be identified with the distance of the spectral points from the zero order, we realize that $\tilde{\psi}(R)dR$ in (69) is the wave function in the diffraction plane, not yet disturbed by aberrations in the imaging system. We call $\tilde{\psi}(R)$ the *object related wave function in the diffraction plane*.

In the diffraction plane, masking of different kinds can be introduced by inserting an aperture diaphragm, or by means of a phase plate causing a phase shift $\tilde{\Phi}_M(S) = \tilde{\Phi}_M(Q + R)$ and having a thin hole for the zero order, or by using zone plates, etc. Masking in the diffraction plane is especially advantageous, since each point of the mask has an influence on one single spatial frequency only. Because of this, the masking can be described by a complex factor $\tilde{F}_M(Q + R)$, depending only on the single object parameter R and on instrumental parameters:

$$\tilde{F}_M(Q + R) = |\tilde{F}_M(Q + R)| \exp[i\tilde{\Phi}_M(Q + R)]. \tag{82}$$

In order to obtain the image-related wave function $\tilde{\psi}'(R)$ in the diffraction plane, the object-related wave-function $\tilde{\psi}(R)$ in the same plane must be multiplied with $\tilde{F}_M(Q + R)$:

$$\tilde{\psi}'(R) = \tilde{F}_M(Q + R)\tilde{\psi}(R);$$

or, as deriving of (75) from (71), we obtain:

$$\tilde{F}'(Q + R) - \tilde{F}_M(Q + R)\tilde{F}(Q + R); \qquad \tilde{F}'(S) - \tilde{F}_M(S)\tilde{F}(S) \tag{83}$$

We call $\tilde{F}(S)$ the *transparency of the object-related*, and $\tilde{F}'(S)$ the *transparency of the image-related wave-function in the diffraction plane*. Eq. (83) is a particular case of (71)–(75). As an example, we take axial illumination; i.e. $Q = 0$ and an aperture hole with the radius R_e. Then we have $\tilde{F}_M(R) = 1$ for $R \leq R_e$ and 0 for $R > R_e$ and we see that frequencies $> R_e$ are filtered out. The same is true for a zone plate. All frequencies, the spectral points of which are covered by the opaque areas of the plate are filtered out. This is why we speak about a filter equation (see Röhler, 1967, especially pp. 22 and 32). The influence of phase plates on the imaging process can be described by the phase $\tilde{\Phi}_M$ in (82).

In addition to the effects due to masking, those caused by the lens aberrations must also be considered, as will be done in Section 5.8. The transition from $\tilde{\psi}'(R)$ to $\psi'(x')$, i.e. from the image-related diffraction pattern to the image, is the inverse to that from $\psi(x)$ to $\tilde{\psi}(R)$. This means that $\psi'(x')$ is obtained from $\tilde{\psi}'(R)$ by the inverse Fourier transform and is thus determined by (69b).

5.7 Use of the Sampling Theorem

In the Fourier expansion we now consider the finite limits $\pm x_e$ of the object field. The lowest spatial frequency contained in the object is $R_e = 1/2x_e$. All higher frequencies $R_k = k/2x_e$, with $k = 2, 3 \ldots$ are multiples of this basic frequency. For this reason, a discrete spectrum of spatial frequencies exists instead of a continuous one. That the object field is not periodically repeated outside its borders is expressed by the fact that the distribution of the wave function in each spectral point has the shape of a $\sin(2\pi Rx)/2\pi Rx$-function. Therefore, the Fourier expansion of $\tilde{\psi}(R)$ can be written in the form

$$\tilde{\psi}(R) = \sum_{k=-\infty}^{+\infty} \tilde{\psi}\left(\frac{k}{2x_e}\right) \frac{\sin[\pi(R2x_e - k)]}{\pi(R2x_e - k)}. \tag{84}$$

This is the sampling theorem, given by Shannon (1949); see the summary by O'Neill (1963) and Röhler (1967). The spectral points with the coordinate values

$$R_S =_{df} k/2x_e; \quad k = 0; \pm 1; \pm 2; \ldots \tag{85}$$

are called *sampling points*.

$$\Delta R_S = 1/2x_e; \qquad \Delta x_{PS} =_{df} \frac{\lambda f}{2x_e} \frac{(M' - M'_Q)}{M'} \tag{86}$$

is the distance between two sampling points in reduced and natural coordinates of the pupil plane. At each sampling point, the wave function has only one spatial frequency, because the contributions of all other frequencies vanish there.

As an illustration, the distance between two sampling points, when using the data of Section 2.2, is $\Delta x_{PS} \approx 0.1$ μm. As a result of the sampling theorem, the wave function within the object field is determined by a countable number of spectral points, i.e. by the wave function at a countable number of points in the diffraction plane.

When the coherence condition was described in (18), information about the object field diameter was needed in order to determine the permissible size of the illumination aperture. Using (86), Eq. (18) becomes now

$$\Delta\varphi \ll \frac{\lambda}{2}\Delta R_S. \tag{87}$$

Accordingly, the illumination aperture should be small enough not to blur the zero-points in the distribution of each sampling point.

5.8 Influence of Lens Aberrations in the Geometric-Optical Treatment

Aberration Disks in the Diffraction Spectrum. Each sampling point represents an image of the electron source, broadened by diffraction at the borders of the object field. All these images are further broadened by the lens aberrations. Because of this blurring, restrictions similar to those used in the last section for the illumination aperture should be made. First we study the broadening of the spectral points by spherical aberration of the objective. The border of the object field acts as an aperture stop for the strongly demagnified imaging of the electron source into the spectral points. On the whole, the ray paths are the inverse of those used in the highly magnified imaging of the object. Therefore, we expect to obtain aberration disks of the spectral points with very small radii δ_P. Detailed calculations, taking into account (5)–(6) and $C_s(M'_Q) \approx C_s(M')/M'^4_Q$ (see Archard, 1958; Hawkes, 1968), lead to the result:

$$\delta_P = \left[\frac{x_e}{f} \frac{M'}{M'_Q - M'} \right]^3 C_s(M'). \tag{88}$$

The absolute value of δ_P should be small compared to the absolute value of $\Delta x_{PS}/2$. Assuming this, we obtain the relation:

$$x_e \ll f \frac{M' - M'_Q}{M'} \sqrt[4]{\frac{\lambda}{4C_s(M')}}. \tag{89}$$

Example. If we insert the instrumental data of Section 2.2 in the right-hand side of the equation, we find $x_e \ll 10$ μm.

The next question is, how far may the masking plane deviate from the plane of diffraction? As an answer we find that the absolute value of the aberration disk δ_D due to defocusing,

$$\delta_D = \Delta z_P \frac{x_e}{f} \frac{M'}{M'_Q - M'}, \tag{90}$$

should be small compared to the absolute value of $\Delta x_{PS}/2$. This leads to

$$x_e \ll f \frac{M' - M'_Q}{M'} \sqrt{\frac{\lambda}{4\Delta z_P}}. \tag{91}$$

Example. If we insert the instrumental data of Section 2.2 and if we allow a mechanical deviation $\Delta z_P = 0.3$ mm of the mask from the diffraction plane, the allowable object field is limited by $2x_e \ll 2500$ Å.

Distortion of the Diffraction Pattern. All other geometric-optical third-order aberrations are omitted, because they do not contribute to an understanding of the main characteristics of optical transfer. We will discuss only the influence of distortion on the diffraction pattern as a particularly illustrative example.

The relative position of the spectral points is shifted by distortion. Therefore in the image-related wave-function of the diffraction plane, we do not find the set of exact whole numbers k given in (85). For this reason, the higher harmonics in the image are out of tune. Because each spatial frequency R_n in the image is built up by two sampling points $k = \pm(n + v)$ with $v \ll 1$, the local modulation of the spatial frequencies is correct in the paraxial region. Only at a larger distance from the axis does it become out of phase. Thus the resulting error leads to a limitation of the image field already known (isoplanatism patch, see Section 5.1). It is easy to show that a ray, diffracted at an angle α_k and having the transverse aberration (12) in the image plane due to spherical aberration, undergoes in the pupil plane a transverse aberration Δx_P which is smaller than (12) by a factor $\approx 1/(M' - M'_Q)$. From (10), the relative distortion of the diffraction pattern is

$$\frac{\Delta x_P}{x_{Pk} - x_{PQ}} = \frac{\Delta R_k}{R_k} = \frac{C_s(M')}{f}R^2\lambda^2\left(\frac{M'}{M' - M'_Q}\right)^2. \qquad (92)$$

Since $C_s[M'/(M' - M'_Q)]^2/f \approx 1$, diffraction by atomic structures ($1/R \approx 50\lambda$) is affected by a relative distortion of less than 10^{-3}. Therefore in the image of an object-field diameter of 1000 Å, as was supposed in Section 2.2, distortion is of minor importance.

Object and Exit Pupil in the Objective Field. Until now the object and the exit pupil have been assumed to be in field-free space. In reality, they are well within the magnetic field of the objective. Thus the rays are deflected by this field before reaching the object plane and will be deflected again after leaving the pupil plane. Both these effects can be expressed by a scaling factor. Since we can disregard scaling problems by using reduced coordinates, these effects are of little or no importance for the systematic theory. However, in designing zone plates for the pupil plane, these effects are of some practical importance. This problem has been solved in principle by Lenz (1964). According to his calculations, zone plates in the back focal plane of the Glaser field with a lens power $k^2 = 0.6$ have radii that are 4% smaller than those calculated by using the equations above.

5.9 The Wave Aberration

The decisive influence of the lens aberrations on the imaging process can only be determined by wave optics. In order to understand it, we need the concept "wave aberration".

In the image space, which is considered field free, we construct reference spheres around each image point. These spheres intersect the axis in the pupil plane. For perfect imaging, the reference spheres would simultaneously be wave-fronts converging into the above image points. The aberrations of the lens system cause distortions of these wave-fronts. The deviation from spheres as a function of the distance x_P from the axis in the pupil plane is the wave aberration $W(x_P)$. It is a function of the upper limit of the integral[8]

$$W(x_P) = -\int_0^{x_P} \frac{\Delta z_B x}{(z_B - z_P)^2}\,dx, \quad x = \text{current pupil coordinate.} \quad (93)$$

Here, Δz_B is the total longitudinal aberration of a ray intersecting the pupil plane at x_P. Eq. (93) relates the ray and the wave concepts. Δz_B should be considered as a function of x_P, which is related to the spatial frequencies as shown in (10). Therefore, we know in principle that the lens aberrations influence the transfer of the spatial frequencies by producing phase shifts. For a more detailed study, $W(x_P)$ must be divided into the individual components of aberration. Since the object field is very small, only spherical aberration and defocusing are important. Therefore, Δz_B can be written as

$$\Delta z_B = \Delta z_s + \Delta z_d. \quad (94)$$

Here, Δz_s and Δz_d are the components of longitudinal aberration caused by spherical aberration and defocusing as defined in (14)–(17). Inserting these quantities in (94), expressing $(z_B - z_P)$ by f and M' according to (4)–(5), and considering that the spherical aberration constant C_s and defocusing are independent of x_P, we integrate (93) and obtain

$$W(x_P) =_{df} W_s(x_P) + W_d(x_P) = \frac{C_s(M')x_P^4}{4f^4}\left(\frac{M'}{M' - M'_Q}\right)^4$$

$$+ \frac{\Delta z x_P^2}{2f^2}\left(\frac{M'}{M' - M'_Q}\right)^2. \quad (95)$$

[8] For the derivation of the differential form of this equation see e.g. O'Neill (1963), especially p. 50, Eq. (4-7; 8).

The same result, expressed in reduced coordinates according to (8) is

$$W(S) = \frac{C_s(M')}{4}\lambda^4 S^4 + \frac{\Delta z}{2}\lambda^2 S^2 \tag{96}$$

Since the objective has rotational symmetry, this function is even.

Generalized Coordinates. In (96) the wave aberration depends on the instrumental parameters f and $C_s(M')$, on the magnifications M' and M'_Q, and on the defocusing Δz. Each electron microscope, therefore, has its own family of wave aberration curves which characterize its imaging properties. It is possible to condense such families into a single curve by introducing the following *generalized coordinates*[9]:

$$\text{generalized wave aberration:} \quad \mathscr{W} =_{df} \frac{W}{\lambda}; \tag{97}$$

generalized pupil coordinate:

$$\mathscr{X}_P =_{df} -\frac{x_P M'}{f(M' - M'_Q)} \cdot \sqrt[4]{\frac{C_s(M')}{\lambda}} = +S\sqrt[4]{C_s(M')\lambda^3}$$

$$= (R + Q)\sqrt[4]{C_s(M')\lambda^3} =_{df} \mathscr{R} + \mathscr{Q}; \tag{98}$$

generalized defocusing:

$$\Delta =_{df} -\frac{\Delta z}{\sqrt{C_s(M')\lambda}}. \tag{99}$$

Then, the generalized equation for the wave aberration has the form

$$\mathscr{W}(\mathscr{X}_P) = \frac{\mathscr{X}_P^4}{4} - \frac{\Delta}{2}\mathscr{X}_P^2. \tag{100}$$

This family of curves (Δ as parameter) is valid for image formation using electrons of arbitrary wavelength, objective lenses of arbitrary focal length and spherical aberration, operating at a high, but not necessarily at an infinite magnification,[10] and with arbitrary defocusing of moderate

---

[9] Corresponding equations given by Hanszen (1966a) contain an error. Due to an oversight $M'/(M' - 1)$ was given instead of $M'/(M' - M'_Q)$. The latter expression approaches 1 as $M'_Q \to 0$, as it should.

[10] It is assumed that the function $C_s(M')$ is known. For calculations see Albert (1966), for experiments Kunath and Riecke (1965).

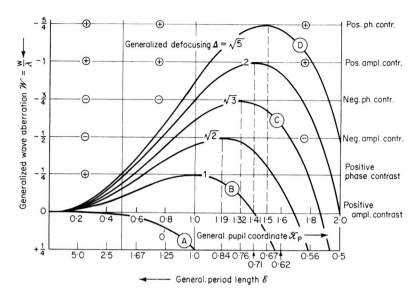

Figure 7 Generalized wave aberrations. The contrast transfer functions of Figs. 11 and 12, belonging to these wave aberrations, are marked with the same letter (Hanszen, 1966a).

value. Moreover, it is valid for light-optical models of the electron microscope; therefore, it is also helpful in describing holographic reconstruction of electron micrographs (see Section 6.10). The family of generalized wave-aberration curves is drawn in Fig. 7. Since $\mathscr{W}(\mathscr{X}_P)$ are even functions, only the positive values are shown. It should be pointed out that the maxima of these curves are of special importance, especially if they have the values $\mathscr{W}(\mathscr{X}_P) = -\frac{n}{4}; n = 1, 2, \ldots$.

In order to complete the system of generalized coordinates, we introduce the following quantities[11] in the object and the image plane:

generalized object and image coordinate:

$$\mathscr{X} = \frac{x}{\sqrt[4]{C_s\lambda^3}}; \qquad \mathscr{X}' = \frac{x'}{\sqrt[4]{C_s\lambda^3}}; \tag{101}$$

generalized period length:

$$\mathscr{E}_k = \frac{1}{\mathscr{R}_k} = \frac{\epsilon_k}{\sqrt[4]{C_s\lambda^3}} = \frac{1}{R_k\sqrt[4]{C_s\lambda^3}} = \frac{1}{(S_k - Q)\sqrt[4]{C_s\lambda^3}}. \tag{102}$$

---

[11] These notations are not identical with those used by Hanszen (1966a).

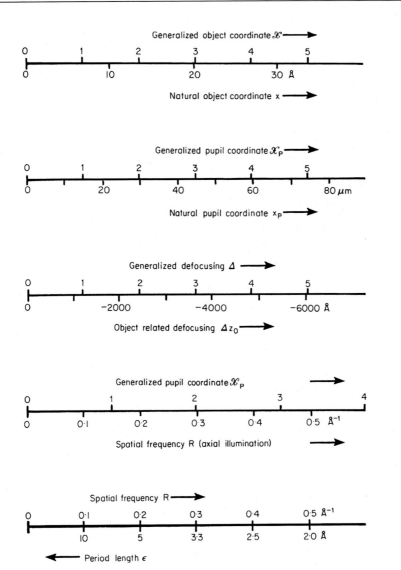

Figure 8 Comparison of the generalized coordinates with the natural coordinates valid for the normal objective.

For the purpose of demonstration the generalized coordinates are compared in Fig. 8 with the natural ones that are based on the numerical data given in Section 2.2. Rewriting Fig. 7 in natural coordinates as in Fig. 8, we see that under practical conditions the wave aberrations may considerably exceed $\lambda/4$.

The Significance of the Wave Aberration for the Imaging Process. The meaning of (96) is the following. The diffracted wave which intersects the diffraction plane at the point S acquires, along its whole path length from the object to the image, a path-length difference $W(S)$ due to lens aberrations. Therefore, the wave arrives at the image plane with a phase difference $\tilde{\Phi}_L = 2\pi W(S)/\lambda$. The effect is the same as if the phase shift occurred in the pupil plane. Thus, the influence of the wave aberration on the imaging process can be described by the factor

$$\tilde{F}_L(S) = \exp[i\tilde{\Phi}_L(S)] = \exp[i2\pi W(S)/\lambda], \qquad (103)$$

by which the object-related wave function $\tilde{\psi}(R)$ in the diffraction plane is to be multiplied. We combine this factor with the factor $\tilde{F}_M(Q+R)$ describing the masking. Thus, we define the *pupil function* $\tilde{F}_P(Q+R)$ as the product of both and obtain, with reference to (10),

$$\begin{aligned}\tilde{F}_P(Q+R) &= \left|\tilde{F}_M(Q+R)\right|\exp\{i[\tilde{\Phi}_L(Q+R) + \tilde{\Phi}_M(Q+R)]\} \\ &= \left|\tilde{F}_M(Q+R)\right|\exp\{i[2\pi W(Q+R)/\lambda + \tilde{\Phi}_M(Q+R)]\}. \end{aligned} \quad (104)$$

Instead of (83) we can write

$$\tilde{\psi}'(R) = \tilde{F}_P(Q+R)\tilde{\psi}(R); \qquad \tilde{F}'(Q+R) = \tilde{F}_P(Q+R)\tilde{F}(Q+R). \quad (105)$$

Comparing this with (71) we see that the pupil function $\tilde{F}_P(Q+R)$ is identical with the transfer function for the wave function $\tilde{k}(R)$:

$$\tilde{k}(R) = \tilde{F}_P(Q+R). \qquad (106)$$

The equation has an imperfection, however. The diffraction pattern is symmetrical to the undiffracted beam, which does not coincide with the axis when oblique illumination is used. $\tilde{F}_L$ is symmetrical to the axis whereas $\tilde{F}_M$ can be of any shape. Therefore, we cannot expect that $\tilde{k}(R)$ has any symmetry with respect to the point $R = 0$. Because of this there are complications, as described in Section 5.4, in a mathematical formulation of the imaging process. In the interest of brevity, only the following special cases will be treated:

a. Axial illumination, i.e. $S = R$. Semitransparent masks are not used. The borders of the aperture are located symmetrically about the axis, i.e. $\tilde{F}_M(+R) = \tilde{F}_M(-R)$; $|\tilde{F}_M| = 0$ or 1. For the phase plates, we have $\tilde{\Phi}_M(+R) = \tilde{\Phi}_M(-R)$.

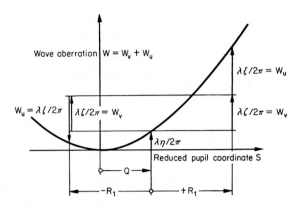

Figure 9 Separation of the wave aberration W into an even component W_v and an odd component W_u (Hanszen & Morgenstern, 1965); $\lambda\eta/2\pi \equiv W(Q)$ in Eq. (108).

b. Oblique illumination, i.e. $S = Q + R$. Either no masking is used, i.e. $|\tilde{F}_M| \equiv 1$, or one side of the diffraction pattern is eliminated, i.e.

$$|\tilde{F}_M| \equiv 0 \quad \text{for } -\infty < R \leqq 0$$

and

$$|\tilde{F}_M| = 1 \quad \text{for } 0 < R < +\infty;$$

phase shifts are not used, i.e. $\tilde{\Phi}_M = 0$.

Even and Odd Components of the Wave Aberration. For convenience, we separate the wave aberration W into an odd component W_u and an even component W_v with respect to $W(R=0)$ (see Fig. 9):

$$W_u(Q+R) = -W_u(Q-R) =_{df} \left[W(Q+R) - W(Q-R)\right]/2$$
$$=_{df} \xi(R)\lambda/2\pi;$$

with

$$\xi/2\pi = \mathscr{R}^3\mathscr{Q} - \Delta\mathscr{R}\mathscr{Q} + \mathscr{R}\mathscr{Q}^3, \tag{107}$$

$$W_v(Q+R) = W_v(Q-R) =_{df} \left[W(Q+R) + W(Q-R)\right]/2 - W(Q)$$
$$=_{df} \zeta(R)\lambda/2\pi;$$

with

$$\zeta/2\pi = \mathscr{R}^4/4 - \Delta\mathscr{R}^2/2 + 3\mathscr{R}^2\mathscr{Q}^2/2. \tag{108}$$

We call ξ the odd, and ζ the even phase shift of the wave aberration.

5.10 Phase Shifts and Masking Effects in the Diffraction Plane

The results of the last two sections are:

(a) Both phase and amplitude modulation of the object can be derived from the object–related wave function in the diffraction plane according to (61) and (63).

(b) The image-related wave function $\tilde{\psi}'(R)$ in the diffraction plane can be obtained from the object related function $\tilde{\psi}(R)$ by multiplying the latter with the pupil function $\tilde{F}_P(R+Q)$. The whole image information is already contained in $\tilde{\psi}'(R)$.

(c) The essential pupil effects are shifts in phase and in the size of the aperture.

How these effects influence the optical transfer of phase and amplitude components of a compound object (79) into the image intensity is shown in Table 2, lines 1 through 4. The results will be discussed in Section 5.12.

5.11 The Amplitude-Contrast and Phase-Contrast Transfer-Functions of the Electron Microscope

Image Modulation. As was done in Eqs. (45) through (47) for the object, we can define the amplitude and phase modulation of the wave function as well as the intensity modulation *for the image.* As we have seen, only the intensity modulation is important in contrast transfer theory. We define this quantity by:

$$K_I'(R) =_{df} \frac{\tilde{A}_I'(R)}{\tilde{A}_I'(R=0)} \exp\left[i\xi_I'(R)\right] \quad \text{with} \quad \frac{\tilde{A}_I'(R=0)}{2 \times 2x_e} = 1. \tag{109}$$

Examples for pure amplitude and phase objects are given in Table 2, line 6a; b.

The Transfer of the Different Object Modulations into the Image. Within the scope of transfer theory, it is necessary and sufficient to know the following:

(a) How the modulation (47), (64) of the intensity of a pure amplitude object is transferred into the image intensity;

(b) How the phase modulation (46) of a pure phase object is transferred into the image intensity.

The information can be obtained from the following quantities:

Table 2 Examples of masking effects and contrast transfer-functions

$F(x) = 1 + \hat{A}_k \cos(2\pi R_k x + \xi_{Ak}) + i\hat{\Phi}_k \cos[2\pi R_k x + \xi_{\Phi k}]$ with $\hat{A}_k; \hat{\Phi}_k \ll 1$: i.e. $K_A(R_k) = \hat{A}_k \exp(i\xi_{Ak}); K_\Phi(R_k) = \hat{\Phi}_k \exp(i\xi_{\Phi k})$

	Object	(A) even phase shift	(B) odd phase shift	(C) even and odd phase shift	(D) elimination of one side phase shift on the other side*	(E) dark field
(2.1)	Object					
(2.2)	Object related diffraction spectrum	$\tilde{F}(R=0) = 1; \tilde{F}(\pm R_k) = \frac{\hat{A}_k}{2}\exp(\pm i\xi_{Ak}) + i\frac{\hat{\Phi}_k}{2}\exp(\pm i\xi_{\Phi k})$				
(2.3)	Masking	$\tilde{F}_P(R=0) = 1;$ $\tilde{F}_P(\pm R_k) = \exp(i\zeta)$	$\tilde{F}_P(R=0) = 1;$ $\tilde{F}_P(\pm R_k) = \exp(\pm i\xi)$	$\tilde{F}_P(R=0) = 1;$ $\tilde{F}_P(\pm R_k) = \exp[i(\zeta \pm \xi)]$	$\tilde{F}_P(R=0) = 1;$ $\tilde{F}_P(-R_k) = 0;$ $F_P(+R_k) = \exp(i\chi)$	$\tilde{F}_P(R=0) = 0;$ $\tilde{F}_P(\pm R_k) = 1$
(2.4)	Image related diffraction spectrum	$\tilde{F}'(R=0) = 1;$ $\tilde{F}'(\pm R_k) = \exp(i\zeta)(\frac{\hat{A}_k}{2}\exp(\pm i\xi_{Ak}) + i\frac{\hat{\Phi}_k}{2}\exp(\pm i\xi_{\Phi k}))$	$\tilde{F}'(R=0) = 1;$ $\tilde{F}'(\pm R_k) = \frac{\hat{A}_k}{2}\exp[\pm i(\xi_{Ak} + \xi)] + i\frac{\hat{\Phi}_k}{2}\exp[\pm i(\xi_{\Phi k} + \xi)]$	$\tilde{F}'(R=0) = 1;$ $\tilde{F}'(\pm R_k) = \exp(i\zeta)[\frac{\hat{A}_k}{2}\exp[\pm i(\xi_{Ak} + \xi)] + i\frac{\hat{\Phi}_k}{2}\exp[\pm i(\xi_{\Phi k} + \xi)]]$	$\tilde{F}'(R=0) = 1;$ $\tilde{F}'(-R_k) = 0; \tilde{F}'(+R_k) = \exp(i\chi)(\frac{\hat{A}_k}{2}\exp(i\xi_{Ak}) + i\frac{\hat{\Phi}_k}{2}\exp(i\xi_{\Phi k}))$	$\tilde{F}'(R=0) = 0;$ $\tilde{F}'(\pm R_k) = \frac{\hat{A}_k}{2}\exp(\pm i\xi_{Ak}) + i\frac{\hat{\Phi}_k}{2}\exp(\pm i\xi_{\Phi k})$
(2.5)	Transparency of the image intensity	$F_I'(x') = 1 + 2\hat{A}_k \cos\zeta \cos(2\pi R_k x' + \xi_{Ak}) - 2\hat{\Phi}_k \sin\zeta \times \cos(2\pi R_k x' + \xi_{\Phi k})$	$F_I'(x') = 1 + 2\hat{A}_k \cos(2\pi R_k x' + \xi_{Ak} + \xi)$	$F_I'(x') = 1 + 2\hat{A}_k \cos\zeta \cos(2\pi R_k x' + \xi_{Ak} + \xi) - 2\hat{\Phi}_k \sin\zeta \times \cos(2\pi R_k x' + \xi_{\Phi k} + \xi)$	$F_I'(x') = 1 + \hat{A}_k \cos(2\pi R_k x' + \xi_{Ak} + \chi) - \hat{\Phi}_k \sin(2\pi R_k x' + \xi_{\Phi k} + \chi)$	$F_I'(x') = 1 + \frac{\hat{A}_k^2}{2}[1 + \cos(2\pi R_k x' + 2\xi_{Ak})] + \frac{\hat{\Phi}_k^2}{2}[1 + \cos(2\pi R_k x' + 2\xi_{\Phi k})]$
(2.6a)	Intensity modulation in the image of a pure amplitude object i.e. $\hat{\Phi}_k = 0$	$K_I'(R_k) = 2\hat{A}_k \cos\zeta \exp(i\xi_{Ak})$	$K_I'(R_k) = 2\hat{A}_k \exp[i(\xi_{Ak} + \xi)]$	$K_I'(R_k) = 2\hat{A}_k \cos\zeta \exp[i(\xi_{Ak} + \xi)]$	$K_I'(R_k) = \hat{A}_k \exp[i(\xi_{Ak} + \chi)]$	$K_I'(2R_k) = \exp(i2\xi_{Ak})$
(2.6b)	Intensity modulation in the image of a pure phase object i.e. $\hat{A}_k = 0$	$K_I'(R_k) = -2\hat{\Phi}_k \sin\zeta \exp(i\xi_{\Phi k})$	$K_I'(R) = 0$	$K_I'(R_k) = -2\hat{\Phi}_k \sin\zeta \exp[i(\xi_{\Phi k} + \xi)]$	$K_I'(R_k) = -\hat{\Phi}_k \exp[i(\xi_{\Phi k} + \chi - \pi/2)]$	$K_I'(2R_k) = \exp(i2\xi_{\Phi k})$
(2.7a)	Amplitude-contrast transfer-function	$D_I(R) = \cos\zeta(R)$	$D_I(R) = \exp[i\xi(R)]$	$D_I(R) = \exp[i\xi(R)]\cos\zeta(R)$	$D_I(R) = (1/2)\exp[i\chi(R)]$	
(2.7b)	Phase-contrast transfer-function	$B(R) = -2\sin\zeta(R)$	$B(R) = 0$	$B(R) = -2\exp[i\xi(R)]\sin\zeta(R)$	$B(R) = -\exp[i\{\chi(R) - \pi/2\}] = \exp[-i\{\chi(R) + \pi/2\}]$	

* The masking $\tilde{F}_P(R=0) = 1; \tilde{F}_P(-R_k) = e^{i\chi}; \tilde{F}_P(+R_k) = 0$ would lead to: $D_I(R) = (\frac{1}{2})\exp[-i\chi(R)]; B(R) = -\exp[-i\chi(R) - \pi/2]] = \exp[-i[\chi(R) + \pi/2]]$.

Amplitude-contrast transfer-function (defined for pure, weak amplitude objects only):

$$D_I(R) =_{df} \frac{K_I'(R)}{K(R)} = \frac{\tilde{A}_I'(R)}{\tilde{A}_I(R)} \exp\left[i\left(\xi_I' - \xi_I\right)\right];$$ (110)

Phase-contrast transfer-function (defined for pure, weak phase objects only):

$$B(R) =_{df} \frac{K_I'(R)}{K_\Phi(R)} = \frac{\tilde{A}_I'(R)}{\tilde{\Phi}(R)} \exp\left[i\left(\xi_I' - \xi_\Phi\right)\right].$$ (111)

The value of these functions for a fixed spatial frequency $R = R_k$ is called the *amplitude-contrast or phase-contrast transfer-factor* for this frequency. Knowing these factors for all frequencies, we know the contrast transfer functions for all weak amplitude and phase objects.

The contrast transfer functions belonging to the pupil functions reviewed in Table 2 are presented on line 7a; b of this table (for particulars see Hanszen & Morgenstern, 1965).

5.12 Discussion of Table 2

Even Phase Shift ζ (Column A). Both the amplitude and the phase structure are transferred into the image. In this case, the transfer is described by the contrast transfer functions given in Table 2.7. The functions also show that there is no phase contrast for $\zeta = n\pi$ and no amplitude contrast for $\zeta = (2n+1)\pi/2$ where $n = 0; \pm 1; \pm 2; \ldots$

It must be pointed out that an even phase shift is capable of producing both a phase and an amplitude contrast. Therefore, the concept "contrast by means of a phase shift" is not identical with the concept "phase contrast".

Odd Phase Shift ξ (Column B). An odd phase shift cannot produce phase contrast. The strength of the amplitude contrast produced is independent of R. Unfortunately, lateral phase shifts ξ occur, the strength of which depends on R.

Even and Odd Phase Shift (Column C). Both the amplitude and the phase contrast are present. Both the amount and the lateral phase of D_I and B depend on R.

Elimination of the Diffraction Pattern on One Side; Phase Shift χ on the Other Side of the Pattern (Column D). By means of a mask for accomplishing the above, amplitude and phase contrast which are independent of R are produced simultaneously. There are lateral phases in both cases, however.

Elimination of the Zero Order, i.e. Darkfield (Column E). We examine the image of the object (Table 2.1) and find the frequency $2R_k$ instead of the object frequency R_k. This is true, because we can no longer overlook the interference intensities between the diffracted beams; for details see Hanszen (1969a). If the object contains several spatial frequencies, the interferences of each diffracted wave with all the others must be taken into account. Thus the spatial frequencies are not transferred independently of each other; hence no transfer functions can be given.

Relations Between D_I, B, and $\tilde{k}$. There are, of course, relations between $D_I(R)$; $B(R)$ and $\tilde{k}(R)$. In the special case of an even phase shift only, i.e. $\tilde{k}(R) = \exp\{i\zeta(R)\}$ we can rewrite (Table 2.7)

$$D_I(R) = Re\,\tilde{k}(R); \qquad B(R) = -2Im\,\tilde{k}(R). \qquad (112a)$$

According to Hauser (1962), we obtain in the general case:

$$D_I(R) = \frac{1}{2|\tilde{k}(Q)|^2}\left[\tilde{k}^*(Q)\tilde{k}(Q+R) + \tilde{k}(Q)\tilde{k}^*(Q-R)\right]; \qquad (112b)$$

$$B(R) = \frac{i}{2|\tilde{k}(Q)|^2}\left[\tilde{k}^*(Q)\tilde{k}(Q+R) - \tilde{k}(Q)\tilde{k}^*(Q-R)\right]. \qquad (112c)$$

5.13 Supplementary Comments on the Imaging of Two-Dimensional Objects

As already pointed out, spatial frequencies masked by an aperture stop, especially by the opaque rings of zone plates, are irretrievably lost in the image, since both the diffraction spectrum and the image intensity of the objects correspond uniquely.

It was questioned (Langer & Hoppe, 1966) whether this statement, valid for one-dimensional objects, would still be valid for two-dimensional ones. To find the answer we study the radiation leaving the object point (x_k, y_k) and entering the image in the point (x'_k, y'_k). According to Fig. 6, this radiation extends over the entire pupil plane. The reduced coordinates in this plane are denoted by R_x and R_y. The intensity in every image point is generated by radiation coming from all points of the pupil plane. Therefore, masking at any point of this plane leads, as a rule, to intensity changes in each image point.

We now study the image points along the line $(y' = 0; x')$ of the image plane. This subset of image points is not only formed by the wave functions of the subset $(R_y = 0; R_x)$ of the pupil points, but also by the wave functions of the complete set of pupil points. It is possible to produce a given intensity distribution (e.g.

$$F'_I(x', y' = 0) = 1 + 2\tilde{A}_k \cos 2\pi R_{xk} x' \quad \text{with } \tilde{A}_k \ll 1)$$

in different ways (in the example by

$$\tilde{F}'(R = 0) = 1; \qquad \tilde{F}'(\pm R_k) = \pm\sqrt{R_{xk}^2 + R_y^2} = \tilde{A}_k/2$$

with an arbitrary value for R_y). Each of these spectra, however, produces a different intensity distribution in the residue set $(x', y' \neq 0)$ outside the line considered. If the points $R = \pm R_{xk}$ of the pupil plane are covered in this case, the period length $1/R_{xk}$ appears even then in the image along the line $(x'; y' = 0)$. However, one should not overlook the loss of information caused by masking, since there may be noticeable disturbances in other parts of the image plane.

In Fig. 10, an illustrative example is given. At the top we see a particular intensity distribution in the pupil and at the bottom the corresponding image intensity. If the line $(R_y = 0; R_x)$ in the diffraction figure is covered (above right), the intensity distribution in the line $(y' = 0; x')$ of the image at the bottom right is altered only imperceptibly (statement by Langer & Hoppe, 1966). Note, however, the spikes starting from the image points into the image field $(y' \neq 0; x')$. The spikes have to be regarded as "spurious structures"; see Section 9.3. Our statement that the relation between diffraction spectrum and Gaussian image is unique is not valid for two arbitrarily selected sub-sets of spectral and image points but only for the complete sets. The information stored in each spectral point does not depend on the information in other points. Therefore, information that is lost by masking cannot be replaced by information stored in the remaining spectral points. It is another matter, whether the information eliminated by the mask is important or not. For example, the information that the electrons are blocked outside the object is unimportant. From this example we learn that information limitations by masking are less serious, if prior information is available.

5.14 Summary and Conclusions

We have succeeded in establishing the imaging properties of the electron microscope from the diffraction spectrum by presenting the contrast transfer-functions (Table 2.7a, b). In the following discussion, therefore, we

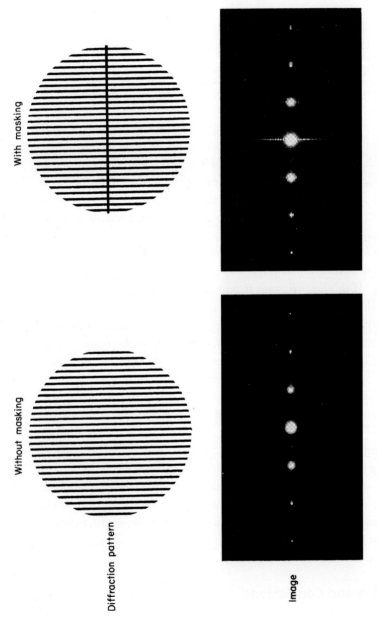

Figure 10 Effect of masking in the pupil plane on the two dimensional image (demonstration-experiment with a light optical model for the electron microscope; see Hanszen, 1968). Due to masking, each image point degenerates into a diffraction spectrum of the opaque strip which serves as a mask.

do not need to study Gaussian image formation by means of interference of the wave functions $\tilde{\psi}'(R)$. In particular, we do not want to deal with the difficult problem of point image formation and "point resolution". On the contrary, we determine the image quality from the properties of the spatial frequency spectrum in the pupil plane. When the contrast transfer functions (Table 2.7a, b) of an electron microscope are known (e.g. by knowing the lens aberrations and pupil conditions) it is easy to predict the properties of a micrograph of a known object, and it is possible to determine the structures of an unknown object from its micrograph apart from the structures which fall within a transfer gap. When the transfer-function is unknown, a micrograph of an object containing all spatial frequencies of interest gives much information about the shape of this function; see Section 6.5.

6. CONTRAST BY MEANS OF PHASE SHIFT: THE CONTRAST TRANSFER-FUNCTIONS FOR AXIAL ILLUMINATION

6.1 The Pupil Function

We can now combine the results of Sections 2.1, 3.3, and 5.8, 5.9, 5.11. In the case of axial illumination, we have $Q = 0$, $R = S$ and therefore $\mathscr{Q} = 0$, $\mathscr{R} = \mathscr{X}_P$ for the generalized pupil coordinates. In this special case, $\mathscr{X}_P$ is identical with the generalized spatial frequency $\mathscr{R}$. The pupil function (104) can now be written in the form

$$\tilde{F}_P(R) = \left|\tilde{F}_M(R)\right| \exp\left\{i\left[2\pi W(R)/\lambda + \tilde{\Phi}_M(R)\right]\right\}. \tag{113}$$

Here $W(R)$ is the wave aberration given in (96) and $|\tilde{F}_M(R)|\exp(i\tilde{\Phi}_M)$ the masking. Using generalized coordinates and considering (96ff), (113) can be rewritten in the form

$$\tilde{F}_P(\mathscr{X}_P) = \left|\tilde{F}_M(\mathscr{X}_P)\right| \exp\left\{i\left[2\pi \mathscr{W}(\mathscr{X}_P) + \tilde{\Phi}_M(\mathscr{X}_P)\right]\right\}. \tag{114}$$

According to (102), we obtain in this particular case

$$\mathscr{X}_P = \mathscr{R} = \frac{1}{\epsilon}\sqrt[4]{C_s\lambda^3}. \tag{115}$$

6.2 The Contrast Transfer-Functions in the Absence of a Mask

In this case, we have $|\tilde{F}_M(\mathscr{X}_P)|\exp[i\tilde{\Phi}_M(\mathscr{X}_p)] \equiv 1$. Therefore, considering (100) and Eqs. (2.7a, b) of Table 2, column A, we obtain:

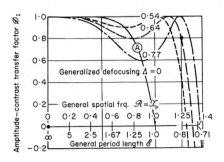

Figure 11 Generalized amplitude-contrast transfer-functions for generalized defocus-ings $0 \leq \Delta < 1$. Transfer-function A of this figure belongs to the wave aberration A in Fig. 7 (Hanszen, 1966a).

Amplitude-contrast transfer-function

$$\mathscr{D}_I(\mathscr{R} = \mathscr{X}_P) = \cos\left[2\pi \mathscr{W}(\mathscr{X}_P)\right] = \cos\left[2\pi\left(\frac{\mathscr{X}_P^4}{4} - \Delta\frac{\mathscr{X}_P^2}{2}\right)\right]; \qquad (116a)$$

Phase-contrast transfer-function

$$\mathscr{B}(\mathscr{R} = \mathscr{X}_P) = -2\sin\left[2\pi \mathscr{W}(\mathscr{X}_P)\right] = -2\sin\left[2\pi\left(\frac{\mathscr{X}_P^4}{4} - \Delta\frac{\mathscr{X}_P^2}{2}\right)\right]. \quad (116b)$$

Only one example for (116a) is given in Fig. 11. Otherwise, we restrict our considerations to (116b), i.e. we deal with transfer conditions for pure and weak phase objects only. For further particulars see Hanszen and Morgenstern (1965) and Hanszen (1966a).

Three families of phase contrast transfer-functions of practical importance are given in Fig. 12. The generalized defocusing Δ is the parameter. The first thing that we recognize is the contrast formation by means of objective aberrations only, i.e. phase contrast formation without any masking in the pupil. Unfortunately, this contrast varies with the spatial frequency. The contrast transfer functions oscillate between ± 2. Therefore, one part of the spatial frequency spectrum is imaged, positive while the other part shows a negative contrast. We call a contrast positive if object details of advancing phase, i.e. of more positive inner potential, appear bright in the image.

The contrast transfer-functions have zero points. Therefore, there are spatial frequencies which are not transferred to the image; i.e. the image has *frequency-gaps*.

Dependent on the sign of the transfer-function between two successive zero points, the corresponding frequency interval is called the *transfer interval* with positive or negative phase contrast. The position and magnitude of these intervals depend on Δ. The maximum values $|\mathscr{B}_{max}| = 2$ of $\mathscr{B}$ are reached at generalized frequencies $\mathscr{R}_{ex}$ for which

$$\mathscr{W} = n/4; \quad n = \pm1; \pm3; \pm5; \ldots \tag{117}$$

Therefore, the transfer properties can also be easily determined from the wave aberration curves in Fig. 7. If the microscope is in focus, i.e. $\Delta = 0$, the transfer intervals are very narrow because the curve of the related wave aberration is steep. If the defocusing counteracts spherical aberration, i.e. $\Delta > 0$, the intervals are broadened. The above is true for a slightly under-excited lens.

Highly favorable transfer functions are obtained when the wave aberration curves have an extremum, the height of which is given by one of the values quoted in (117). The transfer interval, including such an extremum, is especially broad. We call it the *main transfer interval*. The extrema in the main transfer intervals are characterized by the following values of $\mathscr{W}$, $\mathscr{X}_P$, and Δ

$$\mathscr{W}_{ex} = -\Delta_n^2/4; \qquad \mathscr{X}_{Pex} = \sqrt{\Delta_n}; \qquad \Delta_n = \sqrt{-n}; \quad n = -1; -3; -5; \ldots \tag{118}$$

The main transfer interval appears at higher spatial frequencies as defocusing is increased. It is impossible, however, to move this interval continuously along the frequency-coordinate by varying Δ.

The three graphs in Fig. 12 show the phase-contrast transfer-functions producing main transfer intervals with $\Delta = 1; \sqrt{3}; \sqrt{5}$. The useful width of the main transfer interval is marked by arrows; for details see Hanszen (1966a). Micrographs showing point distances below 4 Å are usually obtained by applying a defocusing $\Delta \approx \sqrt{5}$. In this case, these structures are transferred by the main transfer interval shown in Fig. 12C.

The essential information range of a micrograph is determined by the limits of the main transfer interval. According to (98), *both* limits depend on spherical aberration. If this aberration decreases, the main transfer interval moves toward higher spatial frequencies. However, in the scale of period lengths, it gets seriously narrower (see the numerical values given in Table 3). The conventional assumption that the "resolution" of an electron microscope is improved by lowering the spherical aberration is, therefore, very dubious.

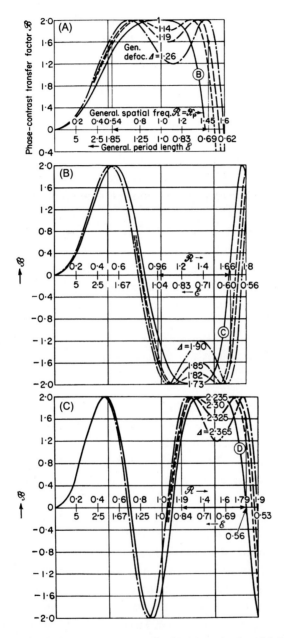

Figure 12 Generalized phase-contrast transfer-functions in the vicinity of $\Delta = +1$, $+\sqrt{3}$, and $+\sqrt{5}$. The effective limits of the main transfer intervals are marked by arrows on the abscissa. The wave aberrations belonging to the transfer-functions B, C, and D are marked by the corresponding letter in Fig. 7 (Hanszen, 1966a).

Table 3 Electron microscopical data for high resolution microscopy at 100 kV

		A	B
Positive phase contrast according to Fig. 12A	Period lengths, transferred with a contrast higher than 80%	12.4–4.6 Å	7.4–2.9 Å
	Radii in the pupil plane, limiting the main transfer interval	≈8.1–21.6 μm	≈6.0–16.2 μm
	Defocusing $\Delta I/I$	$-3.5 \cdot 10^{-5}$	$-3.6 \cdot 10^{-5}$
	Smallest defocusing steps required	$1.5 \cdot 10^{-6}$	$1.5 \cdot 10^{-6}$
Negative phase contrast according to Fig. 12B	Period lengths, transferred with a contrast higher than 80%	7.0–4.0 Å	4.2–2.4 Å
	Radii in the pupil plane, limiting the main transfer interval	≈14.3–24.7 μm	≈10.7–18.5 μm
	Defocusing $\Delta I/I$	$-5.4 \cdot 10^{-5}$	$-5.6 \cdot 10^{-5}$
	Smallest defocusing steps required	$0.9 \cdot 10^{-6}$	$0.9 \cdot 10^{-6}$
Positive phase contrast according to Fig. 12C	Period lengths, transferred with a contrast higher than 80%	5.6–3.8 Å	3.4–2.2 Å
	Radii in the pupil plane, limiting the main transfer interval	≈17.7–26.7 μm	≈13.3–19.9 μm
	Defocusing $\Delta I/I$	$-6.8 \cdot 10^{-5}$	$-7.1 \cdot 10^{-5}$
	Smallest defocusing steps required	$0.75 \cdot 10^{-6}$	$0.75 \cdot 10^{-6}$

Column A: Using a "normal" objective, the data of which are given in Section 2.2.
Column B: Using a high-efficiency objective, $f' = 1.2$ mm; $C_s = 0.5$ mm; $C_{ch} = -0.7$ mm.

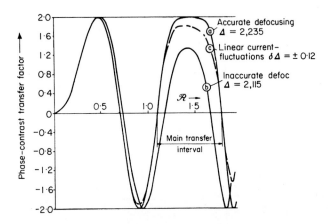

Figure 13 Influence of an inaccurately adjusted defocusing, of electric current fluctuations and of a finite illumination aperture on the phase contrast transfer function with $\Delta = \sqrt{5}$. (a) Accurate adjustment; (b) "underfocus" with $\Delta I/I = 3.5 \cdot 10^{-6}$ in the case of the normal objective; (c) like (a) but with linear current fluctuations of maximal value $\Delta I/I = \pm 3.5 \cdot 10^{-6}$ (Hanszen, 1966b). $\delta\Delta$ in the figure equals δ_I in Eq. (121).

The defocusing steps available on an electron microscope should be small enough to give a wave aberration necessary for obtaining a main transfer interval. Fig. 13, curve b, demonstrates the effect of a small deviation from the nominal defocusing value. In Table 3, the values for the desired minimum steps are given. Compare these values to the relative energy width of 1×10^{-5} or greater which is usually present in the electron beams because of the thermal velocity distribution and Boersch effect; for details see Section 6.4.

The results of the present section can be summarized as follows: The portion of the electron microscope from the electron source to the exit pupil can be considered as a spectroscopic instrument for spatial frequencies. The lens aberrations act as a filter for those frequencies. Some frequencies are completely missing. The modulation of others has decreased or even changed sign. The filter properties depend on the amount of defocusing.

6.3 The Effect of Axial Astigmatism

When axial astigmatism is present, defocusing depends on the azimuth $\theta = \arctan(R_y/R_x)$, if the axes of astigmatism are located in the R_x and R_y directions of the diffraction plane and $R = \sqrt{R_x^2 + R_y^2}$ is the spatial fre-

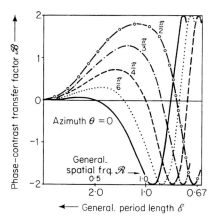

Figure 14 Phase contrast transfer functions in dependence on the azimuth θ; the astigmatic defocusing difference is assumed to be $2\Delta_a = 0.72$ ($\Delta_m = +0.48$).

quency in question. We can write

$$\Delta(\theta) = \Delta_m - \Delta_a \cos 2\theta \tag{119}$$

where $2\Delta_a$ is the defocusing difference due to astigmatism and Δ_m the mean defocusing. Experiments and calculations on the effect of astigmatism on the frequency spectrum of the image have been published by Thon (1968b). Transfer theory characterizes this aberration by means of a family of transfer functions with θ as a parameter. An example is given in Fig. 14.

6.4 Current and Voltage Fluctuations and the Energy Width of the Illuminating Beam

According to what has already been explained about transfer theory, the electron microscope image has frequency gaps, and the transfer intervals following the main transfer intervals become smaller and smaller. The number of these intervals is not limited. The maximum contrast $|\mathcal{B}_{max}| = 2$ should be reached at a certain frequency within each interval.

The situation is different if we consider fluctuations of voltage or current in the instrument. According to (16) they cause variable defocusing resulting in temporal fluctuations of the contrast transfer functions. The effective transfer function is the mean value of the function during the time of recording. In order to obtain this mean value, we have to know the temporal variation of the accelerating voltage U and lens current I. We give here some results only and refer for details to Hanszen and Trepte (1970, 1971a).

When the fluctuations are symmetrical with respect to the mean value, the effective phase contrast transfer function is given by

$$\overline{\mathscr{B}} = G \cdot \mathscr{B} = -2G\sin(2\pi\mathscr{W}),\tag{120}$$

where $G = G(\delta, \mathscr{R})$ is the envelope of the transfer function. G depends only on the generalized spatial frequency $\mathscr{R}$ and on a quantity δ which characterizes the fluctuation. For linear current fluctuations (e.g. sawtooth fluctuations) with the maximum generalized deviation $\pm\delta_l$ we have, for instance,

$$G_l = \sin(\pi|\delta_l|\mathscr{R}^2)/\pi|\delta_l|\mathscr{R}^2.\tag{121}$$

For sinusoidal fluctuations with a maximum deviation $|\delta_s|$, we obtain

$$G_s = J_o(\pi|\delta_s|\mathscr{R}^2),\tag{122}$$

where J_o is the zero order Bessel function; and for fluctuations obeying the *Gaussian error function* we obtain

$$G_g = \exp\left[-(\pi\delta_g/4)^2\mathscr{R}^4/\ln 2\right]\tag{123}$$

where δ_g characterizes the half-width of the fluctuations in generalized coordinates.

In each case, the zero points of the ideal transfer functions (Table 2.7) remain unchanged. An example of the influence of (121) on the phase contrast is shown in Fig. 13, curve c. Comparing this curve with curve b, which characterizes an imperfect defocusing, we see that fluctuations between the defocusing values $\pm\delta_l$ have less influence on contrast than a defocusing difference δ_l. This is easy to understand, because the fluctuations are characterized by a *mean* value and the imperfect defocusing by an *extreme* value.

Looking back at the special case given in (123), a family of envelopes for the contrast transfer function, with δ_g as parameter, are drawn in Fig. 15. When G_g drops below a certain value, no contrast is produced in practice. If this value is $1/e$, the $\mathscr{R}$-value for which $G_g = 1/e$, i.e.

$$\mathscr{R}_e = 2\sqrt[4]{\ln 2/(\pi\delta_g)^2}\tag{124}$$

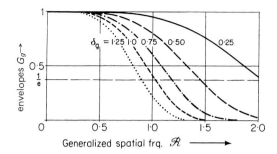

Figure 15 Envelopes G_g of the contrast transfer functions when current or voltage fluctuations, having a Gaussian distribution, occur. The half width of this distribution is expressed by the generalized defocusing δ_g.

gives the practical limit of the contrast transfer function. In terms of the smallest transferred period length ϵ_e we obtain the following equation using Eqs. (16) and (99); (102) with a mean current fluctuation $\Delta I/I$:

$$\epsilon_e = \sqrt{\frac{\pi}{2(\ln 2)^{\frac{1}{2}}}\lambda |C_{ch}| \left|\frac{\Delta I}{I}\right|}. \tag{125}$$

From this we see the importance of chromatic aberration as it affects the transfer limit.

Example. Using the above mentioned data for the normal objective and assuming current fluctuation of the order 1×10^{-5}, we obtain $\epsilon_e = 3.8$ Å. This value is too large for resolving atomic distances.

Since the energy distribution in the illuminating beam is asymmetric, its influence on the optical transfer is more difficult to calculate. It can, however, in most cases be described approximately by (123) using adapted values for δ_g; see Hanszen and Trepte (1971a).

6.5 Experimental Confirmation of the Transfer Theory

Thon's experiments (1965–1968) can be regarded as proof of contrast theory. He used carbon foils 100 Å thick as objects. As known from electron diffraction these foils have a continuous frequency spectrum, the intensity of which, although decreasing with $1/\mathscr{R}^4$, is still clearly noticeable for $\mathscr{R} = 2$. It is a necessary but not sufficient criterion for perfect imaging that the spectrum of a micrograph is the same as that of the object; the differences between object spectrum and micrograph spectrum give evidence about optical transfer.

Since electron micrographs are highly magnified, the structures on the photographic plate are large enough to be studied by diffraction of light waves. For this purpose a small aperture is put into a laser beam, in order to spread the beam by diffraction and illuminate a sufficiently large area of the micrograph. The diffraction pattern of this area—called a diffractogram—appears around the image of the aperture produced by a lens and can be photographed.

We assume that the transparency for the wave function of the object and the transparency for the image intensity are

$$F(x) = 1 + i \int_0^\infty \tilde{\Phi}(R) \cos\left[2\pi Rx + \xi_\Phi(R)\right] dR; \tag{126}$$

$$F_I'(x') = 1 + \int_0^\infty B(R)\tilde{\Phi}(R) \cos\left[2\pi Rx' + \xi_\Phi(R)\right] dR$$

$$= 1 - 2 \int_0^\infty \tilde{\Phi}(R) \sin\left[2\pi \mathscr{W}(R)\right] \cos\left[2\pi Rx' + \xi_\Phi(R)\right] dR$$

$$= 1 - \int_{-\infty}^{+\infty} \tilde{\Phi}(R) \sin\left[2\pi \mathscr{W}(R)\right] \exp\left\{i\left[2\pi Rx' + \xi_\Phi(R)\right]\right\} dR, \tag{127}$$

since $2\pi \mathscr{W}$ is even in R.

We assume further that the density on the photographic plate is roughly given by the same expression. Local variations in thickness occurring in photographic emulsions can be compensated by using an appropriate immersion liquid. Then, the Fourier transform of the light-optical transparency for the wave function in the plate is

$$\tilde{\tilde{F}}(R = 0) = 2x_e;$$

$$\tilde{\tilde{F}}(R \neq 0) = \tilde{\Phi}(R \neq 0) \sin\left[2\pi \mathscr{W}(R)\right] \exp\left[i\xi_\Phi(R)\right] \tag{128}$$

and the intensity-transparency of the diffractogram is

$$\left(\frac{\tilde{\tilde{F}}(R = 0)}{2x_e}\right)^2 = 1; \quad \left(\frac{\tilde{\tilde{F}}(R \neq 0)}{2x_e}\right)^2 = \tilde{\Phi}^2(R \neq 0) \sin^2\left[2\pi \mathscr{W}(R)\right]/4x_e^2$$

$$= \tilde{\Phi}^2(R \neq 0) B^2(R)/16x_e^2. \tag{129}$$

In other words, the diffraction pattern shows the product of the object spectrum and the square of the phase contrast transfer function. In Fig. 16 this

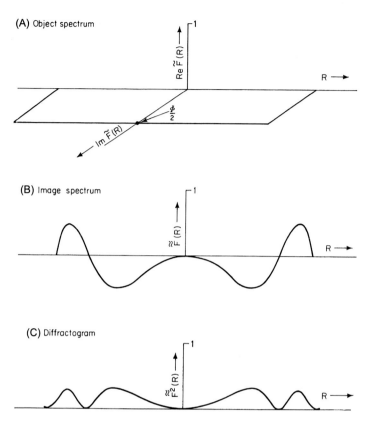

(A) Object spectrum

(B) Image spectrum

(C) Diffractogram

Figure 16 The diffractometer method. Object: $F(x) = 1 + i\tilde{\Phi} \int \cos(2\pi Rx)dx$. (A) Fourier representation $\tilde{F}(R)$ of the object, identical with the object related diffraction transparency; (B) Fourier representation $\tilde{\tilde{F}}(R)$ of the micrograph, identical with the transparency of the wave function occurring in the diffraction plane of the diffractometer; (C) Transparency $\tilde{\tilde{F}}^2(R)$ of the intensity in the recording plane of the diffractometer, similar to the density-distribution in the diffractogram.

behavior is demonstrated for a particular object (126) under the assumption $\tilde{\Phi} = \text{const}$; $\xi_\Phi = 0$. Fig. 16C represents a photometric curve recorded diagonally through the diffractogram. If the objective has axial astigmatism, these curves depend on the azimuth θ. This means that the diffractogram method can also be used to determine astigmatism; compare Fig. 17 with Fig. 14. Fig. 18 shows the different appearance of the same object area at different degrees of defocusing. Some characteristic diffractograms are also given. A detailed discussion of the structures appearing in different micrographs is given by Hanszen (1967). There is no "sharp" image within the

(A)   (B)

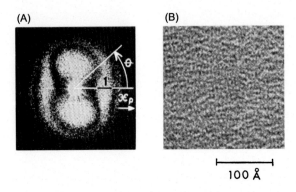

100 Å

Figure 17 Diffractogram (A) of a micrograph (B) having astigmatism. The diffractogram describes the same facts as Fig. 14 (Thon, 1968b, Fig. 18).

micrographs of a focal series. There is no way of knowing how such an image could be distinguished from the others (Thon, 1967).

In order to evaluate the diffractograms of a focal series, the generalized defocusing is related to an arbitrary zero point; as a function of this parameter, diameter and width of the diffraction rings are drawn in a diagram. By displacing this diagram along the Δ-axis of another diagram showing the theoretical curves, the experimental data can be fitted to the theoretical functions, as is done in Fig. 19. In this manner, the absolute defocusing can be determined from the entire focal series. The experimental values fit only to the theoretical curves characterizing phase contrast, but not to the curves characterizing amplitude contrast. Therefore in Fig. 19, we have an experimental proof that a thin carbon foil is in a good approximation a pure phase object.

Obviously, the diffractograms do not give the sign of the contrast. The fact that the neighboring rings belong to transfer intervals of opposite sign is an additional piece of theoretical information.

Contrast inversions dependent on defocusing can also be directly recognized in the micrographs. Especially when there is a considerable amount of defocusing, i.e. $(\Delta/2)\mathscr{X}_p^2 \gg \mathscr{X}_p^4$, the phase contrast transfer functions are odd functions of Δ. Changing the sign of Δ, therefore, results in micrographs that are negatives of the previous ones, although each micrograph is composed of transfer intervals with alternating sign (see Fig. 20). Since amplitude-contrast transfer functions are even, Fig. 20 gives a further proof to the fact that a thin carbon foil is a pure phase object.

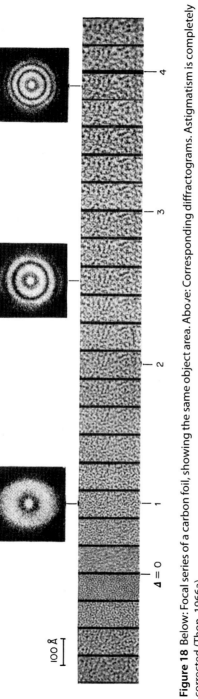

Figure 18 Below: Focal series of a carbon foil, showing the same object area. Above: Corresponding diffractograms. Astigmatism is completely corrected (Thon, 1966a).

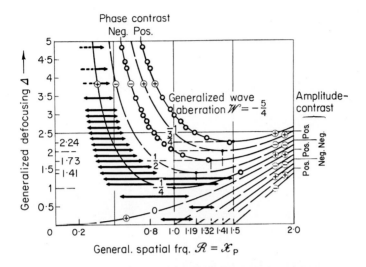

Figure 19 Maximum contrast (O) and width (←→) of the first transfer interval, data plotted in a $\Delta(\mathscr{R})$-diagram; experimental values given by Thon (1968b).

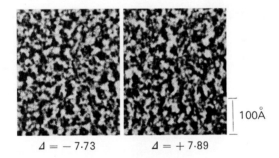

$$\Delta = -7.73 \qquad \Delta = +7.89$$

Figure 20 Total change of phase contrast by inverting the sign of a strong defocusing (Thon, 1967).

6.6 Limitation of the Objective Aperture

If an aperture with the generalized diameter $\mathscr{X}_{Pe} = \mathscr{R}_e$ is used, we have:

$$|\tilde{F}(\mathscr{X}_P)| = 1; \quad \tilde{\Phi}_M(\mathscr{X}_P) = 0 \quad \text{for } |\mathscr{X}_P| \leq |\mathscr{X}_{Pe}|,$$

and

$$\tilde{F}(\mathscr{X}_P) = 0 \quad \text{for } |\mathscr{X}_P| > |\mathscr{X}_{Pe}|.$$

Taking into consideration (104), we obtain the same contrast-transfer functions (Table 2.7) in the interval $|\mathscr{X}_P| \leq |\mathscr{X}_{Pe}|$ as would be obtained without using an aperture. However, for $|\mathcal{X}_P| > |\mathcal{X}_{Pe}|$, the contrast transfer functions vanish. Unlike the case of incoherent illumination (see Section 10.2) the aperture stop now merely acts as a low-pass filter. It has, however, no influence on the contrast transfer of the remaining spatial frequencies and consequently does not raise the contrast of weak objects. This is true because higher terms in the $(R = 0)$ part of (62) could be omitted. Therefore, in the image of a weak object, amplitude or phase contrast cannot be produced by using an aperture stop. On the other hand, in the image of strong objects, aperture contrast can be produced. In this case, however, the linear transfer theory fails, so that no transfer function can be defined. In weak objects, the aperture can eliminate only transfer intervals with the wrong sign.

6.7 Absorption Plates[12]

The improvement of transfer when using zone plates for amplitude objects (Hoppe, 1961, 1963) and phase objects (Lenz, 1963)[13] can be described by transfer theory as follows. If the transfer intervals with negative sign are eliminated by opaque zone rings, a contrast transfer function is obtained which has only positive values. There are, however, wide frequency gaps (see Fig. 21A). These theoretical predictions have been confirmed by diffractometer experiments of Möllenstedt et al. (1968) (see Fig. 22). It is not possible to replace the lost information by other information stored in the same micrograph, as discussed in Section 5.5. If objective apertures, especially zone plates, are used, the image points deteriorate into diffraction patterns of the aperture. Maskings of this kind are therefore advantageous only if the resulting disks are smaller than the original disks due to lens aberrations.

6.8 Weakening of the Undiffracted Beam

If the zero order is totally removed, the phenomena described in Table 2.E occur. In this case, there is no linear transfer. Thus, the contrast depends on the object and no contrast transfer functions can be defined; for further investigations see Hanszen (1969a).

In order to increase the modulation (i.e. the "contrast") in the image intensity, the wave function of the zero diffraction order can be weakened, as long as the wave functions of all diffraction orders are small compared to

---

[12] Compare with Sections 6.7–6.10, also Tsujiuchi (1963).
[13] For a correction of the radii given, see Hanszen et al. (1963), especially p. 485.

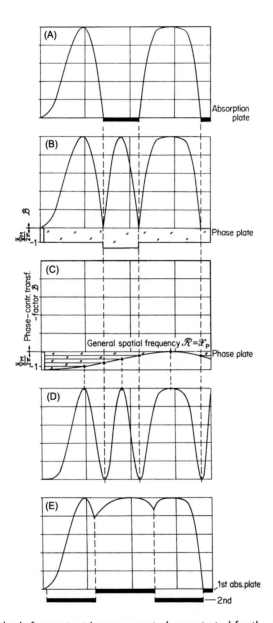

Figure 21 Methods for contrast improvement, demonstrated for the case of curve D in Fig. 12C. (A) Eliminating the transfer intervals with negative sign (Lenz, 1963); (B) Inversion of the sign in these intervals (Hanszen & Morgenstern, 1965); (C) Correction of the wave aberration in order to obtain the value $(n - 1)\lambda/4$ by a phase plate having locally variable thickness (Hanszen & Morgenstern, 1965); (D) Light optical reconstruction (Gabor, 1949); (E) Double exposure of a micrograph using proper defocusing and complementary zone plates (Hanszen, 1966b).

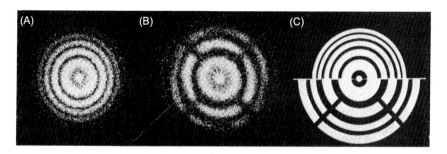

Figure 22 Diffractograms of a carbon foil by Thon (1968a, 1968b) demonstrating the action of a Lenz plate: (A) using no zone plate, (B) using a zone plate, (C) schematic diagram explaining the filter effect; see Hoppe et al. (1969), Fig. 12.

the zero order. There is an experimental complication, however, because every weakening is combined with phase delays. A suggestion by Le Poole, for performing the weakening without any perturbations in the image field due to diffraction in the absorption plate, is mentioned by Hanszen (1969a, p. 126).

6.9 Phase Plates

The possibilities of correcting micrographs by means of phase plates are discussed by Hanszen and Morgenstern (1965) on the basis of transfer theory. From (113)–(114) we see that the pupil function can be made unity, if one introduces in the pupil plane a non-absorbing, phase shifting foil with a hole for the zero diffraction order. The thickness of this foil should vary locally in such a manner, that the wave aberration of the objective is compensated up to a remainder $2\pi n$; this means that

$$2\pi \mathscr{W}(\mathscr{X}_P) + \tilde{\Phi}_M(\mathscr{X}_P) = 2\pi n; \quad n = 0; \pm 1; \pm 2; \ldots \quad (130)$$

A foil of this kind would produce an amplitude-contrast transfer-function $\mathscr{D}_I(\mathscr{R}) \equiv 1$. In this case, the image of an amplitude object would be perfect; i.e. all spatial frequencies would be transferred with an equal and optimal positive amplitude contrast factor.

In order to obtain equal and optimal positive contrast for phase objects, i.e. $\mathscr{B}(\mathscr{R}) \equiv 2$, the phase shift $\tilde{\Phi}_M$ of the foil should be given by

$$2\pi \mathscr{W}(\mathscr{X}_P) + \tilde{\Phi}_M(\mathscr{X}_P) = \begin{cases} 2\pi(n - 1/4); & \text{for } \mathscr{X}_P \neq 0 \\ 2\pi n; & \text{for } \mathscr{X}_P = 0 \end{cases} \quad (131)$$

Since non-absorbing phase plates are realizable in a sufficient approxima-tion, the manufacturing of plates according to (131) should be possible in spite of technical difficulties. The periodic potential of the atoms does not noticeably disturb the image field as long as the decisive period lengths are small compared with the diameter of the sampling points; for in this case the beams diffracted by the foil fall outside the image field.

Until zone plates of this kind are produced we have to be satisfied with a discontinuous zone plate. According to Fig. 21B such plates should have the same radii as the absorption plate in Fig. 21A. The transparent zones of the absorption plate should be replaced by zones which shift the phase by $2\pi n$, the opaque zones by zones which shift the phase by $2\pi(n + 1/2)$. Plates of this kind do not suppress the frequency intervals of negative contrast, but rather invert the contrast in these intervals. According to Fig. 21B, however, considerable transfer gaps still remain. Although manufacturing of such plates is not more difficult than that of absorption plates, they have not yet been employed to improve imaging.

6.10 Reconstruction of Electron Micrographs by Light Optical Methods

The reconstruction method proposed by Gabor (1948, 1949, 1951) can be explained by transfer theory as follows (Hanszen, 1970a). The electron mi-crograph of the object (126), with the transparency (127), is illuminated in the same way as is done with monochromatic light in the diffractometer. Except for a fixed scaling factor, the light optical system used for recon-struction should have the same generalized transfer function as the electron microscope. In this case, the reconstructed image has the intensity distribu-tion

$$F_I''(x'') = 1 + \text{const} \int_0^\infty B^2(R \neq 0) \cos\left[2\pi R x'' + \xi_I(R)\right] dR \qquad (132)$$

and therefore, a positive definite transfer function $B^2(R)$ or $\mathscr{B}^2(\mathscr{R})$: see Fig. 21D. The effect of the unavoidable frequency gaps of the resulting transfer functions should, however, not be underestimated.

The micrograph on the photographic plate has an amplitude and a phase structure. If the amplitude structure is bleached, a pure light-optical phase object remains, and the light-optical step can be directly described by one of the phase-contrast transfer functions of the family $\mathscr{B}(\mathscr{R})$. The only exper-imental requirement is to adjust the generalized light-optical defocusing to the generalized defocusing in the electron microscopical step. If the phase structure in the unbleached plate is suppressed by means of appropriate immersion, a pure amplitude object remains, the imaging of which is described by an amplitude-contrast transfer function of the family $\mathcal{D}_I(\mathcal{R})$ if no other measures are taken. This function, however, can be converted into the corresponding phase-contrast transfer function $\mathcal{B}(\mathcal{R})$ by means of a $\lambda/4$-plate, inserted into the zero order of the diffraction spectrum. In this case the reconstruction, which leads to contrast of uniform sign, is also described by (132); for details see Hanszen (1970a).

Better results should be obtained by *single-sideband holography*. When one side of the diffraction pattern in the microscope is cut off by an opaque mask, the only imperfections of the micrograph are lateral displacements (136) of the spatial frequencies (see transfer function (2.7b) in column D of Table 2). The image will be perfectly reconstructed, when these displacements are compensated in the light optical process. This can be done by a masking which leads to the conjugate complex transfer function (Table 2; D, 7b) (see the model experiments by Hanszen, 1969b, 1970b).

6.11 Information Gain by Evaluating Focal Series: Using Complementary Objective Diaphragms

If the image is not improved by masking methods, there will be frequency gaps in the transfer functions of Fig. 12 which in many cases may be serious. In order to interpret unknown structures, the evaluation of only a single micrograph may therefore lead to misinterpretations.

The frequency gaps move with defocusing; this means that the information, missing in one micrograph, can be substituted by information from other micrographs of a focal series. Conventional micrographs are less appropriate for information replenishment since, according to Fig. 12, contrast alternations create intricate conditions. Reconstruction should rather be based on micrographs like Fig. 21A, B or on reconstructions as in Fig. 21D. Starting from micrographs like those in Fig. 21A was proposed by Hanszen (1966b) and verified (1970a) by light-optical model experiments; see Fig. 23. The image improvement of a phase object is interpreted in Fig. 21E: the first micrograph is produced using zone plate I and the defocusing $\Delta = 1$. The frequency gap is filled, in this micrograph, by a second micrograph in which zone plate II and the defocusing $\Delta = \sqrt{5}$ are used.

By varying Δ only, it is impossible to obtain two phase-contrast transfer functions such that all minima of the first function coincide with the maxima of the second one. Therefore, it is impossible to obtain an image containing the whole frequency spectrum by reconstructing it from

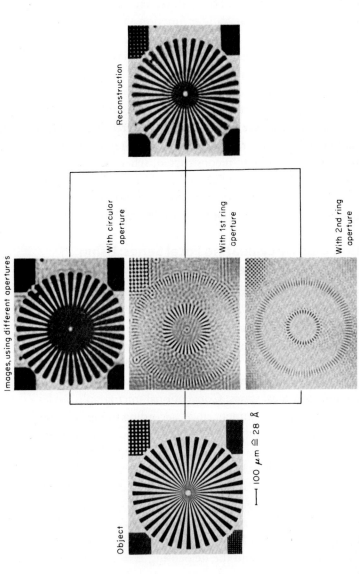

Figure 23 Transfer improvement by complementary objective diaphragms, demonstrated by means of a light-optical amplitude object. In order to eliminate the spatial frequencies suffering contrast alterations, the objective aperture has to be severely limited (image, shown above); the frequencies lost in the first image can be restored by the method described in Section 6.11 (images at the center and the bottom) (Hanszen, 1968, where the captions of Figs. 1 and 3 have to be exchanged).

only two original micrographs. At worst, the number of required plates photographed at carefully calculated Δ-values is equal to the number of the frequency gaps to be removed. Unfortunately, the contrast in the reconstruction decreases with the number of micrographs used, because the undiffracted beam which produces the background is needed for constructing *each* image (see Section 5.4).

7. CONTRAST BY MEANS OF PHASE SHIFT: CONTRAST TRANSFER FUNCTIONS FOR OBLIQUE ILLUMINATION

7.1 Without Masking

Calculations for axial illumination are easy, because the spatial frequency R can be identified with the reduced pupil coordinate S related to the axis point of the pupil plane, and therefore $\mathscr{R}$ with $\mathscr{X}_P$. This is not true for oblique illumination. In this case the wave aberration must be separated, according to (107) and (108), into an odd component $W_u(R)$ and an even one $W_v(R)$. Writing $\xi = 2\pi \mathscr{W}_u(R)$ and $\zeta = 2\pi \mathscr{W}_v(R)$, it can be seen from Table 2.7a,b.C that the contrast transfer functions are complex. The even component of the wave aberration determines the amount and the odd component the phase of the contrast transfer function. Therefore, in addition to the effects known from axial illumination, lateral displacements of the spatial frequencies occur in the image. For this reason, it is not advantageous to operate with oblique illumination without aperture limitation.

7.2 Elimination of One Side of the Diffraction Pattern

If a symmetrical objective aperture is used and the illumination is oblique enough to place the zero diffraction order near the edge of the objective aperture, the simple relations of Table 2.7.D exist. We give the results in generalized coordinates. The coordinate $\mathscr{Q}$ of the undiffracted beam coincides with the coordinate $\mathscr{X}_{Pe}$ of the edge, and the generalized spatial frequency, as defined in (98), is

$$\mathscr{R} = \mathscr{X}_P - \mathscr{X}_{Pe} = \mathscr{X}_P - \mathscr{Q}. \tag{133}$$

For the phase shifts χ, we can write

$$\chi = 2\pi \left[\mathscr{W}(\mathscr{X}_P) - \mathscr{W}(\mathscr{X}_{Pe}) \right] \tag{134}$$

and, according to (2.7.D) for the contrast transfer functions,

$$\mathcal{D}_I = (1/2)\exp\{2\pi i[\mathcal{W}(\mathcal{X}_P) - \mathcal{W}(\mathcal{X}_{Pe})]\};$$

$$\mathcal{B} = \exp\{2\pi i[\mathcal{W}(\mathcal{X}_P) - \mathcal{W}(\mathcal{X}_{Pe}) + (1/4)]\}. \tag{135}$$

The absolute value of the transfer factor, therefore, is the same for all spatial frequencies and independent of defocusing. Using (27), the lateral displacement is $\delta\mathcal{X}'$ of the image structures with respect to the object structures is

$$\delta\mathcal{X}'(\mathcal{R}) = -[\mathcal{W}(\mathcal{X}_P) - \mathcal{W}(\mathcal{X}_{Pe})]/\mathcal{R};$$

for amplitude transfer or

$$\delta\mathcal{X}'(\mathcal{R}) = -[\mathcal{W}(\mathcal{X}_P) - \mathcal{W}(\mathcal{X}_{Pe}) - (1/4)]/\mathcal{R} \tag{136}$$

for phase transfer respectively. This is dependent on spatial frequency and defocusing. From (136) we learn that the essential image properties can already be derived from the wave aberration curves.

The phase-contrast transfer properties will now be discussed with the aid of Fig. 24. We assume defocusing to be $\Delta = \sqrt{2}$, by which the shape of the wave aberration curve is given, and the coordinates of the diaphragm edge to be $\mathcal{X}_{Pe} = \pm1.56$. Furthermore we assume the undiffracted beam to intersect the pupil plane at $\mathcal{X}_{Pe} = -1.56$. As can be seen from the diagram, the highest transferred spatial frequency is twice that in the case of central illumination. Spatial frequencies, for which $\mathcal{W}(\mathcal{X}_P) - \mathcal{W}(\mathcal{X}_{Pe}) = n - 1/4$ with $n = 0, \pm1, \pm2, \dots$ holds, do not suffer lateral displacements.

In order to reduce the lateral displacements of the other frequencies as far as possible, tilting of the illumination and defocusing should be adjusted in such a way:

(1) that the most important frequency interval is covered by an extremum of the wave aberration,

(2) that this extremum has the height $(n - 1/4)$, compared with the wave aberration at the zero order.

For further details see Hanszen and Morgenstern (1965), Hanszen and Trepte (1971b, 1971c); for details about single-sideband holography see Hanszen (1969b, 1970b).

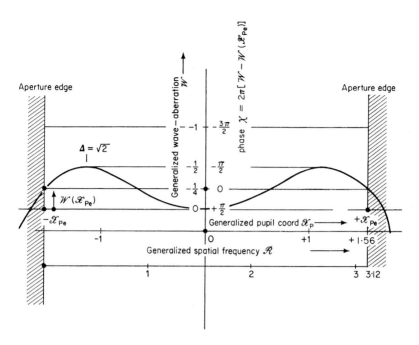

Figure 24 An example showing the phases χ occurring with oblique illumination. The zero diffraction order touches the aperture edge.

7.3 Practical Applications[14]

The discussion of contrast transfer under oblique illumination was brief because of its limited range of application. As we have seen in the last section, the transfer limit can be raised by the factor of 2 in one coordinate direction. Unfortunately, however, transfer is drastically reduced in the perpendicular direction.

One-sided oblique illumination is therefore only suitable for investigating very special objects, e.g. one or a small number of lattice planes having spacings so small that it is impossible to separate them by axial illumination (see Dowell, 1963; Komoda, 1964, 1966, 1967; Komoda & Otsuki, 1964). In order to obtain more information about crystal lattices from electron micrographs than from electron diffraction patterns, it should be expected that lattice images indicate the position of each lattice plane in a reliable way. It has been established, however, that lateral displacements occur depending on spatial frequency (i.e. reciprocal lattice spacing) and defocusing. Only if (136) vanishes for the considered lattice spacing $1/\mathscr{R}$ is the information

---

[14] For single-sideband holography, see Section 6.10.

obtained on the lattice position correct (especially with respect to the crystal boundaries). Unfortunately, it is difficult to find out if this requirement is fulfilled. Lateral displacements always occur when the rotational symmetry of the optical system is disturbed, as is the case for one-sided oblique illumination. The symmetry can be restored by using conical oblique illumination; see the general considerations of Hanszen and Morgenstern (1965). This illumination, however, is a particular case of partial coherent illumination (Hanszen & Trepte, 1971b, 1971c); see also Section 10.1.

8. THE PROBLEM OF POINT RESOLUTION

8.1 The Concepts of Contrast Transfer and Resolution Limit

So far the concept of "resolving power" has been avoided. If the imaging properties of an electron microscope are described by a "resolution limit", one may be led to think that all structures larger than the "smallest resolved distance" are transferred perfectly and all smaller structures not at all. By means of transfer functions a very detailed description of the imaging properties can now be given.

Contrast transfer theory is based on "spatial frequencies" as the object elements to be analyzed, while older theory is based on "object points" as elements; see Section 4.3. The two concepts differ from each other in a similar way as do the corpuscular and wave interpretations of matter. An infinite spectrum of spatial frequencies is related to an object point. For this reason, in accurate discussions, the reciprocal spatial frequency should not be identified with the distance between two object points as is often done. In fact, a similar complementarity between point-image and transfer function can be given as is known from quantum theory (see e.g. Röhler, 1967, p. 34f).

8.2 Definition of Point Resolution

Two object points are said to be resolved when their Airy disks overlap in such a way, that the intensity in the saddle between the two maxima is 75% or less of the intensity in the maxima. In most calculations of the intensity distribution in the image of the two disks, incoherent illumination is assumed (see Glaser, 1956, p. 371). Fig. 25 shows, however, that in electron microscopy, where illumination is practically coherent, the overlapping is more complicated. Thus we should not identify the half-width of

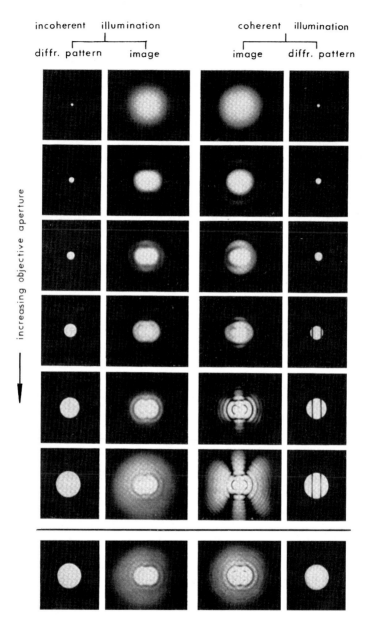

incoherent illumination   coherent illumination

diffr. pattern   image   image   diffr. pattern

increasing objective aperture

Figure 25 Light-optical model experiments demonstrating point resolution. Object: diaphragm with two holes of 9 μm diameter at a distance of 18 μm: $\lambda_c = 6328$ Å; $C_S = 1660$ mm; $\Delta = 0$. Left column: using incoherent illumination, the intensities are to be added. Right column: using coherent illumination, the wave functions are to be added. The Fresnel fringes appearing are *spurious structures*. Row at the bottom: double exposure; in each case, one of the holes was covered.

the diffraction disk of a *single* atom with the "point resolution" of the electron microscope. Point resolution can only be determined by calculating the overlapping of the wave function in *both* disks and taking interference effects into account. Shape and size of the diffraction disks, and therefore also the resolution, depend on aperture, spherical aberration, and defocusing. In a paper by Scherzer (1949) the influence of these parameters was treated, at least in all fundamentals.

8.3 Transfer-Theoretical Treatment of Point Resolution

We now study the image formation of an object having a homogeneous background interrupted by two small, equal phase details (shortly called "object points") at a distance x apart. In the diffraction pattern, the shape of these details is described by the wave function at the pupil points $R_s = k/x$ with $k = \pm 1; 2; \ldots$. That the object detail repeats only twice is due to the fact that each spectral point is a "sampling point" with a *finite* width; see Section 5.7. In order to resolve the point distance x in the micrograph, it is sufficient to use only the zero diffraction order and the spectral points with $k = \pm 1$. In order to produce positive phase contrast, it is necessary to have a wave aberration $\mathcal{W} = n - 1/4$ with $n = 0; \pm 1; \ldots$ at these points. Furthermore, it is necessary to maintain this value of the wave aberration in an extended interval around the pupil points considered, because the information of having only two object points, which is stored in the width of the sampling points, must also be transferred with phase contrast into the image. This condition is best fulfilled for the generalized pupil coordinate $\mathcal{X}_P = 1$ and a defocusing $\Delta = 1$; then n equals zero. Relating $\mathcal{X}_P$ to the distance x of the object points according to (115), we obtain the condition for resolving this distance,

$$x = \sqrt[4]{C_s \lambda^3} \tag{137}$$

which complies with the resolution limit of the older theory up to a factor of about $1/2$. The new description is more reliable, because (137) was derived under the assumption of coherent illumination and under the requirement of producing phase contrast. Its statements, however, are much more restricted than the statement of older resolution theory because there is no doubt that phase contrast transfer of two object details having distances smaller *or* larger than (137) is made questionable by the new theory. This means that even in the simple case of weak objects it is impossible to describe the image quality by a single numerical value, e.g. the resolution limit. Instead, we need the contrast transfer functions as evaluated in previous chapters.

8.4 Practical Consequences

At best, a resolution limit x_{min} can be given if a transfer function exists for which all period lengths $\epsilon \geqq x_{min}$ are transferred with sufficient contrast. Since an electron microscopist is not interested in imaging the "coarse" structures known from light microscopy, the unavoidable frequency gap of the phase-contrast transfer functions in the vicinity of $\mathscr{R} = 0$ (see Fig. 12) raises no problems. Thus, the resolution limit of the electron microscope is given in practice by the high-frequency end of the main transfer-interval in Fig. 12A. Consequently, in generalized coordinates, $\mathscr{E}_{min} = 1/\mathscr{R}_{min} = 1/1.45 = 0.69$. The resolution limit of the normal objective defined in this sense would be $x_{min} = 4.6$ Å.

As has already been pointed out, an alternating transfer function such as given in Fig. 12C is generally used to obtain high resolution micrographs. In this case, the resolution limit is usually quoted to be $\mathscr{E}_{min} = 1/1.79 = 0.56$ (i.e. $x_{min} = 3.8$ Å for the normal objective). In light optics, however, resolution of this kind would be called "spurious resolution". Röhler (1967) writes in his monograph (p. 61): "We should, speak of spurious resolution in this case, as such micrographs "normally cannot be used for imaging problems because the shape of non-periodic objects is incorrectly reproduced". This criterion of light optics illustrates the problematic situation of present electron microscopy as well as the need to develop reconstruction methods and microscopes with better transfer functions. In order to elucidate the present situation, we refer to the experimental results of Thon (1967), especially p. 450, showing that all micrographs of a focal series show the same resolution. The indication of a resolution limit by means of point resolution tests is, therefore, a very unreliable criterion for the quality of an image.

9. APERTURE CONTRAST

9.1 Comparison Between Contrast by Phase Shift and Aperture Contrast

We have seen, that in the case of *weak objects* a symmetric objective aperture serves only the purpose of eliminating those frequencies which are transferred to the image with the wrong sign of contrast; it has, however, no

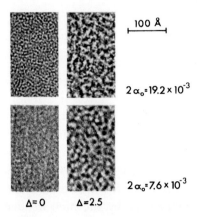

$100 \ \text{Å}$

$2\alpha_0 = 19.2 \times 10^{-3}$

$2\alpha_0 = 7.6 \times 10^{-3}$

$\Delta = 0 \qquad \Delta = 2.5$

Figure 26 Influence of the objective aperture $2\alpha_0$ on the image of a weak object (carbon foil). Micrographs taken with the "normal" objective described in Section 2.2. The objective aperture only limits the spatial frequency spectrum in the image without influencing the transferred frequencies.

influence on the contrast transfer factor of the spatial frequencies that appear in the image (see Fig. 26). With coherent illumination, therefore, no enhancement of resolution can be achieved by giving the objective aperture an optimal size, as is possible with incoherent illumination.

The aperture stop, however, does raise the contrast of strong objects. This effect is well known as "scattering-absorption contrast". Unfortunately, the aperture diaphragm eliminates the highest spatial frequencies from the image. Instead of them, phase structures of larger dimension become visible in the image. The contrast transfer conditions for imaging a phase edge are discussed in detail by Hanszen et al. (1963; p. 483).

In the case of *weak objects*, no aperture contrast is possible and the contrast caused by phase shifts can be described by linear theory. In the case of *strong objects*, aperture contrast is possible, but it does not follow the linear theory. Since in the latter case the imaging depends on the object, only characteristic examples can be discussed (see Hanszen et al., 1963, p. 481ff; and Hanszen & Morgenstern, 1965, pp. 219f and 223). In the following section some supplementary considerations about the strong phase object (49) are given.

9.2 An Example for Aperture Contrast

In (49) the diffraction spectrum of a strong single-spatial-frequency phase object was given. We saw that the Fourier coefficients for $|k| > 1$ are significant only for the amount of modulation, but not for the value of the frequency in the object. For this reason, we want to find out whether we can eliminate the diffraction orders $|k| > 1$. If we do so by using an aperture stop, the image-related transparency in the diffraction plane is

$$\tilde{F}'(R = 0) = J_o(\tilde{a});$$
$$\tilde{F}'(R = \pm R_1) = iJ_{+1}(\tilde{a}) \exp[2\pi i \mathscr{W}(R_1)], \qquad (138)$$

where $\mathscr{W}(R_1)$ describes the influence of spherical aberration and defocusing. From this, we obtain the wave function and the intensity transparency in the image:

$$F'(x') = J_o(\tilde{a}) + 2iJ_1(\tilde{a}) \exp[2\pi i \mathscr{W}(R_1)] \cos 2\pi R_1 x' \qquad (139)$$

$$\left.\begin{aligned} F_I'(x') = \quad & J_o^2(\tilde{a}) + 2J_1^2(\tilde{a}) - 4J_o(\tilde{a})J_1(\tilde{a}) \sin 2\pi \mathscr{W} \cos 2\pi R_1 x' \\ & \qquad\qquad\qquad + 2J_1^2(\tilde{a}) \cos 2\pi 2R_1 x' \\ =_{df} \quad & A(\tilde{a}) \qquad + \qquad\qquad B(\tilde{a}, \mathscr{W}, R_1) \\ & \qquad\qquad\qquad + C(\tilde{a}, 2R_1) \end{aligned}\right\} \qquad (140)$$

The background A of the image is modulated by two terms B and C. Even in the case $\mathscr{W}(R_1) \equiv 0$—where no contrast formation by phase shifts is possible—the modulation term C remains. It therefore expresses the aperture contrast. Unfortunately, the image intensity does not contain the object frequency R_1, but rather the double frequency $2R_1$ (see Fig. 27, curve a). The decisive requirement of contrast theory, that each frequency of the image has to correspond to the same frequency in the object, is now violated. Moreover, the intensity modulation in the image is not proportional to $\tilde{a}$; thus the contrast transfer depends on the object. The modulation in the image completely disappears whenever $J_1(\tilde{a}) = 0$; that is when $\tilde{a} = 3.83, 7.02$, etc.

As a result of the wave aberration the basic frequency can be transferred into the image as well (see term B in (140) and Fig. 27, curve b). This transfer also depends on the object. It disappears for $J_o(\tilde{a}) = 0$ and for $J_1(\tilde{a}) = 0$. Consequently, there are object thicknesses for which the modulation in the image vanishes completely. Nevertheless, at some specific values of defocusing only the frequency $2R_1$ is present in the image. It is impossible, however, to have only the basic frequency in the image.

If a phase object consists of numerous spatial frequencies $R_1; R_2; \ldots R_k$, it may happen that a low diffraction order of one frequency coincides with a higher order of another frequency. If the corresponding area of the pupil plane is masked, the transfer of *several* frequencies is affected and image interpretation is difficult.

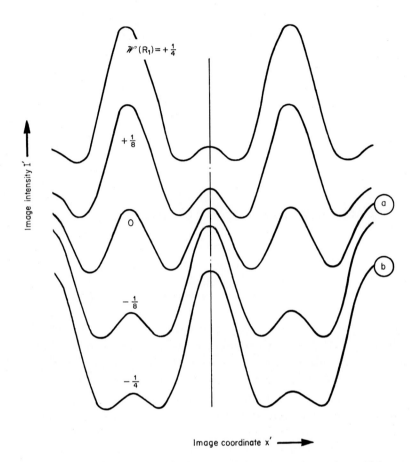

Figure 27 Intensity distribution in the image of the strong phase object $F(x) = 1 + i\tilde{a}\cos 2\pi R_1 x$, limiting the diffraction pattern to $-R_1 < R < +R_1$: defocusing, expressed by $\mathscr{W}(R_1)$ varies. (Transfer-theoretical interpretation of a diagram given by Komoda, 1964.) The micrographs corresponding to curves a and b are shown in Fig. 28.

9.3 Spurious Structures

The theoretical results of (140) and Fig. 27 have been confirmed by experiment (Komoda, 1964). Fig. 28A shows a micrograph which is "in focus", according to geometrical optics. It does not show the lattice spacings 12.6 Å actually present in the object, but "half spacings" of 6.3 Å. Structures of this kind (sometimes inaccurately called "defocusing artifacts") are *spurious structures*. The defocused micrograph in Fig. 28B, on the other hand, shows natural lattice spacings.

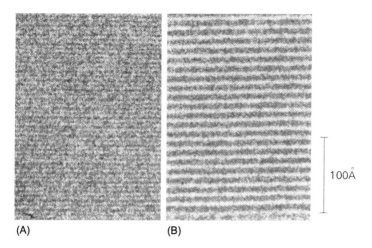

(A)   (B)

Figure 28 Electron micrographs showing the same portion of a Cu-phthalocyanine crystal, taken with axial illumination and symmetric objective aperture but slightly different focusing. The corresponding curves in Fig. 27 are marked by a and b (Komoda, 1964).

It is difficult to define the concept "spurious structures" in electron microscopy; for particular considerations see Hanszen (1969a). It is convenient to reserve this term for smaller details, i.e. for spatial frequencies appearing in the image that are higher than those existing in the object. Knowing the radius R_e of the object aperture, it is easy to say that all structures $x' < 1/R_e$ appearing in the micrograph of a phase object are spurious structures.[15]

Dark field images may also lead to objectionable spurious structures in the micrograph; for a detailed discussion see Hanszen (1969a). A drawing, given by Hanszen (1967; Fig. 4c and b), illustrates how spurious structures are produced when imaging strong phase objects with aperture contrast or in dark field.

10. CONTRAST TRANSFER FOR PARTIALLY COHERENT AND FOR INCOHERENT ILLUMINATION

10.1 Partially Coherent Illumination

As previously mentioned, coherent illumination, i.e. illumination by a point source, is an idealization. In reality, sources of finite diameter are used, and

---

[15] This statement does not hold for amplitude objects, where the object *intensity* is usually the reference quantity, see Hanszen and Morgenstern (1965), p. 223, footnote 1.

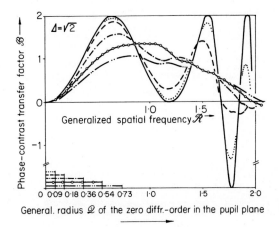

Figure 29 A family of phase-contrast transfer-functions $\mathscr{B}(\mathscr{R})$. Parameter is the generalized illumination aperture $\mathscr{Q}$ (see the scale at the abscissa); defocusing is $\Delta = \sqrt{2}$. The frequency gap disappears when the illumination aperture is increased. $\mathscr{Q} = 0.18$ corresponds to an illumination semi-aperture of $1 \cdot 10^{-3}$ for the normal objective.

different points of the source radiate incoherently. If the source is not too large, illumination is *partially coherent* and the effective transfer function can be calculated by taking the mean value of the transfer functions for each point of the source. Numerical calculations of these transfer functions and experiments with a light optical model take a long time; a publication on them is under way (Hanszen & Trepte, 1971b, 1971c). Here we give only some general results. For small illumination apertures, the zero points and the maximum contrast values of the main transfer intervals remain unchanged. For larger apertures, the zero points move and the contrast always decreases. As shown in Fig. 29, there are values of defocusing for which contrast transfer functions can be produced with a uniform sign, but with a rather low contrast transfer factor, by widening the illumination aperture. In the case of a normal objective, the semi-illumination angles needed are not much larger than $1 \cdot 10^{-3}$.

A ring condenser gives a special kind of partially coherent illumination. Such a ring-shaped source can be regarded as composed of pairs of point sources located under every azimuth at equal distances from the axis and radiating incoherently. It was shown by Hanszen and Morgenstern (1965) that, contrary to one-sided oblique illumination, this illumination eliminates lateral displacements. On the basis of the formulae given there, contrast transfer functions for conical illumination are obtained by integrating over the azimuth θ. Under optimal conditions with this type of illumination, transfer functions having uniform sign up to high spatial frequencies are obtained. However, contrast is strongly reduced for all frequencies; see Hanszen and Trepte (1971c).

10.2 Incoherent Illumination

An illumination is said to be incoherent, when the illumination aperture is so large that the undiffracted beam fills the whole objective aperture. With incoherent illumination, phase structures cannot be shown. Hence, such an illumination is advantageous when the phase component of a mixed object is to be suppressed. In order to increase the amplitude component, a low voltage microscope combined with an energy filter has to be used. For such an instrument, see Wilska (1965).

With incoherent illumination, the intensity transparency of the object is linearly transferred into the intensity distribution of the image regardless of the modulation depth of the object. Here it may be sufficient to give only a few details pertinent to electron microscopy; for particulars see Hanszen, Rosenbruch, and Sunder-Plassmann (1965).

When the undiffracted beam fills the entire pupil the beams diffracted by the spatial frequency R_k in the object have also the cross section of the pupil. However, their centers are displaced by $\pm R_k$. Only those portions of these beams which are hatched in Fig. 30A pass through the aperture and participate in image formation. For coherent illumination the frequency R_k is no longer transferred, as in the case drawn in Fig. 30A at the right, because the center of one circle lies outside the radius of the other. For incoherent illumination this frequency appears, although weakly, in the image, because the hatched area in the figure is not yet reduced to zero. The amplitude contrast transfer function for an aberration free objective (Fig. 30C; curve $h_i/h_a = 0$) is nearly triangular; it ends at $\mathscr{R}_e = 2$, which is double the generalized radius of the aperture. When we use the same aperture and incoherent illumination, therefore, the highest transferred spatial frequency in the image of an amplitude object, with arbitrary modulation depth, is twice as high as in the image of a weak object using coherent illumination.[16]

If a circular aperture is used, and the diameter of it is so small that the spherical aberration gives no wave aberration larger than $\lambda/4$ at any point inside the aperture, the true transfer function coincides with that for perfect

---

[16] For incoherent illumination, there are no objections to covering the central portion of the objective aperture.

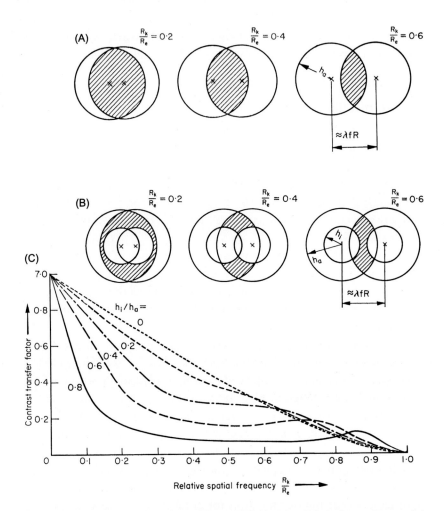

Figure 30 Contrast formation for incoherent illumination. (A) Relative positions of the undiffracted beam and of the beam which is diffracted at the spatial frequency R_k in the pupil plane; R_k is increasing from left to right. (C) Contrast transfer-functions of the aberration-free objective, using ring apertures in the objective having the same outer radius h_a but different inner radii h_i, as shown in (B). The area of the pupil plane hatched in (B) is responsible for contrast transfer (Linfoot, 1957).

imaging; see Fig. 31A, curve a. There are no practical focusing problems in this case. If a larger aperture is used the influence of the spherical aberration should be compensated by a slight underfocusing in order to keep the wave aberration within the hole as small as possible. In this manner, transfer functions can be obtained which are extended to higher spatial fre-

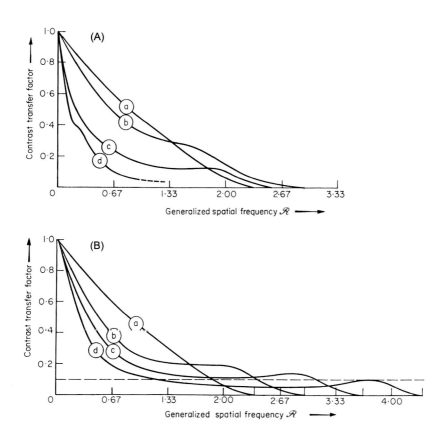

Figure 31 Contrast transfer-functions of a real objective for incoherent illumination. The generalized radii of the objective aperture are denoted by $\mathscr{H}$ (Hanszen et al., 1965). (A) Circular objective apertures: (a) $\mathscr{H}_a = 1.17$; $\Delta = 0.70$. (b) $\mathscr{H}_a = 1.50$; $\Delta = 1.08$. (c) $\mathscr{H}_a = 1.84$; $\Delta = 1.30$. (d) $\mathscr{H}_a = 2.17$; $\Delta = 1.90$. (B) Ring apertures: (a) $\mathscr{H}_a = 1.17$; $\mathscr{H}_i = 0.00$; $\Delta = 0.70$. (b) $\mathscr{H}_a = 1.50$; $\mathscr{H}_i = 0.68$; $\Delta = 1.45$. (c) $\mathscr{H}_a = 1.84$; $\mathscr{H}_i = 1.12$; $\Delta = 2.50$. (d) $\mathscr{H}_a = 2.17$; $\mathscr{H}_i = 1.74$; $\Delta = 4.00$.

quencies without appreciable loss of contrast at low frequencies; see curve b, Fig. 31A. The improvement declines rapidly if the aperture is increased further, see curves c and d. It is possible, however, to raise the contrast transfer factor close to the resolution limit by using ring apertures in the objective. Unfortunately the contrast transfer factor decreases for low frequencies in this case. Figs. 30B, C demonstrate the improvement of transfer by ring apertures, using an aberration-free objective and a fixed outer radius of the aperture ring. Fig. 31B demonstrates the same for a lens with aberration, using apertures with adapted ring radii and adapted defocusing, in order to obtain optimal transfer functions. The primary condition is to keep the changes of the wave aberration below $\lambda/4$ within the free opening of the ring. No appreciable improvement in transfer can be expected when using zone plates with several rings.

11. TEST OBJECTS

In order to determine the contrast transfer properties of an electron microscope, test objects are needed having continuous spatial frequency spectrum in the interval of interest. In light optics, radial gratings (as in Fig. 23) are mainly used for this purpose. There are no such test objects for electron microscopy.

We have already seen in Section 6.5 that a thin carbon foil is often a suitable test object, because it has a "white" spatial frequency spectrum with sufficiently intense diffracted waves (Thon, 1965, 1966a, 1966b, 1966c, 1967, 1968a, 1968b). The diffractogram of an electron image of such a foil gives far-reaching information about the contrast transfer functions (see for example (129) and Fig. 18). Unfortunately, a carbon foil has no regular structures and contrast transfer properties can hardly be determined visually from the random image structure.

An example of an object having a regular structure and containing all spatial frequencies is given by a phase edge. The spectrum of such an object is weak, however, and therefore it is difficult to record and to evaluate. Information on contrast transfer in this case can be obtained from the Fresnel fringes appearing in the image of the object. It has been shown (Hanszen & Morgenstern, 1966), that Fresnel fringes are less characteristic for electron microscopical contrast transfer functions when the objective aperture is limited. In this case, the shape of the fringe system depends on the aperture, but the contrast transfer functions do not. Better information can be expected in experiments without an aperture stop.

If a biprism interference field (see Möllenstedt & Düker, 1956) replaces the object in an electron microscope, wave aberration and contrast transfer cannot be obtained from the fringe *contrast* in the image (Lenz, 1965b). However, one can derive them from the *lateral displacements* (Hanszen, 1967). Experiments of this kind appear to be very difficult to carry out.

Artificial objects such as proposed by Boersch (1943) can serve as test objects also (see Hanszen, 1967). Such experiments have not yet been done. For information concerning transfer tests by means of an enlarged light optical model for the electron microscope, see Hanszen (1968, 1969a, 1969b, 1970a, 1970b); an example is given in Section 6.11.

12. HISTORICAL REMARKS

The references in the preceding chapters concern only papers contributing further details; historical aspects were not considered. Therefore, a short historical evaluation is given in the following.

The results of light optical imaging and transfer theory can be assumed to be known; see the summary by Born and Wolf (1964) and the monograph by Röhler (1967) with numerous bibliographical references. Special mention should also be made of the publication of Zernike (1942) about image formation with special relation to phase contrast, and its contrast theoretical interpretation by Menzel (1958, 1960) and Hauser (1962), which extended the theory to problems of particular interest in electron microscopy.

The wave optical imaging theory of the electron microscope began with the papers of Boersch (1936, 1938, 1943) on the importance of Abbe's diffraction pattern and his recognition of the fact that electron microscopical objects behave as phase objects (Boersch, 1947). Proposals were already made in these papers for phase shifting foils in the diffraction plane, and the phase shifting influence of spherical aberration was discussed.

The theory of electron microscopical resolution (Glaser, 1943) turned only slowly away from incoherent illumination. Focusing conditions were first included by Scherzer (1949); Figs. 3 and 4 of his paper gave examples, although unrecognized ones, of phase-contrast and amplitude-contrast transfer-functions. Further details about imaging theory were presented by von Borries (1949), Uyeda (1955), and Haine (1957), considering partially the concept of phase objects. These articles as well as some more recent ones (Vyazigin & Vorobev, 1963; Eisenhandler & Siegel, 1966b; Reimer, 1966, 1969; Heidenreich & Hamming, 1965; Heidenreich, 1966, 1967) are more concerned with the problem of resolving atomic structures than with instrumental transfer. Moreover, the distinction between "contrast formation by phase shift" as an instrumental method and "phase-contrast imaging" (i.e. imaging of phase structures in the object as intensity structures in the image) as a requirement for image formation is not always strictly observed. The proposal for and the realization of zone plates as objective apertures (Hoppe, 1961, 1963; Langer & Hoppe, 1966, 1967; Lenz, 1963; Eisenhandler & Siegel, 1966a; Siegel, Eisenhandler, & Coan, 1966; Möllenstedt et al., 1968) spurred new studies about image formation. The question, how such a masking influences contrast formation, could be answered by transfer theory (Hanszen et al., 1963,

1965; Lenz, 1965a, 1965b; Morgenstern, 1965). This theory closely connects the contrast and resolution problems, which earlier were treated rather separately. For several years it has been known (von Borries & Lenz, 1956; Lenz & Scheffels, 1958; Faget, Fagot, & Fert, 1960; Fert, 1961; Fagot, Ferré, & Fert, 1961; Farrant & McLean, 1965; Dowell & Farrant, 1965; Hahn, 1965) that object structures the size of which depends on defocusing can be emphasized in the image. In investigations of this kind Albert, Schneider, and Fischer (1964) and Thon (1965) also considered the influence of spherical aberration, which was neglected in the previous papers. Independently of transfer theoretical considerations they came to nearly the same results as that theory. Thon (1965, 1966a, 1966b, 1966c, 1967, 1968a, 1968b) introduced the carbon foil as electron microscopical test object and used successfully the diffractometer method, known from X-ray investigations.

To quote all publications on the problem of resolving lattice spacings in crystals would lead us too far. The papers of Komoda (1964, 1966, 1967) should be mentioned, however. A theoretical treatment of these experiments brought results which are discussed in transfer theory from a general point of view.

The possibility of reconstructing electron micrographs by holography was first recognized by Gabor (1948, 1949, 1951); see also Haine and Mulvey (1952). Transfer theory is capable of outlining the possibilities of this method by describing it in terms of the spatial frequency concept and comparing the method with other procedures for image improvement.

Contrast theory enjoys increasing popularity among the designers of electron microscopes, who need an imaging theory which is independent of the object. Electron microscopists have been more reserved, however. It is important to them, of course, to know which misinterpretations of electron micrographs are possible, under which conditions they may occur, and how we can avoid them. It was the author's desire to work out these points for the readers of this article.

ACKNOWLEDGMENTS

I wish to express my gratitude to Professor Alvar P. Wilska, Professor Jan B. Le Poole, and Professor Kurt W. Just in Tucson, Arizona, as well as to Dr. Rolf Lauer and Dr. Lutz Trepte in Braunschweig, Germany, for reading the manuscript and for many useful remarks.

REFERENCES

Albert, L. (1966). *Optik, 24*, 18.

Albert, L., Schneider, R., & Fischer, H. (1964). *Z. Naturforsch.*, *19a*, 1120.

Archard, G. D. (1958). *Rev. Sci. Instrum.*, *29*, 1049.

Boersch, H. (1936). *Ann. Phys. (5)*, *26*, 631. *Ann. Phys. (5)*, *27*, 75.

Boersch, H. (1938). *Z. Tech. Phys.*, *19*, 337.

Boersch, H. (1943). *Phys. Z.*, *44*, 202.

Boersch, H. (1947). *Z. Naturforsch.*, *2a*, 624.

Boersch, H. (1948). *Optik*, *3*, 24.

Born, M., & Wolf, E. (1964). *Principles of optics* (2nd edition). Oxford: Pergamon Press.

Dowell, W. C. T. (1963). *Optik*, *20*, 535.

Dowell, W. C. T., & Farrant, J. L. (1965). In *International conference on electron diffraction and crystal defects* (p. I P-2).

Eisenhandler, C. B., & Siegel, B. M. (1966a). *J. Appl. Phys.*, *37*, 1613.

Eisenhandler, C. B., & Siegel, B. M. (1966b). *Appl. Phys. Lett.*, *8*, 258.

Faget, J., Fagot, M., & Fert, Ch. (1960). In *Electron microscopy: Proceedings of the Delft conference: Vol. I* (p. 18). Delft: Nederlandse Vereniging voor Electronenmicroscopie.

Fagot, M., Ferré, J., & Fert, Ch. (1961). *C. R. Hebd. Séances Acad. Sci.*, *252*, 3766.

Farrant, J. L., & McLean, J. D. (1965). In *Conference on electron diffraction and crystal defects* (p. I P-5).

Fert, Ch. (1961). *J. Phys. Rad.*, *22*, 26S.

Gabor, D. (1948). *Nature*, *161*, 777.

Gabor, D. (1949). *Proc. R. Soc. A*, *197*, 454.

Gabor, D. (1951). *Proc. Phys. Soc. B*, *64*, 449.

Glaser, W. (1943). *Z. Phys.*, *121*, 647.

Glaser, W. (1956). In S. Flügge (Ed.), *Handbuch der Physik: Vol. 33*. Berlin: Springer. 123 pp.

Hahn, M. (1965). *Z. Naturforsch.*, *20a*, 487.

Haine, M. E. (1957). *J. Sci. Instrum.*, *34*, 9.

Haine, M. E., & Mulvey, T. (1952). *J. Opt. Soc. Am.*, *42*, 763.

Hanszen, K.-J. (1966a). *Z. Angew. Phys.*, *20*, 427.

Hanszen, K.-J. (1966b). In *Electron microscopy: Proceedings of the Kyoto conference: Vol. 1* (p. 39). Tokyo: Maruzen.

Hanszen, K.-J. (1967). *Naturwissenschaften*, *54*, 125.

Hanszen, K.-J. (1968). In *Electron microscopy: Proceedings of the Rome conference: Vol. 1* (p. 153). Rome: Tipografia Poliglotta Vaticana.

Hanszen, K.-J. (1969a). *Z. Angew. Phys.*, *27*, 125.

Hanszen, K.-J. (1969b). *Z. Naturforsch.*, *24a*, 1849.

Hanszen, K.-J. (1970a). *Optik*, *32*, 74.

Hanszen, K.-J. (1970b). In *Microscopie électronique, 7me congrès international: Vol. I* (p. 21).

Hanszen, K.-J., & Lauer, R. (1965). *Optik*, *23*, 478.

Hanszen, K.-J., & Morgenstern, B. (1965). *Z. Angew. Phys.*, *19*, 215.

Hanszen, K.-J., & Morgenstern, B. (1966). *Optik*, *24*, 442.

Hanszen, K.-J., Morgenstern, B., & Rosenbruch, K. J. (1963). *Z. Angew. Phys.*, *16*, 477.

Hanszen, K.-J., Rosenbruch, K. J., & Sunder-Plassmann, F. A. (1965). *Z. Angew. Phys.*, *18*, 345.

Hanszen, K.-J., & Trepte, L. (1970). In *Microscopie électronique, 7me congrès international: Vol. I* (p. 45).

Hanszen, K.-J., & Trepte, L. (1971a). *Optik*, *32*, 519.

Hanszen, K.-J., & Trepte, L. (1971b). *Optik*, *33*, 166.

Hanszen, K.-J., & Trepte, L. (1971c). *Optik*, *33*, 182.

Hauser, H. (1962). *Opt. Acta, 9*, 121.

Hawkes, P. W. (1968). *Br. J. Appl. Phys. (2), 1*, 131.

Heidenreich, R. D. (1966). *Bell Syst. Tech. J., 45*, 651.

Heidenreich, R. D. (1967). *J. Electron Microsc., 16*, 23.

Heidenreich, R. D., & Hamming, R. W. (1965). *Bell Syst. Tech. J., 44*, 207.

Hopkins, H. H. (1961). In K. J. Habell (Ed.), *Proc. Conf. Opt. Instr. Technol.* (p. 480). London: Chapman and Hall.

Hoppe, W. (1961). *Naturwissenschaften, 48*, 736.

Hoppe, W. (1963). *Optik, 20*, 599.

Hoppe, W., Katerbau, K.-H., Langer, R., Möllenstedt, G., Speidel, R., & Thon, F. (1969). *Siemens Rev., 36*, 24.

Jeschke, G., & Niedrig, H. (1970). *Acta Crystallogr. A, 26*, 114.

Komoda, T. (1964). *Optik, 21*, 93.

Komoda, T. (1966). *Jpn. J. Appl. Phys., 5*, 603.

Komoda, T. (1967). In *Proceedings of the 25th EMSA meeting* (p. 234). Baton Rouge: Claitor.

Komoda, T., & Otsuki, M. (1964). *Jpn. J. Appl. Phys., 3*, 666.

Kunath, W., & Riecke, W. D. (1965). *Optik, 23*, 322.

Langer, R., & Hoppe, W. (1966). *Optik, 24*, 470.

Langer, R., & Hoppe, W. (1967). *Optik, 25*, 413, 507.

Lenz, F. (1963). *Z. Phys., 172*, 498.

Lenz, F. (1964). *Optik, 21*, 489.

Lenz, F. (1965a). *Labor Invest., 14*, 808.

Lenz, F. (1965b). *Optik, 22*, 270.

Lenz, F., & Scheffels, W. (1958). *Z. Naturforsch., 13a*, 226.

Linfoot, E. H. (1957). *Opt. Acta, 4*, 12.

Linfoot, E. H. (1964). *Fourier methods of optical image evaluation* (p. 15). London: Focal Press.

Lohmann, A., & Wegener, H. (1955). *Z. Phys., 143*, 413.

Menzel, E. (1958). *Optik, 15*, 460.

Menzel, E. (1960). *Optics in metrology* (p. 283). Oxford: Pergamon Press.

Möllenstedt, G., & Düker, H. (1956). *Z. Phys., 145*, 377.

Möllenstedt, G., Speidel, R., Hoppe, W., Langer, R., Katerbau, K.-H., & Thon, F. (1968). In *Electron microscopy: Proceedings of the Rome conference: Vol. 1* (p. 125). Rome: Tipografia Poliglotta Vaticana.

Morgenstern, B. (1965). *Z. Naturforsch., 20a*, 972.

Nagendra Nath, N. S. (1939). *Akust. Z., 4*, 263.

O'Neill, E. L. (1963). *Introduction to statistical optics.* Reading, MA: Addison Wesley Publishing Corp.

Reimer, L. (1966). *Z. Naturforsch., 21a*, 1489.

Reimer, L. (1969). *Z. Naturforsch., 24a*, 377.

Röhler, R. (1967). *Informationstheorie in der Optik.* Stuttgart: Wiss. Verl. Ges.

Scherzer, O. (1949). *J. Appl. Phys., 20*, 20.

Shannon, C. E. (1949). *Bell Syst. Tech. J., 27*, 379, 623.

Siegel, B. M., Eisenhandler, C. B., & Coan, M. G. (1966). In *Electron microscopy: Proceedings of the Kyoto congress: Vol. 1* (p. 41). Tokyo: Maruzen.

Thon, F. (1965). *Z. Naturforsch., 20a*, 154.

Thon, F. (1966a). *Z. Naturforsch., 21a*, 476.

Thon, F. (1966b). In *Electron microscopy: Proceedings of the Kyoto congress: Vol. 1* (p. 23). Tokyo: Maruzen.

Thon, F. (1966c). *Physikertagung München* (p. 101) (Fachberichte).

Thon, F. (1967). *Phys. Bl.*, *23*, 450.

Thon, F. (1968a). In *Electron microscopy: Proceedings of the Rome conference: Vol. 1* (p. 127). Rome: Tipografia Poliglotta Vaticana.

Thon, F. (1968b). Thesis. Tübingen.

Tsujiuchi, J. (1963). *Prog. Opt.*, *2*, 131.

Uyeda, R. (1955). *J. Phys. Soc. Jpn.*, *10*, 256.

von Borries, B. (1949). *Z. Naturforsch.*, *4a*, 51.

von Borries, B., & Lenz, F. (1956). In *Electron microscopy: Proceedings of the Stockholm conference* (p. 60). Stockholm: Almqvist and Wiksell.

Vyazigin, A. A., & Vorobev, Y. V. (1963). *Bull Akad. Sci. USSR Phys. Ser.*, *27*, 1103.

Wilska, A. P. (1965). *Labor Invest.*, *14*, 825.

Zernike, F. (1942). *Physica*, *9*, 686, 974.

INDEX

Printed in the United States
By Bookmasters